AF553349

TEXTBOOK OF SOLID STATE AND NUCLEAR PHYSICS

By

Dr. Ajit Kumar Sharma

Deptt. of Physics

D.A.V. (P.G.) College

Bulandshahr

(U.P.)

DISCOVERY PUBLISHING HOUSE PVT. LTD.

NEW DELHI-110 002

Published by:
DISCOVERY PUBLISHING HOUSE PVT. LTD.
4383/4B, Ansari Road, Darya Ganj
New Delhi-110 002 (India)
Phone: +91-11-23279245; 23253475; 43596065
+91 9811179893 / +91 9871656464
E-mail: discoverybooksindia@gmail.com
orderdphbooks@gmail.com
namitwasan9@gmail.com
web: www.discoverypublishinggroup.com

First Edition: **2011**
Reprinted: **2023**

ISBN: 978-81-8356-850-0

Textbook of Solid State and Nuclear Physics

Printed at:
Infinity Imaging Systems
Delhi

Preface

The book has been written to meet the requirements of graduate students of all Indian Universities. The matter of this book has been detail theory and sufficient solved examples. We have been selected sufficient problems from various Universities examination papers.

One of salient course syllabi features of the present edition of the book is the inclusion of a large number of typical worked out problems which will elucidate various abstract principles and theories discussed in the text. It covers the complete syllabus of physics prescribed by Technical Universities. We have tried to our best to keep the book free from the misprint. The author shall be grateful to the readers who point out errors and omissions which inspite of all care might have been there.

We shall indeed be very thankful to our colleague for their recommendations this book of for their students. The present book will be warmly received by the students and teachers.

—Author

Preface

[illegible] book has been written to meet the requirement of [illegible] students of [illegible]. All the matter of the book has been [illegible] theory and sufficient [illegible] and problems [illegible] examination [illegible].

[illegible] the book [illegible] and [illegible] of [illegible] the [illegible] syllabus of physics prescribed by [illegible] Universities. We have tried our best to keep the book free from the misprints. The author shall be grateful to [illegible] who point out errors and omissions which in spite of all care might have been there.

We shall indeed be very thankful to our [illegible] who recommend this book for their students. The present [illegible] will be warmly received by the students and teachers.

Contents

1

CLASSIFICATION OF NUCLEAR REACTORS AND REACTIONS

INTRODUCTION

Most of the nuclear reactors are of the heterogeneous type, in which fuel is concentrated in plates, rods, or hollow cylinders, which are distributed in a regular pattern (lattice) within the moderator. Reactors are further classified according to the energy of the neutrons that cause fission.

Reactors may be classified in a wide variety of ways. For example they may be classified by the manner in which the fuel and the moderator are mixed. In a *homogeneous reactor* the fuel and moderator arc mixed in the form of a solution. These reactors ire known as *thermal, intermediate* or fast according to the thermal neutrons (0.025 eV), intermediate neutrons (1 to 10 keV) and fast neutrons (above than 10 keV) respectively. Reactors are also classified according to their purpose. The types to be examined here are:

(a) Research Reactors,

(b) Production Reactors,

(c) Power Reactors.

RESEARCH REACTORS

Research reactors are used primarily to supply neutrons for physical research and radioisotope manufacture. In these reactors the total energy liberated is comparatively small, since they operate at a low level of reactivity. In these cases cooling is required only to prevent over heating. We shall describe here five main types of research reactors :

Graphite-Moderated Research Reactor

The graphite-moderated, heterogeneous, natural uranium reactor has the distinction of being the first reactor built and operated. This first reactor, now called CP-1 (Chicago Pile-1), was put into operation in 1942. This was operated without coolant flow at a power level of 200 watts, Similar assemblies with air as coolant, like the GLEEP (graphite low energy experimental pile) and BEPO (British experimental pile) at Harwell (England) and the X—10 Clinton pile at Oak Ridge and BNL (Brookhaven National Laboratory) are well known examples as they have been the work horses of reactor technology and research. All these reactors use fuel in the shape of cylindrical rods, or variations on this shape, clad usually in aluminium. "Several hundred tons of graphite are used as moderator and reflector, with several hundred cylindrical channels passing right through the graphite to allow the introduction and positioning of the fuel elements. Cooling gas is made to pass over the elements through the channels at low pressure.

WATER BOILER TYPE REACTOR

Perhaps one of the roost misleading names of a reactor type is the water boiler. The water boiler is usually a homogeneous mixture of a highly enriched uranium salt dissolved in ordinary water. The name water boiler often leads to a mistaken belief that these reactors are the boiling water type of reactor. Actually the boiling water reactors arc quite different and are under the bead "power reactors". The name arises from the fact that a sudden increase in power will cause the formation of steam bubbles in the solution, which in turn will shut down the reactor quickly. Under normal conditions, the solution does not boil as its temperature is kept below 80°C by circulating water through the coils inside of the core vessel.

The first reactor of this type was built as Los Alamos in early 1944. It was named LOPO (low power) because of its low power operation. It was improved in December for high power operation and was named HYPO (high power). It was further improved for superpower, named SUPO (Super power), and is in operation since 1951.

Another water boiler of note is the reactor built at the North Carolina State College in 1953. It is the first reactor owned and operated by a University anywhere in the world. Similar assemblies, like WBNS (water boiler neutron source) California and ARR (Armour research reactor) Chicago and KEWB (kinetic experimental water boiler) California are well known examples. Few properties of these reactors are given in Table 1.

Table 1 : Water Boiler, Water Coolant, Water Moderator, Homogeneous Research Reactor

Name	Size	Fuel	Reflector	Power (kW)/ Neutron flux
LOPO	Sphere 1'dia 15 litre	Enriched uranyl sulphate solution; 15 litres; U^{238} (580 gm.), U^{238} (3380 gm.).	BeO and Graphite	5×10^{-5}
HYPO	"	Enriched U-nitrate solution, U^{235} (897 gm.) U^{238} (5341 gm.).	BeO and Graphite	63×11^{11}
SUPO	"	90% enriched U-nitrate, 0.8 kg. in 13 litres.	Graphite	35-45 1.7×10^{12}
WBNS	"	90% enriched uranyl nitrate solution, U^{235} (700 gms.).	Graphite	0.005 2×10^{5}
North Carolina	Cylinder $(D=10\frac{3}{4}$, H=11')	90% enriched UO_2 SO_4 solution,. U^{235} (848 gm.) in 14 litres.	Graphite	10 5×10^{11}
ARR	15 litre spherical core	88% enriched uranyl sulphate.	Graphite	50 10^{12}
KEWB	Sphere (1,1/4, dia)	90% enriched uranyl sulphate, U^{235} (1750 gm.).	Graphite	50 9×10^{11}

The water boiler reactor has proved to be versatile research tool providing intense sources of neutrons and γ-rays. It has a great merit of simplicity of design and construction. Its use is somewhat limited because of its low power rating.

Swimming Pool Reactor

The turn swimming pool reactor is the popular name of light water moderated heterogeneous reactor which consists of a lattice of enriched uranium fuel immersed in a large pool of water. The water in the pool plays the role of moderator, coolant and shield. The chief advantages of this type of reactor are simplicity, low cost, safety, flexibility and accessibility.

The first swimming pool reactor was completed in 1950 at the Oak Ridge National Laboratory. It was named BSR because was designed with the view of allowing radiation effects on bulky materials to be observed by immersing them in the water surrounding the core.

The Indian Reactor *Apsara* belongs to this type. This reactor was completed on August 4, 1956 and was the first reactor to go into operation in Asia. It was named Apsars on the name of Legendary sea mermaid because it is half under the tea and half outside. It was designed and built by the Indian engineers and scientists under the guidance of the late Dr. H.J. Bhaba. This reactor is designed for a maximum power level of 1 mw and contains large quantity of light water in a concrete pool. The fuel assembly is attached to steel frames and is suspended from above from an adjustable bridge placed on rails. This bridge spanes the pool. The fuel elements are arranged in the rectangular assembly. These are made of enriched uranium alloy and arc held inside aluminium cans. Few control rods of Cd metal are also introduced in the fuel assembly and are adjusted automatically so that reactor does not become super critical. Holes are provided through graphite reflector through which thermal neutrons are available.

Table 2 : Swimming Pool Water Coolant, Water Moderator, Heterogeneous Research Reactor

Name	Size	Fuel	Reflector	Power (kW) Neutron flux
BSR	20'×20'×40'	MTR type 3.5 kg. U^{235}, Fuel core 1'×1'×2'	H_2O or BeO	100-1000 2×10^{13}
NRLR (Naval)	$49' \times 32' \times 28\frac{1}{2}$	MTR type, Fuel core 21"×21"×24'	H_2O or Graphite	100-1000 10^{12}
Baitelle RR		MTR type 2.8 kg. U^{235}	H_2O or Graphite	1000 1.4×10^{13}
Swiss Reactor	10' dia 2' deep	23 MTR type fuel assemblies.	H_20	10-100 10^{11}
Melusine (France)		Enriched uranium	H_2O	1000 10^{13}
Apsara (India)		Enriched uranium alloy	Graphite	100-1000 10^{12}

When the reactor becomes critical the water surrounding the core emits a brilliant Cerenkov light glow, due to the passage of energetic electrons through water, can be seen from above the pool in the broad day light as well. The first fuel charge was removed from the reactor in

Dec. 1964 since it had attached its maximum burn up. Maintenance of its various components/improvements in cooling and control systems were carried out in the first quarter of 1965. It was started up with the second stage of fuel elements, obtained from U.K., in July 1965. The standard element of second fuel charge contains 140 gm of U^{235} (80% enriched uranium).

This reactor has been used for the production of isotopes, irradiation of biological samples, and for testing the various types of reactors and health physics instruments. It has been used for fundamental research in nuclear physics and structure of materials. A large number of isotopes such as P^{32}, S^{35}, I^{131} and Au^{198}, produced in this reactor, are used in research instruments and hospitals. There are about two dozen swimming pool reactors available in the world. Few properties of some reactors are given in table 2.

Light-Water-Moderator, Tank type Reactor

This reactor is very similar to pool reactor, in principle, except that the core is suspended in a deep cylindrical tank of water and cannot be moved and it preferred over 'one of the pool type when a large number of holes's for neutron beams in many different directions is an important requirement. A reactor with increased neutrons flux was designed and built in 1952 at the National Reactor Testing Station in Arco, Idaho and was named Material Testing Reactor (MTR). The MTR fuel elements are heterogeneous plate type elements. These plates, often referred to as sandwich type plates, contain an inner core of uranium-aluminium alloy which Is 0.02" thick. This core is clad (bread) on either side by 0.02" thick aluminium. Eighteen of the plates are brazed together in aluminium side plates with a spacing of 0.117" between plates to form which is generally known as the MTR type element. Each one of the MTR elements contains 140 gm. of U^{235} and the critical mass for the reactor is approximately 3 kg.

Heavy-Water-Moderator, Tank type Reactor

The desirability of obtaining heavy water for use as a neutron moderator lies in the extremely small thermal absorption cross section deuterium. Because of this property the neutron economy in a heavy water system is greatly improved. The first heavy water moderator reactor was the Chicago Pile 3 (CP-3) which went into operation on may 1944 at the Argonne National Lab. and was dismantled in January 1950. It is a simple tank type of reactor. Canada has specilized in heavy water natural uranium reactor because of the abundant supply of natural uranium and no facilities for enrich ment of natural uranium. The first reactor constructed in Canada was completed in Sep. 1945.

Cirns : formerly known as CIR, is India's second research reactor, built as a joint Indo-Canadian project under the Colombo Plan at Trombay. This reactor was completed on 10 July, 1960. The name CIR stands for Canadian Indian Reactor and the new name cirus is the name of the star close to our solar system. It is a natural uranium heavy water moderated and light water cooled high flux research reactor with a terminal power of 40 MW.

It is Operated round the clock. The standard fuel rod contains about 55 kg. of natural uranium. For normal operation the reactor core contains about 185 fuel rods and 23 tonnes of heavy water. This reactor is used for basic and applied research in physics, chemistry, biology, metallurgy and reactor technology, for the production of large quantities of high activity radio isotopes as well as plutonium-239 and uranium-233.

Zerlina : Is India's third research reactor. It attained criticality on Jan. 14, 1961. The name zerlina stands for *zero energy lattice investigation nuclear assembly* as it was made to study the lattice assemblies of different kinds. In this reactor natural uranium is used as a fuel and heavy water as a moderator. The standard fuel rod contains about 45 kg. of natural uranium. The standard core contains 80 fuel rods and about 8 tonnes of heavy water. It is operated on a two shift basis upto a maximum power of 100W and is being extensively used for physics and studies of uranium—heavy water lattice.

Purnima : It is India's first experimental zero energy fast reactor. It attained criticality on May 22, 1972. It uses plutonium oxide as fuel. It is used for the training of personnel in the operation of plutonium fuelled fast reactors. Experiments were carried out by scientists to measure parameters used in the construction of the core of the pulsed fast reactor at Kalpakham.

R-5 : It is a thermal research reactor with a nominal power of 100 MW and is located adjacent to 40 MW cirus reactor. The natural uranium fuel elements of cirus and zerlina and the plutonium oxide for purnima are fabricated in Trombay. This fabrication facility went into operation in June 1959. The present capacity of this plant is about 30 to 44 tones of fuel elements per year.

PRODUCTION REACTORS

The purpose of a production reactor to convert fertile into fissile material. The fissible materials are used as a fuel in other reactors. In Hanford works production reactors, the natural uranium is used as a fuel, graphite as a moderator and water as a coolant. Because of the relatively large number of neutrons captured by a hydrogen nuclei present in water, the proportion available for capture by U^{238} is reduced considerably. The efficiency of conversion of U^{238} into Pu^{239} is thus appreciably less than unity. The ratio is always smaller than 0.8 in this type of reactor.

The efficiency of the conversion process can be increased by decreasing resonance escape probability. It is mainly by capture of neutrons in the resonance region. U^{238} is converted into Pu^{239}. Since maximum value of multiplication factor in uranium graphite reactor is 1.07, hence any decrease in the resonance escape probability can make the reactor impossible to attain criticality. Multiplication factor would be a large as 1.25 if heavy water is used as a moderator. A substantial decrease in the resonance escape probability is thus permissible without making reactor impossible to attain criticality. Thus we see that conversion efficiency of a production reactor using heavy water as a moderator is greater than the reactor using graphite as a moderator.

POWER REACTORS

The primary purpose of a power reactor is the utilization of the fission energy produced in the reactor core and to convert it into useful power. Heat generated in fission process is used, either directly, or indirectly, to produce steam at high temperature and pressure or to heat a compressed gas. The steam or heated gas is used as a working substance in a turbine. The turbine can be connected to an electric generator or can be used as a source of mechanical power. Choice of the heat exchange material is governed by three important considerations.

(d) It must have a high thermal conductivity so as to carry away the heat efficiently and give it up in a heat exchanger.

(b) It must have a low neutron capture cross-section so as not to upset the reactor characteristic.

(c) It must not be decomposed by intense radiation.

Now we shall discuss some of the reactor types which appear to show the greatest promise for power production.

1. Pressurized Water Reactor

This is a heterogeneous reactor that uses slightly enriched uranium (1.4 to 2% of U^{235}) as fuel and light water as a moderator and coolant. This type of reactor has received much attention in the United Slates. The fuel elements are in the form of rods or plates. The core is contained in a vessel under a pressure of 1000 to 2000 pounds per square inch. The pressurized water is circulated through the reactor core, from which it removes heat, and then through on external heat exchanger. Heat exchanger is used to abstract heat energy from the coolant in order to make it available for conversion into other forms of energy. High pressure is maintained in the core in order to raise the boiling point of the coolant to high enough temperature only because a consequence of the laws of thermodynamics is that heat can be more coefficiently convened into useful energy if it is released at a high temperature.

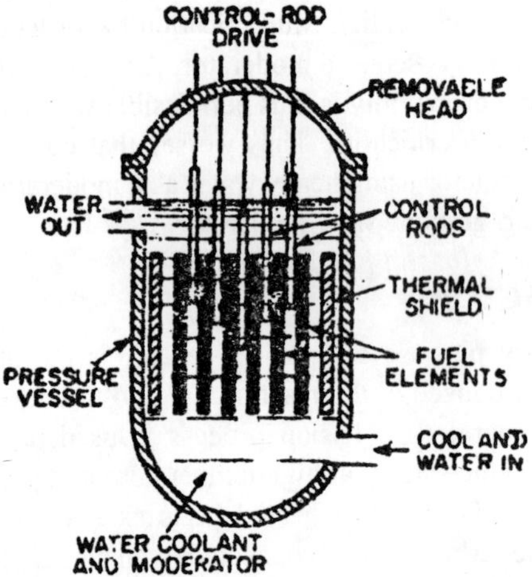

Fig. 1 : Schematic diagram of the pressure vessel and core of a pressurized water reactor

This first full scale nuclear power plant in the United States at Shipping port, Pennsylvania, was completed at the end of 1957. The first one produced originally 60 mega watts of electric power. This power was latter increased to 100 megawatts with a capability of going to 150 megawatts of electricity Several other reactors of this type have been (or are being) constructed in the United States and in other countries e.g., France, Italy and U.S.S.R.

2. Boiling water Reactor

The boiling water reactor is a small nuclear power plant designed to allow steam to be generated directly in the reactor core. This uses light water as moderator coolant. This reactor system is more advantageous than pressurized water reactor. The pressure in the reactor vessel does not have to be much greater than the required steam pressure while in PWR the pressures used to prevent boiling are significantly larger. External heat exchangers are also not required. Enriched uranium is used as a fuel.

The experimental boiling water reactor at Argonne National Laboratory was the first reactor of this type. It was designed in the end of the year 1956 and electricity at the power of 5 megawatts was produced. The first boiling water reactor to produce substantial amount of electric power was started up in 1960 at the nuclear power station, near Chicago, Illinois.

Central station power plants utilizing boiling water reactors are also in operation in other parts of U.S.A. as well as in Italy and Germany. The maximum power produced by most recent plants is 1000 megawatts of electricity and the steam produced is at a temperature of 285° and a pressure of about 1000 psi.

The 420 MW energy station, at Tarapur (India), consists of two enriched uranium reactors of the boiling water type. These reactors were built with the help of the Central Electric Company of the U.S.A. under a turn key contract, on the basis of specifications provided by the Tarapur Atomic Power Project of the Department of Atomic Energy. Nuclear power began to flow from the station when it became operational at 8.15 p.m. on April I, 1969. These reactors operated extremely well in the first few years. However, some problems were raised in the years thereafter. Now since 1975 the performance is rather good.

This type of reactors are particularly adapted to integral super-heating. The steam is produced in one part of the core and super-heated in another part. The first such plant, completed in 1964, was the Boiling Nuclear Superheat (BONUS) reactor in Puerto Rico. In this reactor water is boiled in the central region of the core and the steam is superheated in the outer region.

The second superheat reactor, completed in 1965, is at the Pathfinder power station at Sioux Falls. Sout Dakota. The arrangement is reverse of that in the first one. The temperature of exist steam is 480° in the first case and 440°C in the second one.

3. Heavy Water Moderated Reactor

There arc two advantages of reactors moderated by heavy water as compared with ordinary water as a moderator. First, the total cost of the core and cost of the consumed fuel arc both less in the heavy water moderated reactor because natural uranium or very slightly enriched material is used as a fuel. Second, the conversion ratio for the regeneration of fissile material can be close to unity. Both factors decrease the cost of the power, produced. As the high price of the heavy water increases the cost of power, this type of reactor cannot be used for large plants with an electrical capacity of 500 MW.

The first proto-type of a heavy water moderated power reactor is the Nuclear Power Demonstration Reactor (NPDR) completed in 1962 at Canada. It is of the pressurized water reactor type with heavy water as moderator and coolant instead of the light water. The coolant leaves the reactor at 277° and produced steam at about 230° in the heat exchanger.

The first boiling water reactor which uses heavy water as moderator is the Halden (Norway) reactor. In this reactor heavy water coolant-moderator boils and produces heavy steam. It is only one of its kinds in existence and has a designed thermal powers of 20 megawatts.

The second nuclear power station (430 MW) near. Kola at Rana Pratap Sagar in Rajasthan employs heavy water moderated reactors using natural uranium as fuel. The first unit of the reactor became critical in 1972 and started feeding power in 1973 into the power system of Rajasthan. In this power station the design and bulk of the equipment have come from Canada, but the construction and installation activities were done by the Department of Atomic Energy. Unfortunately there have been failures in two stages due to deficiencies in design and manufacture of the turbine generator; designed in U. K. and supplied from Canada. For the second unit of this station under construction, a major step in indigenisation has been taken. The reactor vessel, shields, boilors and fuelling machines have been made in and India and the responsibility has been taken over by Indian Scientists and Engineers.

The third nuclear power station (470 MW) is under construction at Kalpakham about 100 km away from Madras. In this instance complete responsibility for design. Engineering and construction of the project rests with the Department of Atomic Energy. The design generally follows the design used at Rajsthan, significant changes have been made in increase the output of the units by about 10%. This is 80% Indian, only

20% components are from aboard. Apart from reactor vessel and shields, boilers and fuelling machines, the equipment manufactured in India include the turbine generators, zirconium alloy components and the entire instrumentation and control system. The total fuel charge as well as heavy water will come from Indian plants.

The fourth Indian Atomic power plant is under construction at Narora. It will have two atomic reactors of 235 MW each and well deliver 440 MW of power to the northern power grid when completed. It will use natural uranium at fuel and heavy water as moderator like the power stations at Ranapratap Bagh and Kalpakham. This power station will be 100% from Indian resources.

Organic Moderated Reactor

Another method for achieving high temperatures at moderate pressures in a thermal reactor is to moderase and cool with an organic liquid containing a high hydrogen content and of his high boiling point. The organic moderated Reactor Testing Experiment (OMRE), was completed in 1957 at the National reactor Testing Station. Arco, Idaho. The organic liquid used in this reactor is a mixture of di and ter-phenyle and has a normal boiling point at atmospheric pressure of about 340°C and 370°C at the operating pressure of some 200 psi. The temperature is limited to a maximum of 325°C because the thermal decomposition of the organic compounds become more than can be tolerated. The experience gained was utilized in the design of a small power plant, called the Piqua Nuclear Power Facility (PNPF) at Piqua Ohia. The coolant used is a mixture of the three terphenyls ($C_{18}H_{14}$) at a pressure of 120 psi. It leaves the reactor at a temperature of 302°1C. Superheated steam at 288°C and a pressure of 440 psi is produced in the heat exchanger. The electric output of this plant is 11.5 megawatts.

Gas-Cooled Reactor

Several of the early experimental reactors in the United States were cooled with forced-air. Until 1957, this type of reactors were not developed in the United States because of various reasons. In the mean time air was employed as the coolant in the graphite moderated production reactors at Windscale in U. K. The graphite moderated power reactor with nitrogen carbon dioxide as a coolant and natural uranium as the fuel have been constructed in England, France and other countries. The carbon dioxide gas at a pressure of about 150 to nearly 400 psi in the later designs is pumped through the reactor leaving at a temperature around 400°C. The superheated steam is generated, at temperatures in the approximate range of 320° to 390° in the heat exchanger About 300 megawatts of electricity is produced by most of the larger plants. Helium gas also used in some of the gas cooled reactors. In High Temperature Gas Cooled Reactor, completed in 1966 at Peach Bottom, Pennsylvania, the helium gas leaves the core temperature of 750°C and produces superheated steam at 538°C at and 1450 psi pressure.

The use of beryllium oxide as a moderator in gas cooled power reactor was proposed by F. Daniels and his associates in 1944. With improvements in fabrication Experimental Beryllium Oxide Reactor was constructed but the work was terminated in 1966 because cost of fabrication is a significant disadvantage in power reactor with this moderator.

Sodium Graphite Reactor

The high boiling point of liquid sodium makes it an excellent cooled for a high temperature reactor. High temperatures can be achieved without high pressures. Sodium Reactor Experiment was developed to its critical state in 1957. In this reactor graphite was used as moderator, sodium as coolant and either natural or slightly enriched uranium as the fuel.

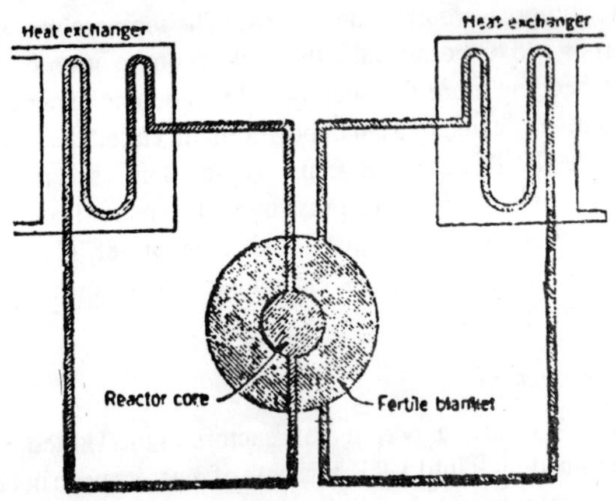

Fig. 2 : Schematic Diagram of homogeneous reactor.

Another reactor based upon SRE was completed in 1962. This is known as Hallam Nuclear Power Facility (HNPF). The temperature of the sodium leaving the reactor was 507°C and the steam was generated at a temperature of 446°C and a pressure of 850 psi. Its design power was 76 megawatts of electricity. It is very advantageous because sodium has excellent heat transfer characteristics and does not require excessive pumping power. After operating at full power for nearly two years, it was found that several of the stainless steel moderator cans had cracked and allowed sodium to leak. The cracking was caused by rupture of the steel due to the extended operation at high temperatures. In August 1965, the decision was made ultimately to close down the reactor.

Liquid Fuel Reactor

Aqueous homogeneous reactors are characterized by having the fissionable material such as uranyl sulphate or nitrate dissolved in light or heavy water and circulating this solution through a spherical vessel of stainless steel, in which chain reaction takes place and through a heat exchanger, in which the heat is transferred from hot solution to a working fluid and back to the vessel. The Homogeneous Reactor Experiment (HRE) at Oak Ridge National Laboratory was built in 1952 to test the feasibility of maintaining a fission chain at high temperatures and pressures. The advantages of an aqueous homogeneous reactor are the high power density, low fuel inventory, continuous removal of fission products and radiation damage products, high degree of nuclear stability, simple mechanical design and absence of structural materials in reactor core. Beside these advantages there are some drawbacks also.

(i) Great care must be taken to prevent leakage of highly radioactive solution at high temperature and pressure.

(ii) High pressures are required to produce high temperatures with the formation of steam in the reactor.

(iii) Radiolytic decomposition of H_2O or D_2O causes a number of problems.

Fast Reactor

The development of power reactors in which the bulk of fissions arc produced by high energy neutrons is of importance in the civilian power reactor development programmes. Such reactors not only produce useful power but also regenerate more fissile material than is consumed. Breeder reaction is prominent that is why it is also known as breeder reactor.

The first Experi-mental Breeder was completed in August 1951 and produced over 100 watts of electric power on 20th December. It produced the first electricity from nuclear energy. In addition to the use of plutonium oxide and carbide, other alloy of plutonium, cerium and cobalt were also used as a fuel in designing of reactors.

KINDS OF NUCLEAR REACTIONS

When a target nucleus is bombarded by means of energetic projectile particles. The bombardment process may give rise to the following events single or jointly.

Radiative Capture

A particle may combine with a nucleus to produce a new nucleus or a compound nucleus which is in an excited state. The excess energy is emitted in the form of γ-ray photons. This type of process is known as radiative capture example

$$_{12}Mg^{26} + {}_1H^1 \rightarrow ({}_{13}Al^{27})^* \rightarrow {}_{13}Al^{27} + \gamma.$$

Disintegration

On striking the target nucleus the incident particle is absorbed and a different particle is ejected. The product nucleus differs from target nucleus. The incident particle may be an α-particle, proton, neutron, etc. The product particle may be a charged particle or a neutron. Thus process is known disintegration as

$$_7N^{14} + {}_2He^4 \rightarrow {}_3O^{17} + {}_1H^1.$$

Photo Disintegration

The γ-rays are absorbed by the target nucleus, exciting it to a higher quantum state. If the energy is high enough, one or more particles may be liberated. An example is

$$_1H^2 + g \rightarrow {}_1H^{'} + {}_0n^1.$$

High Energy Reactions

In the energy range about 150 MeV, spallation process merges into new kinds of reactions in which new kinds of particles (mesons, strange particles) are produced along with neutrons and protons.

Direct Reactions

A collision of an incident particle with the nucleus may immediately pull one of the nucleons out of the target nucleus by the so called *pick-up reaction.* In the inverse process, a bombarding particle composed of more than one nucleon may loss one of them to the target by the so-called *stripping reaction, e.g.*

$$_{29}Cu^{63} + {}_1d^2 \rightarrow {}_{29}Cu^{64} + {}_1H^1.$$

Heavy Ion Reactions

Nuclear reactions induced by heavy ions exhibit the characteristics both of compound nucleus and of the stripping and pick up mechanisms. Example

$$N^{14} + Pb^{207} \rightarrow N^{13} + Pb^{203}.$$

Spontaneous Decay

β– and α – decay processes may be regarded as this type of nuclear reactions. In these reactions the total energy of the system is not under the experimenter's control.

Spallation Reactions

On capture a incident particle, a heavy nucleus has sufficient energy for the ejection of several particles. Such a reaction is known as spallation reaction, *e.g.,* the nuclear fission, in which heavy nucleus splits mainly in two nuclei,

Example : $_{92}U^{235} + {}_0n^1 \rightarrow {}_{40}Zr^{98} + {}_{52}Te^{136} + 2{}_0n^1.$

Elastic Scattering

The incident particle strikes the target nucleus and leaves without energy loss but, in general, with altered direction of motion. Scattering of α-particles in gold is a good example of this process.

(i) $e^- + {}_6C^{12} \rightarrow {}_6C^{12} + e^-$

(ii) $_2He^4 + {}_{79}Au^{197} \rightarrow {}_{79}Au^{197} + {}_2He^4$

(iii) $_0n^1 + {}_{92}U^{238} \rightarrow {}_{92}U^{238} + {}_0n^1$

Inelastic Scattering

The scattered particle may loss K E in excess of that required for an elastic collision with the nucleus. This being corresponding increase in the internal energy of the nucleus which is excited to a higher quantum state. This inelastic scattering can be represented by the well known example.

(i) $_0n^1 + {}_{82}Pb^{208} \rightarrow {}_{82}Pb^{208*} + {}_0n^1$

(ii) $_3Li^7 + {}_1H^1 \rightarrow ({}_3Li^7)^* + {}_1H^1$

(iii) $_1H^1 + {}_7N^{14} + 7N^{14*} + {}_1H^1.7N^{14*} \rightarrow 7N^{14} + y.$

The star (*) is used to indicate that, after scattering, nucleus is left in an excited state. In this present example the excess energy is latter radiated away in the from of a γ-quantum.

KINDS OF THE CONSERVATION LAWS

We shall merely list the various conservation laws that appear to be valid in ordinary nuclear interactions.

Conservation of Momentum

The total linear momentum of the products must be equal to the linear momentum of the bombarding particle (the target nucleus is ordinarily taken to be at rest).

Conservation of Isotopic Spin

The invariance of the nuclear Hamiltonian function towards the charge character of the nucleons can be expressed analytically as an invariance towards rotational shifts of the axes in isotopic spin space, and there should correspondingly exist a conservation law for the isotopic spin of a nuclear system.

Conservation of Parity

The parity of the system determined by the target nucleons and bombarding particle must be conserved throughout the reaction. The total parity of the system is the product of intrinsic-parities of the target nucleus and bombarding nucleus. No violation of parity has been observed in a nuclear reaction (strong nuclear forces). Although parity does not appear to be conserved in weak interactions.

Conservation of Spin

The important conservation law asserts that the spin character or a closed system cannot change, *i.e.*, the statistics remains same as that existed before reaction.

Conservation of Nucleons

The law of conservation of nucleons states that the nucleons can neither be created nor can be destroyed so that the number of nucleons minus the number of anti-nucleons in the universe remains constant.

Conservation of Charge

The total electric charge of the products must be equal to the total electric charge of initial particles.

Conservation of Angular Momentum

The total angular momentum *l* comprising the vector sum of the intrinsic angular momentum s and relative orbital momentum *l* of the products must be equal to the total angular momentum of the initial particles.

Conservation of Energy

The total energy of the products, including both mass energy and kinetic energy of the particles plus the energy involved must be equal to the mass energy of the initial ingredients plus the kinetic energy of the bombarding particle.

NUCLEAR REACTION KINEMATICS

There are certain restrictions in the conservation of energy and momentum. These restrictions are called kinematic restrictions and this mathematical method is known as *kinematics*. Consider a nuclear reaction

$$X + x \rightarrow Y + y, \quad ...(1)$$

where X, x, Y and y are the target nucleus, bombarding particle, product nucleus and product particle, respectively. It will be assumed that target nucleus is at rest so it has no kinetic energy. Since total energy is conserved in a nuclear reaction, therefore, we get

$$Mxc^2 + (E_x + m_x c^2) = (E_Y + M_Y c^2) + (E_y + m_y c^2 \quad ...(2)$$

where m_x, m_y, M_Y. My all represent respective masses of incident particle, target nucleus, product particle and product nucleus. We now introduce a quantity Q where

$$Q = E_Y + E_y - E_x. \quad ...(3)$$

From eqns (2) and (3), we have

$$(M_x + m_z - M_Y - m_y)c^2 = Q. \quad ...(4)$$

The quantity Q is called the energy balance of the reaction or more commonly *Q-value of the reaction*. If this is +ve the reaction is said to be *exoergic*. This occurs if sum of the masses of incident particle and target nucleus is greater than that of masses of the product nuclei. The

K.E. of the product nuclei being greater than that of the incident particle. If Q is – ve the reaction is said to be *endoergic, i.e.,* energy must be supplied usually as a K. E. of the incident particle.

The term E_Y in eqn. (2) represents the recoil energy of the product nucleus. It is usually small and hard to measure but can be eliminated by taking into account the conservation of momentum. In an experiment to measure a Q value, the bombarding energy E_x and the energy of the ejected particle E_y at some specified angle θ are measured. Thus by applying the laws of conservation of momentum, we have

$$m_{xyx} = M_Y V_Y \cos\phi + m_y y_y \cos\theta \qquad ...(5)$$

$$\text{and } M_Y V_Y \sin\phi = m_y y_y \sin\theta, \qquad ...(6)$$

where v_x, v_y and V_Y are the velocities of incident particle, ejected particle and of product nucleus respectively.

From eqns (5) and (6) eliminating φ, we have

$$M_Y{}^2 V_Y{}^2 = mM_Y^2 Y_Y^2 = m_x^2 + m_y^2 y_y^2 - 2m_x m_y v_x v_y \cos\theta. \qquad ...(7)$$

$$\text{Since } E_x = \frac{1}{2} m_x v_x^2,\ E_y = \frac{1}{2} m_y v_y^2 \text{ and } E_Y = \frac{1}{2} m_Y V_Y^2$$

Hence after eliminating v_x, v_y and V_Y, we get

$$2E_Y M_Y = 2E_x m_x + 2E_y m_y - 4(m_x m_y E_x E_y)^{1/2} \cos\theta$$

$$\text{or} \quad E_Y = E_x \cdot \frac{m_x}{M_Y} + E_y \frac{m_y}{M_Y} - \frac{2}{M_Y}\left(m_x m_y E_x E_y\right)^{1/2} \cos\theta \qquad ...(8)$$

Substituting the value of E_Y in equation (3), we get

$$Q = E_x \frac{m_x - M_Y}{M_Y} + E_y \frac{m_y + M_Y}{M_Y} - \frac{2}{M_y}\left(m_x m_y E_x E_y\right)^{1/2} \cos\theta \qquad ...(9)$$

This is known as Q-value reaction. It gives the desired relation between the energy released and the measured quantities E_x, E_y and θ. For a special case when we are observing the outcoming particle y at 90° to a collimated beam of projectile, the above relation reduces to

$$Q = E_y\left(\frac{1 + m_y}{M_Y}\right) - E_x\left(\frac{1 - m_x}{M_Y}\right). \qquad ...(10)$$

With the modern techniques, it is possible to measure Q values, for nuclear reaction producing charged particles, to an accuracy of 1 part in a thousand or better.

General Solution of the Q-equation

It is clear from eqn. (9) that the variation of E_y with E_x, for a fixed Q, can also be shown by regarding the Q equation as a quadratic in $\sqrt{E_y}$. Thus its solution can be put in the form

$$\sqrt{E_y} = u \pm \sqrt{(u^2 + v)}, \qquad \text{...(11)}$$

$$\text{where } u = \frac{\sqrt{(m_x m_y E_x)}}{M_Y + m_y}\cos\theta \text{ and } v = \frac{M_Y Q + E_x(M_Y - m_x)}{M_Y + m_y}.$$

The energetically possible reactions are those for which $\sqrt{E_y}$ is real and positive. Nuclear reactions are divided energetically as follows

1. Exoergic Reactions

These reactions are possible even for $E_x = 0$. Thus for $E_x \to 0$, equation (11) gives

$$E_y = Q\frac{M_Y}{(m_y + M_Y)}(Q > 0). \qquad \text{...(12)}$$

The kinetic energy E_y is the same for all angles θ. When $E_x \to 0$.

$$Q = E_y + E_Y \text{ and } \theta + \phi = 180°.$$

Commonly projectile x is lighter than the product nucleus Y. Thus ν is +ve for all values of the bombarding energy and only one of the two solutions of eqn., (11) will be positive. Thus E_y to be single valued $Q > 0$ and $M_y > m_x$. E_y depends on cos θ and is smallest in the backward direction θ = 180°.

2. Endoergic Reactions

All endoergic reactions have negative Q-values. When $E_x \to 0$, eqn. (11) gives $u^2 + \nu = -$ve and hence $\sqrt{E_y}$ is imaginary. It means that these reactions are not possible. The smallest value of E_x at which reactions can take place is called the *threshold energy*. The reaction first becomes possible when E_x is large enough to make $u^2 + \nu = 0$, hence

$$E_x = -Q\left[\frac{m_y + M_Y}{m_y + M_Y - m_x - \left(\frac{m_x m_y}{M_Y}\right)\sin^2\theta}\right].$$

At $\theta = 0$, E_x is minimum and is the threshold energy.

$$\therefore \qquad (E_x)_{th} = -Q\frac{(M_Y + m_y)}{(M_Y + m_y - m_x)} \qquad ...(13)$$

By using relation (4) and an excellent approximation $M_x \geq \frac{Q}{c^2}$, we have

$$(E_x)_{th} = -Q\frac{(m_x + M_X)}{M_X} \qquad ...(14)$$

Thus we see that at the threshold or the reaction particles first appear in the $\theta = 0$ direction with the K.E.

$$E_y = u^2 = (E_x)_{th}\left(\frac{m_x m_y}{m_y + M_Y}\right)^2. \qquad ...(15)$$

As the bombarding energy is raised, particle y begins to appear at $\theta > 0$.

NUCLEAR CROSS SECTION

It is usually expressed in terms of an effective area presented by a nucleus towards the beam of bombarding particles, such that the number of incident particles that would strike such an area, calculated upon a purely geometrical basis, is the number observed to lead to the reaction in question. This effective area is called the cross-section for that reaction. Thus the probability of occurrence of a particular nuclear reaction is described by the effective cross-section for that process. The cross-section may also be denned as :

1. The probability that an event may occur when a single nucleus is exposed to a beam of particles of total flux one particle per unit area.
2. The probability that an event may occur when a single particle is shot perpendicularly at a target consisting of one particle per unit area.

Partial Cross-section

The total nuclear cross-section is the effective area possessed by a nucleus for removing the incident particles from a collimated beam by all possible processes. This can be written as a sum of several partial cross-sections which represent the contributions of the various distinct, independent processes which can remove particles from the incident beam. Thus we have

$$\sigma_t = \sigma_s + \sigma_r, \quad ...(1)$$

where σ_l is the total cross-section, as the cross-section that produces measurable scattering and σ_r, the reaction cross-section. Scattering cross-section can be classified as : inelastic scattering and elastic scattering processes, Thus, we get :

$$\sigma_s = \sigma_{cl} + \sigma_{incl.} \quad ...(2)$$

These partial cross-sections can still be sub-divided. In the case of elastic cross-section separate partial cross-sections cannot be written because of the possibility of interference between them. On the other hand all inelastic scattering processes are incoherent and their cross-sections are additive,

$$\sigma_{incl.} = \sigma_s + \sigma_a + \sigma_n + \cdots \quad ...(3)$$

Similarly or the partial cross-section for capture may be sub-divided into component parts associated with the various possible nuclear reactions. Hence we can write

$$\sigma_r = \sigma' + \sigma'' + \sigma''' + \ldots \quad ...(4)$$

Differential Cross-section

The cross-section which defines a distribution of emitted particles with respect to the solid angle is called differential cross-section. It is defined as $d\sigma/d\Omega$ and the partial cross-section for a given process is given-by as follows

$$s = \int \frac{d\sigma}{d\Omega} d\Omega. \quad ...(5)$$

Determination of Cross-section

Experimental values of a can be obtained by the measurement of the attenuation of an incident beam in a known thickness of material. If a monoenergetic beam consisting of *I* particles per unit time distributed uniformly over an area A is allowed to pass through a thin sheet of target material of thickness Δx and having n nuclei per unit volume then we have the available nuclear target area which is the product of the cross-section per target nucleus (σ) and the number of target nuclei lying with in the area A.

$\therefore$ Probability for one incoming particle to hit a nuclear target = Nuclear target area $\sigma nA\,\Delta x$/Total target area A. If the nuclear reaction produces N light product particles per unit time, then the reaction probability is defined as the ratio N/I. Hence we have

$$\frac{N}{I} = \frac{\sigma n A\,\Delta x}{A} \text{ or } \sigma = \frac{N}{\left(\frac{I}{A}\right)(nA\Delta x)}. \qquad ...(6)$$

Hence the nuclear cross section can be defined as the number of light product particles per unit time per unit incident flux and per target nucleus. If a bombarding particle is producing a variety of light reaction products N_1, N_2,... per unit time, then the total cross-section is defined as

$$\sigma_{total} = \frac{N_1 + N_2 + ...}{\left(\frac{I}{A}\right)(nA\,\Delta x)\cdots} \qquad ...(7)$$

If the incident flux (beam intensity) is I_0 and the flux after penetrating a distance x is I, it will become I – dI after a further penetration of distance dx. This reduction in flux is caused by the various collision events occurred with in this distance dx.

$$\therefore \qquad -dI = N = n\sigma\, dxI \text{ or } -\frac{dI}{I} = \sigma n\, dx. \qquad ...(8)$$

This expression contains the product of σ, the cross-section per nucleus or the *microscopic cross-section* and n, the number of nuclei per unit volume. The product is called the *macroscopic cross-section* and is denoted by Σ. Thus we have

$$-\frac{dI}{I} = \Sigma dx \text{ or } I = I_0 e^{-\Sigma x} \qquad ...(9)$$

The experimental arrangement consists of a source and a detector between which can be placed a slab of the experimental material. If I, is the particle intensity reaching the detector in the absence of the absorbing material and I is the value when the slab of thickness x, containing n atoms per unit volume is present, the total cross-section o can be calculated.

In many nuclear reactions the light product particles are not produced in an isotropic manner. We thus define differential cross-section $d\sigma/d\Omega$ in terms of the number of light products dN emitted per unit time in a small solid angle $d\Omega$ at some angle θ with the direction of incident beam. From eqn, we have

$$\frac{1}{I}\frac{dN}{d\Omega}=\frac{nA\,\Delta x\left(\frac{d\sigma}{d\Omega}\right)}{A} \quad \text{or} \quad \frac{d\sigma}{d\Omega}=\frac{\frac{dN}{d\Omega}}{\left(\frac{1}{A}\right)(nA\,\Delta x)}.$$

Macroscopic Cross-section and Mean Free Path

The equation (9) shows that Σ corresponds to linear absorption coefficient and has the dimensions of reciprocal length. The penetration distance λ can be defined as the path length over which the neutrons can travel without suffering a collision with a target. This distance is known as *mean free path.* Since beam intensity is proportional to the beam particle density q, hence we can write eqn. (9) as

$$q = q_0 e^{-\Sigma x}.$$

$$\text{The average path length } \lambda = \frac{\text{Total path length}}{\text{Total no. of particles}} = \frac{\int_0^{\infty} x\,dq}{q_0}$$

$$= \frac{1}{\Sigma}\int_0^{\infty}(-\Sigma x)e^{-\Sigma x}\,d(\Sigma x) = \frac{1}{\Sigma} \quad \text{...(10)}$$

Thus the average path length λ between successive collisions is equal to the reciprocal of the macroscopic cross-section Σ.

CLASSICAL ANALYSIS OF CROSS-SECTION

If the bombarding particle has an impact parameter (the distance of closest approach) b with respect to the nucleus. The angular momentum of the system will be normal to the relative momentum p of the two particles and of magnitude

$$L = pb = \frac{\hbar b}{\lambda}.$$

If $b > R$, the bombarding particle should not have much effect on the target nucleus as it is outside the range of nuclear forces. Thus we expect the most important nuclear interactions to occur with those bombarding particles which have angular momentum less than or equal to the maximum value $\frac{\hbar R}{\lambda}$. Quantum mechanics requires that the angular momentum is quantized and can adopt the following discrete values $\hbar l$ with $l = 0, 1, 2...$ If we now consider the associated wave λ we see that it may be resolved into partial waves, each corresponding to a value of l.

Blatt and Weisskopf have provided the significance of the angular momenta in a plane wave by dividing up the incident beam into cylindrical zones, such that each zone is associated with a different impact parameter. In the central zone there will only be the particles with an impact parameter b less than $\lambda = \left(= \frac{\hbar}{p}\right)$. Actually it is not possible to visualize an impact parameter less than the associated wavelength (*Uncertainly principle*). The angular momentum is zero throughout this zone. The next zone contains all particles with impact parameters between λ and 2λ. Similarly the lth zone contains particles with impact parameters between $l\lambda$ and $(l + 1)\lambda$. The change in angular momentum (*i.e.*, momentum × impact parameter) on passing from zone l to $l + 1$ is $\hbar l$ to $h(l + 1)$.

The area of the lth zone is obtained from the difference between the areas due to radii $(l + 1)\lambda$ and $l\lambda$. This is the cross-sectional area σ_r^l of the lth zone. If it is assumed that each particle hitting the nucleus causes a reaction then the reaction cross-section may be obtained by summing σ_r^l, over all values of l from 0 to l_m

$$\therefore \qquad \sigma_r = \sum_0^{l_m} \pi\lambda^2 (2l+1) = \pi\lambda^2 (l_m + 1)^2. \qquad ...(1)$$

The maximum value of l, corresponds to maximum impact parameter R, is given by $l_m = R/\lambda$. Substituting this value of l_m in the above relation we have maximum reaction cross-section

$$\sigma_r = \pi (R + \lambda)^2. \qquad ...(2)$$

This result suggests the possibility of nuclear reaction cross-sections which are several orders of magnitude larger than the geometrical cross-section of the nucleus.

However, it will be recalled that the incident wave may also be partially reflected and that a transmission factor T_l, depending on the value of l, must be taken into account. The quantum mechanical treatment of the problem gives

$$\sigma_r = \pi\lambda^2 \sum_{l=0}^{\infty} (2l+1) T_l. \qquad ...(3)$$

An interesting case is that of a nucleus which has a radius R very much greater than $\lambda\!\!\!^{-}$, *i.e.,* when the high energy particles strike the nucleus. The nucleus is considered as completely absorbent (black) which corresponds to $T_l = 1$. It is for all values of l upto and including $l_m (= R/\lambda\!\!\!^{-})$. In these conditions the cross-section obtained is same as shown by eqn. (2). This is close to the geometrical cross-section when $\lambda\!\!\!^{-}$ is small is comparison to R. At very high energies, the nucleus exhibits a certain transparency to particles which may penetrate it without interaction. For the weaker energies, $\lambda\!\!\!^{-}$ is too large compared with R for the simplifying hypothesis $T_l = 0$ for large l to hold, since even for l = 0, l > R.

The elastic cross-section depends upon the condition under which the incident wave is reflected. The interference with the incident wave gives rise to amplitudes four times that of the penetrating wave.

$$\sigma_{el}(\max) = 4\pi(R + \lambda\!\!\!^{-})^2. \qquad ...(4)$$

In contrast, when $T_l = 1$ for all the values of l, in the case of the black nucleus previously examined, the minimum value is given as

$$\sigma_{el}\,(\min) = \pi(R + \lambda\!\!\!^{-})^2 = \sigma_{inel}. \qquad ...(5)$$

The effective total cross-section is, therefore, equal to

$$\sigma_l = \sigma_{el} + \sigma_{inel} = 2\pi\,(R + \lambda\!\!\!^{-})^2. \qquad ...(6)$$

Thus we see that for black nucleus the total cross-section is twice the geometrical cross-section. The opaque disc of the nucleus projects a shadow and diffraction occurs as in wave optics. If the nucleus is replaced by a disc of radius (R + $\lambda\!\!\!^{-}$) this shadow may be shown to exist. The disc then emits particle waves of the same intensity in the direction of the incident beam but opposite in phase to those of the beam.

In the derivation of above theory we have not considered the effect of Coulombs repulsion. The Coulomb repulsion will bring the relative K.E. of the charged particle from T (when particle is far apart from nucleus) to T – B, when the particle is just touching the nucleus of radius R. If z and Z are the atomic numbers of incident particle and target nucleus respectively, then $B = zZe^2/4\pi\epsilon_0 R$. Thus the Coulomb repulsion slows down the particle so that the initial moment $(2MT)^{1/2}$ is decreased to $[2M\,(T - B)]^{1/2}$, where M is the reduced mass of the system. Hence the maximum angular momentum.

$$L_m = R[2M(T - B)]^{1/2} = R(2MT)^{1/2} (1 - B/T)^{1/2}$$

$$\therefore \quad b_m = \frac{L_m}{p} = \frac{L_m}{(2MT)^{1/2}} = R\left(1 - \frac{B}{T}\right)^{1/2}.$$

Thus the Coulomb repulsion diminishes the l_m by a factor of $\left(1 - \frac{B}{T}\right)^{1/2}$. The upper limit for the capture of charged particles can be estimated as the area of the disc of radius b_m.

$$\sigma_r = \pi R^2 \left(1 - \frac{B}{T}\right). \qquad ...(7)$$

COMPOUND NUCLEUS

According to the Bohr, the nuclear reaction takes place in two distinct and independent stages :

(i) *formation of a compound nucleus C and*

(ii) *the disintegration of the compound nucleus into the products of the reaction.* The compound nucleus, which is a many body system of strongly interacting particles, is formed by the amalgamation of an incident particle x with a target nucleus X.

$$X + x \rightarrow C^*. \qquad ...(1)$$

The incident particle captured by a nucleus gives up its energy to few nucleons and, as the results of the interaction of these nucleons with all the others, the energy is quickly distributed among all the nucleons of the compound nucleus. The new nucleus thus formed is in excited state. If E_x is the kinetic energy of the incident particle of mass m, the excitation energy of the compound nucleus will be given by

$$E^* = \frac{E_x M}{(m+M) + E_B}, \qquad ...(2)$$

where M is the mass of the target nucleus and E_0 the binding energy of the particle in the compound nucleus. The mode of disintegration of compound nucleus ($C^* \rightarrow Y + y$) is independent of the mode of formation and depends only on its energy, angular momentum and parity. If the different processes lead to the same compound nucleus the decomposition is identical. A compound nucleus, once formed, can decay in a number of different ways, each with its own-intrinsic probability. For example a compound nucleus Al^{27*}, formed by the bombardment of

Na^{23} with α-particles or by other methods, can decay at least in following ways : $Al^{27*} \rightarrow (Na^{23} + \alpha)$, $(Mg^{25} + d)$, $(Mg^{26} + p)$, $(Al^{26} + n)$ or $(Al^{27} + \lambda)$. In each case the residual nucleus can usually be left in one of its excited states, leading to still more modes of decay.

The compound nucleus, has a life time which is long ($10^{-14} - 10^{-18}$ sec) compared to the time for a nucleon to traverse a nucleus ($10^{-20} - 10^{-23}$ sec.). This finite life time is because there can always be a statistical fluctuation in the energy distribution which concentrates enough energy on a nucleon to allow to escape.

ENERGY LEVELS OF NUCLEI

Resonances in the yield curve of a nuclear transmutation will occur whenever the energy of the incident particle together with its B.E. in the compound nucleus is equal to and coincides with an energy level of that nucleus. An energy level that corresponds to an excitation energy greater than the B.E. of the particle in the compound nucleus is a *virtual energy level* and the nucleus is than said to be in a *quasi-stationary state*. The most stable energy state of nucleus is the lowest energy level or *ground state* (*zero potential energy*). Corresponds to this state the energy of the nucleus is the B.E. calculated from the mass defect. For excitation energies less than the B.E., the corresponding energy levels are *bound energy levels*. In the normal state of the nucleus, the lowest bound levels are all occupied. The nucleon will be in one of the normally unoccupied bound states, if the excitation energy is less than about 8 MeV. The region of the virtual levels, contains those energy levels that can be occupied during a nuclear reaction while the nucleus exists in an intermediate state as a compound nucleus.

De-excitation of bound energy levels can take place only by emission of γ-rays, whereas de-excitation from a virtual energy level can occur in a variety of ways, either by particle emission or γ-emission. For example, in certain (α, p) reactions discrete energy values are observed amongst the ejected protons, although monoenergetic α-particles are being used in the bombardment of the target. These experimental results can be understood if one attributes to each energy group a corresponding energy level in the nuclei of their origin. Thus, the largest of the group energies E_0 is associated with the formation of residual nucleus in its lowest or ground state. The smaller energy values (E_1, E_2, E_3) of the groups are associated with residual nuclei which have been left in higher excited energy states. The energy differences between the various groups can be related to the respective energy level spacing in the residual nucleus. This conclusion is supported by the fact that the lower energy groups are accompanied by simultaneous γ-ray emission, whereas the highest energy group is not accompanied by any γ-emission from the residual nucleus

Level Width and De-excitation

The excited state of the compound nucleus has a definite mean life time before it decays by one of the possible modes of de-excitation. We know that

Mean life time τ = 1/disintegration constant λ, ...(1)

where λ is connected with the level width Γ by the relation

$$\Gamma = \hbar\lambda \qquad ...(2)$$

$$\therefore \qquad \Gamma = \frac{\hbar}{\tau} \text{ or } \Gamma\,\tau = \hbar. \qquad ...(3)$$

According to the principle of Heisenberg uncertainty

$$\Delta\, E.\Delta t = \hbar. \qquad ...(4)$$

If we equate the uncertainty in the time measurement Δ t with the mean life time of the excited state, we obtain level width Γ for the uncertainty in the energy of the excited state Δ E. *Hence the level width of the excited state is a spread in energy of the excited level due to the uncertainty in the level.* As the level width is inversely proportional to the mean life time τ, it is clear that a long life time means a very fine and narrow energy level, whereas a short life time means a broad or diffuse energy level of larger level width.

For each individual mode of decay, there is a different probability of decay and, therefore, a different partial width for each decay product. The total width of an energy level is then the sum of the individual partial widths, *i.e.,* $\Gamma = \Gamma_1 + \Gamma_2 + \Gamma_3 + ...$ For bound levels, neglecting internal conversion $\Gamma = \Gamma_\gamma$. But for virtual level $\Gamma = \Gamma_\gamma + \Sigma\Gamma_{particles}$, where the summation includes the widths corresponding to the emission of certain energetically allowed particles. The widths Γ_γ and $\Gamma_{particles}$ are partial widths. For the ground state $\Gamma = 0$.

RECIPROCITY THEOREM

Let us consider a reversible process X + α = Y + y, in which X, x, Y and y occur in arbitrary numbers in a large box of volume V. We are interesting in the relation between the total cross-section $\sigma(x \rightarrow y)$, most generally $\sigma(\alpha \rightarrow \beta)$ of the reaction with entrance channel α and reaction channel β and the total cross-section $\sigma(\beta \rightarrow \alpha)$ of the inverse reaction. For this we use the fundamental theorem of statistical mechanics *(the principle of overall balance),* which states that when the system is in dynamical equilibrium all energetically permissible states are occupied with equal probability. Here we are interested in two particular states, the reaction channels α and β. The theorem is then equivalent to stating that in a given energy range the number of possible channels in the box is proportional to the number of possible channels into the box. The latter is given by

$$N_\alpha = \frac{4\pi p_\alpha^2 \, V \, dp_\alpha}{h^3} = \frac{p_\alpha^2 \, V \, dp_\alpha}{2\pi\hbar^3}.$$

Since $\quad v = \frac{dE}{dp}$, hence $N_\alpha = \frac{p_\alpha^2 \, VdE_\alpha}{2\pi^2\hbar^3 v_\alpha}$. ...(1)

Similarly, we have $N_\beta = \frac{p_\beta^2 \, VdE_\beta}{2\pi^2\hbar^3 v_\beta}$. ...(2)

The energy range for the two channels must of course be the same, *i.e.*, $dE_\alpha = dE_\beta$, hence

$$\frac{\text{No. of channels } \alpha \text{ in the box}}{\text{No. of channels } \beta \text{ in the box}} = \frac{N_\alpha}{N_\beta} = \frac{p_\alpha^2 \, v_\beta}{N_\beta \, v_\alpha}. \quad ...(3)$$

The system is in dynamical equilibrium when the number of the transitions $\alpha \rightarrow \beta$ per second is equal to the number of transitions $\beta \rightarrow \alpha$ per second. The condition usually holds and is known as the *principle of detailed balance.* Further

No. of transitions $\alpha \rightarrow \beta$ per sec $= N_\alpha \times \omega\,(\alpha \rightarrow \beta)$,

where $\omega(\alpha \rightarrow \beta)$ is the transition probability for the reaction $(\alpha \rightarrow \beta)$.

Hence $\quad p\alpha^2 \, v_\beta \, \omega(\alpha \rightarrow \beta) = p\beta^2 \, v_\alpha \, \omega(\beta \rightarrow \alpha)$. ...(4)

The transition probability measures the chance that one particle moving with velocity v in volume V is scattered per sec. Hence the cross-section σ which corresponds to unit incident flux is given by the relation

$$\sigma = \frac{\omega V}{v}. \qquad ...(5)$$

Combining relations (4) and (5) and using $k = \frac{p}{\hbar}$, we have

$$k\alpha^2 \, \sigma(\alpha \to \beta) \, k\beta^2 \, \sigma(\beta \to \alpha) \qquad ...(6)$$

or
$$\frac{\sigma(\alpha \to \beta)}{\lambda\!\!\!^{-}\alpha^2} = \frac{\sigma(\beta \to \alpha)}{\lambda\!\!\!^{-}\beta^2}. \qquad ...(7)$$

We have assumed zero intrinsic angular momenta for the particles so far. If I is the intrinsic angular momentum of any one of the particles, the corresponding density of states then must be multiplied by 2I + 1. This if there are intrinsic momenta for X, x, Y and y, we may write

$$(2I_X + 1)\,(2I_x + 1)\, k_\alpha^2 \sigma(\alpha \to \beta) = (2I_Y + 1)\,(2I_y + 1)\, k_\beta^2 \sigma(\beta \to \alpha). \qquad ...(8)$$

If the initial and final states have definite angular momenta then the above equation must be employed.

NUCLEAR REACTION OF CROSS-SECTION BY BOHR'S

Bohr assumed that the mode of disintegration of a compound system depends only on its energy, angular momentum and parity, but not on the specific way in which it has been produced. Therefore, to calculate the cross-section of a nuclear reaction, it is necessary to determine the cross-section of the two processes, the formation of the compound nucleus and its decay.

$$X + x \to C \to Y + y. \qquad ...(1)$$

Since the break up process of the compound nucleus is independent of the formation process, hence we can write in general

$$\sigma(\alpha, \beta) = \sigma_C(\alpha) \, G_C(\beta), \qquad ...(2)$$

where $\sigma(\alpha, \beta)$ is the cross-section corresponding to a specific entrance channel α and specific exit channel β, $\sigma_C(\alpha)$ is the cross-section for the formation of a compound nucleus C with particle x through channel α and $G_C(\beta)$ is the probability that the compound nucleus C will disintegrate by emitting a particle y through channel β.

For the sake of simplicity we shall neglect the dependence of $G_C(\beta)$ on angular momentum and parity. $G_C(\beta)$ then only depends on the excitation energy (E_C) of compound nucleus. Because of the Heisenberg uncertainty principle this energy cannot be sharply defined. The uncertainty in excitation is measured by a quantity Γ_β, the level width associated with the decay constant λ_β of compound system.

$$\Gamma_\beta \text{ (For } E_C) = \hbar\lambda_\beta = \hbar\tau_\beta \text{ (for } E_C), \qquad ...(3)$$

where τ_β is the decay time of the compound nucleus through the channel β. If we add all the decay constants for the possible exist channels then we get the total decay constant for the compound nucleus. Therefore, the total width of the compound level with an excitation energy E_C is given by

$$\Gamma(\text{for } E_C) = \sum_\beta \Gamma_\beta(\text{for } E_C). \qquad ...(4)$$

The *branching probability,* the relative probability for the compound nucleus at excitation energy E_C to break up by emitting particle y, in terms of decay rates is given by

$$G_C(\beta) = T_\beta \text{ (for } E_C)/T(\text{For } E_C). \qquad ...(5)$$

The assumption of the independence of two processes (the creation and the disintegration of the compound system) gives a relation, which depends only in the $k\alpha$ and $k\beta$. It is derived earlier and can be rewritten as

$$k\alpha^2\, \sigma(\alpha, \beta) = k\beta^2\, \sigma(\beta, \alpha). \qquad ...(6)$$

Using eqns (2) and (5), this relation becomes

$$k\alpha^2 \frac{\sigma_C(\alpha)\,\Gamma_\beta(\text{for } E_C)}{\Gamma(\text{for } E_C)} = k\beta^2 \frac{\sigma_C(\beta)\,\Gamma_\alpha(\text{for } E_C)}{\Gamma(\text{for } E_C)}$$

or

$$\frac{k\alpha^2 \sigma_C(\alpha)}{\Gamma_\alpha(\text{for } E_C)} = \frac{k\beta^2 \sigma_C(\beta)}{\Gamma_\beta(\text{for } E_C)} = U(E_C). \qquad ...(7)$$

This proportional coefficient $U(E_C)$ is a function depending upon the excitation energy E_C of the compound nucleus only, but not on the particular channel used. This relation shows that there is a simple relationship between the probability for breakup and the cross-section for forming the compound nucleus through the same channel. The probability of decay through a channel β

$$G_C(\beta) = \frac{\Gamma_\beta}{\Gamma} = \frac{k\beta^2 \sigma_C(\beta)}{[\Gamma_\alpha + \Gamma_\beta + ...]U(E_C)} = \frac{k\beta^2 \sigma_C(\beta)}{\sum_\gamma k_\gamma^2 \sigma_C(\gamma)}, \qquad ...(8)$$

where the sum is extended over all channels γ into which compound system can decay. It is, therefore, possible to derive the decay probability through a given channel of the compound nucleus if the cross-section $\sigma_C(\gamma)$ for the formation of the compound nucleus by all possible channels are known, and if the Bohr assumption is valid. The validity of Bohr's hypothesis of the compound nucleus has been elegantly demonstrated by several experiments. Let us discuss the experiment of Indian Scientist S.N. *Ghoshal* (1950) in which he produced the same compound nucleus Zn^{64*} with the bombardment of $_{28}Ni^{60}$ by α-particle and $_{29}Cu^{52}$ by protons. Additional 7 MeV was added to the K.E. of protons to produce the same excitation in Zn^*. The reactions observed were

Ni^{60} (α, n) Zn^{63}	Cu^{63} (p, n) Zn^{63}
Ni^{60} (α, 2n) Zn^{62}	Cu^{63} (p, 2n) Zn^{62}
Ni^{60} (α, pn) Cu^{62}	Cu^{63} (p, pn) Cu^{62}.

Since the excitation produced through these two processes are the same, the disintegration probability $G_C(\beta)$ is the same, because it depends upon the excitation produced in the compound nucleus and not upon the mode of formation. If the compound nucleus theory is true, then one should expect.

$$\sigma(p, n) : \sigma(p, 2n) : \sigma(p, pn) = \sigma(\alpha, n) : \sigma(\alpha, 2n) : \sigma(\alpha, pn). \quad ...(9)$$

Ghoshal, here six cross-sections are plotted as functions of K.E. of the β-particle and protons. The proton-energy scale has been shifted by 7 MeV with respect to the α-particle energy scale in order to make the peaks of the proton curves correspond with those of the α-particle curve. It can be seen from figure that the results are
accurate with in 10%, and thus providing evidence for the validity of the compound nucleus theory. The small difference between experimental and theoretical values can be explained as under : Ghoshal's experiment was performed at high bombarding energies. The levels excited in the compound nucleus are more closely spaced and broader and may overlap partially. This can be explained by the *continuum theory.* Thus we see that the compound nucleus theory given here is applicable to low bombarding energy.

CONTINUUM THEORY OF NUCLEAR REACTION

At higher bombarding energies the individual levels of the compound nucleus become broader and also more closely spaced. The total width

Γ becomes much greater than D, the spacing of the levels. The width and number of levels are such that even though the interval is small, many levels may be found in it. Sharp resonances are no longer observable when $\Gamma \geq D$. The spacing between levels is completely occupied, so the *space is described as continuum.* The cross-section for the formation of a compound nucleus is larger for neutrons than for charged particles because Coulomb repulsion between the incident charged particle and the target nucleus is important in the latter case. In the medium and heavy nuclei, the individual level becomes broader and levels become more closer when the energy of the incident particle is large. The continuum theory of nuclear reaction cross-sections treats the individual level not separately but as an average over many resonances.

We shall derive the value of cross-section on the basis of classical mechanics. If the incident particle is a neutron (with J = 0), the cross-section for reaching the nuclear surface is given by the classical target area such as,

$$\sigma_C(n) = \pi R^2. \quad ...(1)$$

We now proceed to a qualitative wave mechanical discussion. This discussion is based upon the fact that there is a sudden change of potential when the particle crosses the boundary of the nucleus, due to the strongly attractive nuclear forces. The sudden change of potential reflects the incoming wave at the nuclear surface and thus decreases the cross-section, especially at low energy.

Let us consider a beam of particles of energy E moving in the direction of x in a potential V(x), which is zero for x < 0 and is $-V_0$ for x > 0. The particles are partially reflected at x = 0. We have to determine now the transmission coefficient T, defined as the ratio of the number of particles transmitted through the region x > 0 to the number of incident particles. Thus

$$\text{For } x < 0, \ \psi = Ae^{ikx} + Be^{ikx}, \quad ...(2)$$

$$\text{where} \quad k = \sqrt{\frac{(2ME)}{\hbar}}. \quad ...(3)$$

$$\text{For } x > 0, \ \psi = Ce^{+iKx} \quad ...(4)$$

$$\text{where} \quad K = \sqrt{\frac{[2M(E+V_0)]}{\hbar}}. \quad ...(5)$$

Here A, B and C are coefficients giving the amplitude of the various wave functions. Since neither the wave function nor the kinetic energy can become infinite at any point, hence both, the wave function and its derivative must be continuous at all points. Using boundary conditions at x = 0, we have

$$A + B = C \text{ and } ikA - ikB = iKC.$$

$$\therefore\ A = \frac{1}{2}\left(\frac{1+K}{k}\right) \text{ and } B = \frac{1}{2}C\left(\frac{1-K}{k}\right) \qquad ...(6)$$

$$\text{Hence } T = \frac{|A|^2|B|^2}{|A|^2} = \frac{4Kk}{(K+k)^2} \qquad ...(7)$$

Hence limiting values of T are of special interest. At high incident energies relation (5) indicates that K = k and thus T approaches 1. On the other hand, as the neutron energy approaches zero,

$$k < K \text{ and } T = \simeq \frac{4k}{K} \simeq 4\left(\frac{E}{V_0}\right)^{1/2}.$$

The position of the particle is uncertain by the amount of its wavelength $\lambda\!\!\!^{-}$. This can be accounted approximately for by replacing πR^2 with $\pi(R + \lambda\!\!\!^{-})^2$. Thus cross-section becomes

$$\sigma_c = \pi(R + \lambda\!\!\!^{-})^2. \qquad ...(8)$$

This equation has to be modified by including the transparency T of the potential barrier. Thus we obtain the cross-section for the formulation of the compound nucleus by S-wave neutrons.

$$\sigma_c = \pi(R + \lambda\!\!\!^{-})^2 4kK/(k + K)^2. \qquad ...(9)$$

For very small neutron energies $k \leq K$ and $\lambda\!\!\!^{-} \geq R$, the above equation becomes

$$\sigma_c = \frac{4\pi}{kK}.\left(\text{Because } k = \frac{1}{\lambda\!\!\!^{-}}\right) \qquad ...(10)$$

At high energies the cross-section approaches the classical value πR^2 and increases with decreasing energy, but less strongly than $\pi(R + \lambda\!\!\!^{-})^2$, because of the reflection at the nuclear boundary.

Similar approach can also be used to obtain an expression for the reaction cross-section of charged particles. In this case the incident beam is deviated by the Coulomb potential $V(r) = Z_1Z_2e^2/2\pi\epsilon_0 r$. The incident particles reaching the nuclear surface thus have a maximum impact parameter $R[1 - (B/E)]^{1/2}$, where B = Coulomb barrier height = V(R). Hence the reaction cross-section can be expressed as

$$\sigma_c = \pi R^2[1 - B/E] \qquad E > B$$
$$= 0 \quad E \leq B. \qquad ...(11)$$

Here we see that the high energy limit of this eqn. is πR^2, identical to that for neutrons. An attempt to obtain a better fit to the continuum theory value of σ_c has been made by Destrovsky *et al.*, in 1959.

BREIT WIGNER DISPERSION FORMULA

The concept of cross-section and level width can be applied to resonances in a quantitative way. In the particularly important case of resonance processes, a theoretical formula for the cross-section was derived by G. Breit and E.P., Wigner in the United States in 1936. In its simplest form, it gives the value of the cross-section in the neighbourhood of a single resonance level formed by an incident particle with zero angular momentum and charge zero so that spin and the Coulomb effects can be ignored. The result is analogous to the theory of optical dispersion, so that the main formula obtained is often called the *dispersion formula.*

Whether or not the level of compound nucleus is bound, excitation by an incident particle can be treated as analogous to the excitation of the oscillations produced in an electric circuit by an electromagnetic wave. We therefore, expect the nuclear cross-section to vary with the incident energy in the same way that the energy in a forced oscillation varies with incident frequency. The classical resonant circuit absorbs energy because of resistive levels. In the nuclear case, damping arises because of the possibility of decay. Because of this possibility the nuclear state has a finite width Γ. The wave function of a decaying state of mean energy E_0 may be written as

$$\psi(r,t) = \psi(r)e^{iE_0 t/\hbar}e^{-\Gamma t/2\hbar}. \qquad ...(1)$$

This corresponds to an exponential decrease of intensity of excitation $|\psi(r, t)|^2$ with a time constant $\tau(=\hbar/\Gamma)$. This wave function also shows that a decaying state is not a state of definite energy E of the form $\Psi(r)e^{-i(E/\hbar)t}$. Never the less, it can be represented by superposition of states of slightly different energies E, each with a different amplitude A(E)

$$\Psi(r, t) = \int_{-\infty}^{+\infty} A(E) e^{-iEt/\hbar} dE. \qquad ...(2)$$

Using the Fourier analysis technique, we can show that the energies E are grouped about a mean energy E_0 with a spread of the order of $\Gamma = \hbar\lambda$. Equating eqns. (1) and (2), we get

$$e^{-\Gamma t/2\hbar} = e^{-\lambda t/2} = \int_{-\infty}^{+\infty} A(E) e^{-iEt/\hbar} dE. \qquad ...(3)$$

According to the Fourier theorem any well behaved function f(t) can be represented as

$$f(t) = \frac{1}{2\pi} \underset{s \to \delta}{\text{Lim}} \int_{-\Omega}^{\Omega} - i\omega t \, d\omega \int_{-\infty}^{\infty} e^{j\omega t'} f(t') dt'. \qquad ...(4)$$

Applying this to the function $e^{-\lambda_t/2}$, we get

$$4(E) = \frac{1}{2\pi\hbar} \int_0^{\infty} e^{\left[i(E-E_0)/\hbar - \lambda/2\right]t'} dt'$$

$$= \frac{1}{2\pi} \frac{1}{(E - E_0) + i\hbar\lambda/2}. \qquad ...(5)$$

Here we have assumed that the decaying system was prepared at the time t = 0. The probability of finding the system with a given energy E is proportional to

$$|A(E)|^2 = \frac{1}{4\pi^2} \frac{1}{(E - E_0)^2 + (\hbar\lambda/2)^2}$$

$$\frac{1}{4} = \frac{1}{4\pi^2} \frac{1}{(E - E_0)^2 + \frac{\Gamma^2}{4}}. \qquad ...(6)$$

This gives the level shape. It is exactly as for pure radiative decay except that particle emission is now included by using the total width Γ instead of the radiative width Γ_r. The cross-section for excitation of the level by collision of particle x width nucleus X is, therefore, expected to have the form

$$\sigma_x = \frac{C}{\left[(E - E_0)^2 + \frac{\Gamma^2}{4}\right]}, \qquad ...(7)$$

where C is constant. For the determination of C, let us suppose that the compound nucleus formation and decay processes take place in a box of volume V which contains one nucleus X and one particle x. If the states are quantised, the no. of states of motion of particles with momentum between p and p + dp = $(4\pi p^2 dpV)/h^3$. The probability of formation of the compound level per unit time = No. of possible states of motion × probability that the nucleus X is contained within the small volume σ_{xv} out by the

$$dP = \frac{4\pi p^2 dp}{h^3} V \cdot \frac{\sigma_{xv}}{V} = \frac{4\pi}{h^3} = v\sigma_x p^2 dp.$$

Substituting the value of σ_x from eqn in the above eqn. and integrating over the energy spectrum we get

$$P = \int_{-\infty}^{+\infty} \frac{4\pi}{h^3} v \cdot p^2 \cdot \frac{C}{(E-E_0)^2 + \frac{\Gamma^2}{4}} dt$$

$$= \int_{-\infty}^{+\infty} \frac{4\pi}{h^3} \cdot \frac{1}{(E-E_0)^2 + \frac{\Gamma^2}{4}} dE \,.$$

If we assume that the variation of the channel wavelength λ of the particle over the level width Γ may be neglected, then the probability of formation per sec

$$= \frac{4\pi C}{h\lambda^2} \cdot \frac{2}{\Gamma} \left[\tan^{-1} \frac{2(E-E_0)}{\Gamma} \right]_{-\infty}^{+\infty}$$

$$= \frac{4\pi C}{h\lambda^2} \cdot \frac{2}{\Gamma} \cdot \pi = \frac{C}{\hbar \pi ƛ^2 \Gamma} \,. \qquad ...(8)$$

Probability of decay per sec $= \dfrac{\Gamma_x}{\hbar}$,

where Γ_x is the partial width of the compound level for the emission of x. In the equilibrium state the probability of formation per sec would equal to the probability of decay per sec of the excited state back into the system X + x. Hence

$$\frac{C}{\pi \hbar ƛ^2 \Gamma} = \frac{\Gamma_x}{\hbar} \text{ or } C = \pi ƛ \Gamma \Gamma_x, \qquad ...(9)$$

Substituting this value in eqn. (7), the cross-section for the formation of the level becomes

$$\sigma_x(E) = \frac{\pi\lambda^2 \Gamma\Gamma_x}{(E - E_0)^2 + \frac{\Gamma^2}{4}} \qquad ...(10)$$

For the process X(x, y) Y, we obtain reaction cross-section as

$$\sigma_r = \sigma_{xy}(E) = \frac{\sigma_x \Gamma_y}{\Gamma} = \pi\lambda^2 \frac{\Gamma_x \Gamma_y}{(E - E_0)^2 + \frac{\Gamma^2}{4}}. \qquad ...(11)$$

Here Γ_x and Γ_y are partial level width defined as

$$\Gamma_{x\tau x} = \hbar \text{ and } \Gamma_{v\tau v} = \hbar$$

where τ_x and τ_v the main life times that the compound nucleus would have if elastic scattering of x or the emission of y were the only possible modes of decay.

If spin is considered, the right band side of eqn. (11) must be multiplied by the factor

$$gc = \frac{(2Ic + 1)}{(2I_x + 1)(2Ix + 1)}, \qquad ...(12)$$

where I_x is the total angular momentum of the incident particle, Ix is that of target nucleus and Ic is of the compound state, which is formed only by those orbital angular momentum l_x which satisfy the conditions

$$Ic + Ix + I_x + I_x \text{ and } \Pi x \Pi_x (-1)^{lx} = \Pi c. \qquad ...(13)$$

Hence for the nuclear reaction in which particles have definite spins, the relation (11) can be written as

$$\sigma_r = \pi\lambda^2 \frac{(2Ic + 1)}{(2I_x + 1)(2Ix + 1)} \frac{\Gamma_x \Gamma_y}{(E - E_0)^2 + \frac{\Gamma_v^2}{4}} \qquad ...(14)$$

This is known as the *Breit Winger resonance formula.* For the (n, γ) reaction in particular

$$\sigma(n, \gamma) = \pi\lambda^2 = \frac{(2Ic + 1)}{2(2Ix + 1)} \frac{\Gamma_x \Gamma_y}{(E - E_0)^2 + \frac{\Gamma^2}{4}} \qquad ...(15)$$

This is maximum when $E = E_0$ and is equal to

$$\sigma_{max}(n, \gamma) = 4\pi\lambda^2 gc\Gamma_n \Gamma_\gamma / \Gamma^2. \qquad ...(16)$$

For $E = E_0 \pm \Gamma/2$, $\sigma = 1/2\ \sigma_{max}$ and hence Γ is the full width at half maximum. A sharp resonance corresponds to a narrow width Γ and thus to an excited state of long life. A largest possible capture cross-section will occur when $\Gamma_n = \Gamma_\gamma = \Gamma/2$. Its maximum possible value $\sigma_{max} = \pi g c \lambda'^2$. The width of the resonance peak affects the cross-section. In a general way, if the peak is broad, covering a large energy range, the cross-sections are likely to be somewhat decreased as compared with the case of sharp and narrow peak. The width of the peak is inversely related with the life of the excited state of the compound nucleus.

All resonance cross-sections except for the elastic re-emission of the incident particle $\sigma(x, x)$, *i.e., elastic resonance scattering.* Besides compound nucleus scattering (*i.e.,* absorption and re-emission of neutron of the same energy), the incident particle wave is scattered by the nucleus as if it were an impenetrable sphere. This type of scattering is known as *potential* or *shape-elastic scattering.* The elastic-scattering cross-section σ_{el} is given by

$$\sigma_{el} = 4\pi\lambda^2 \left[gc \left| \frac{\Gamma_{n/2}}{\left(E - E_0 + \frac{i\Gamma}{2}\right)} + e^{i\phi_l} \sin\phi_l \right|^2 + (1 - gc)\sin^2 f_l \right], \qquad ...(17)$$

where ϕ_l is an energy dependence quantity known as the hard sphere phase shift. For $l = 0$ or S-wave neutrons $\phi = R/\lambda = Rk$. Hence for S-wave neutrons on a spinless target

$$\sigma_{el} = 4\pi\lambda^2 \left| \frac{\Gamma_{n/2}}{(E - E_0) + i(\Gamma/2)} + e^{kR} \sin kR \right|^2 \qquad ...(18)$$

Here the coefficient $4\pi\lambda^2$ is the maximum possible S-wave scattering cross-section. The first term between bars is called the resonance scattering amplitude, which if present alone would lead to eqn. (11). The second term is called the *potential scattering amplitude,* which if present alone would lead to

$$\sigma_{el} = 4\pi\lambda^2 \sin^2 kR. \qquad ...(19)$$

This term varies smoothly with energy. For small energies $kR \leq 1$. As $E \to 0$, we get

$$\sigma_{el} \simeq 4\pi R^2, \qquad ...(20)$$

which is the potential scattering cross-section for an impenetrable sphere when $R \leq \lambda\!\!\!^{-}$.

The resonance term rises to large values near $E = E_0$, but is small elsewhere. For $E < E_0$ the two terms interfere destructively yielding a low value of σ_{el}. For $E > E_0$ the interference is constructive. For charged particles, resonance scattering involves coherence between Coulomb potential scattering, nuclear potential scattering and resonance scattering.

Let us consider the lowest energy resonance of the compound nucleus. For $E \leq E_0$, the denominator of eqn. (20) does not change much with E and Γ_γ is independent of E but Γ_n does depend on E. Hence $\sigma(n, \gamma) \propto \lambda\!\!\!^{-2}\, \Gamma_n$. The probability of elastic emission of a neutron of energy E (momentum p) is proportional to the density of states in momentum space around p. Hence

$$\Gamma_n \propto \frac{p^2 dp}{dE} \alpha p \left\{ \begin{array}{c} E = \dfrac{p^2}{2m} \\ dE = \left(\dfrac{p}{m}\right) dp \end{array} \right\}$$

Thus we have $\sigma(n, \gamma) \propto \lambda\!\!\!^{-2} p$

$$\propto \frac{1}{p} \qquad \left(\because \lambda\!\!\!^{-} = \frac{\hbar}{p}\right)$$

$$\text{or} \qquad \propto \frac{1}{v}, \qquad ...(21)$$

where v is the velocity of neutron. *It is known as $1/v$ – law.* Thus Breit-Wigner formula leads to the conclusion that at low neutron energies the cross-section should be inversely proportional to the neutron speed. Following $1/v$ *region* there occurs the *resonance region* in which the cross-sections rise sharply to high values. The cross-sections are low with neutrons of very high energy.

CONTINUUM THEORY WITH OPTICAL MODEL

The continuum theory does not stand upto experimental tests satisfactorily. Barschall plotted the measured total neutron cross-sections in a three dimensional graph against the neutron energy E and the mass number A, and found that the cross-sections did not decrease smoothly with increasing E, as predicted by the theory, and that the trend of those maxima and minima with energy was a smooth function of A. He showed that the disagreement with the theory was not due to unexpected resonances in individual nuclides, but to a general flow in our theory. The widely spaced shallow maxima and minima were expected from scattering by a potential well, as in the shell model. Thus we see that *our compound nucleus theory must be modified in the light of shell model idea.*

A mathematical mode, known as *optical model,* pictures the interaction among the nucleus as being intermediate to that predicted by the continuum and shell models of the nucleus. It is the model of a complex nuclear potential, the real part produces a potential scattering like the scattering by a hard sphere and imaginary part corresponds to the cross-section for compound nucleus formation. The nucleus can thus be viewed as a *cloudy crystal ball,* which is able to transmit, refract and absorb the incident particle waves, analogous to the transmission, reflection, refraction and absorption of a light beam. In optics we require an index of refraction which is a complex quantity to describe optical phenomena. Due to this analogy with optics, this nuclear model is named as *optical model.*

Let us first discuss the optical model of Fernbach, Serber and Taylor, which involves a nuclear potential of complex form. For the sake of simplicity, we consider the case that the real and imaginary parts of the nuclear potential have the same radial dependence, such as

$$V = V(r)(1 + i\xi) \qquad \text{for } r \leq R$$

$$V = 0 \qquad \text{for } r \geq R. \qquad ...(1)$$

The Schrodinger equation of the neutron with this potential is

$$\nabla^2 \psi + \frac{2M}{\hbar^2}[E - V(r)(1 + i\xi)]\,\psi = 0. \qquad ...(2)$$

For the special case of $l = 0$ neutrons and a square well of depth V_0, for which

$$V(r) = -V_0 \ (r < R),\ V(r) = 0 \ (r > R),$$

the radial equation is given by

$$\frac{du^2}{dr^2} + K_c^2 u = 0, \quad r < R \qquad ...(3)$$

$$\text{and} \quad \frac{d^2u}{dr^2} + k^2 u = 0 \quad r > R, \qquad ...(4)$$

where

$$Kc = \frac{\sqrt{[2M(E+V_0+i\xi V_0)]}}{\hbar} = \left(\frac{2M}{\hbar^2}\right)^{1/2}(E+V_0)^{1/2}\left(1+\frac{i\xi V_0}{E+V_0}\right)^{1/2}$$

$$= \sqrt{\frac{[2M(E+V_0)]}{\hbar + i\xi V_0\left[\frac{M}{2\hbar^2(E+V_0)}\right]^{1/2}}} = K + ik'$$

Hence the solution of eqn. (110) is of the form

$$\psi(r) = \left(\frac{A}{r}\right)\sin Kcr$$

or $$|r\psi|^2 = A^2[\sin^2 Kr + \sinh^2 k'r), \text{ for } r \le R. \quad ...(5)$$

The wave is absorbed as it penetrates into the potential step. The characteristic length known as penetration depth for $E \le V_0$ is given by

$$L = \frac{1}{k'} = \frac{2\hbar(E+V_0)^{1/2}}{\xi V_0\sqrt{(2M)}} = \frac{2\hbar}{\xi(2MV_0)^{1/2}} = \frac{2}{\xi K} \quad ...(6)$$

Substitution of experimental values ($x \approx 0.06$, $V_0 \approx 50$ MeV), we get L of the order of nuclear diameter. This model has explained discontinuities in the variation of the elastic cross-sections for slow neutrons as a function of the mass nu⌐ber of nuclei. The maxima has been observed for radii $R = \left(n+\frac{1}{2}\right)\lambda_{inside}$. The study of elastic scattering has provided valuable information on nuclei radii. The radii so calculated were not corresponding to the relation $R = R_0A^{1/3}$. The above potential form gives too large a cross-section for the elastic scattering process in backward directions.

A much better fit to the experimental cross-sections is the potential well proposed by Woods and Saxon. In it V(r) varies smoothly with r, the distance from the centre of the nucleus, and is of the form

$$V(r) = -(V_0 + iW_0)\,[1 + e^{(r-R)/a}] \quad ...(7)$$

This model gives little idea about the nuclear surface. A better fit to the experimental results is obtained if a surface absorption potential of Gaussian shape, centred around r = R is included. In recent years other forms of optical potentials have been introduced to take into account the deformation of the nuclei, the spin-orbit coupling and also the energy region at which the experiments are being performed. Additional terms are required to take into account these effects. *The real part the potential function accounts for refraction and imaginary part for absorption of the incident nucleons. The spin-orbit term describes changes in spin orientations through the coupling of spin and angular momenta.* This orientation is known as polarization. In addition to the theory of the neutron-nuclear interaction, the optical model has been successively applied to the scattering of protons, deuterons, alpha particles and heavier ions. We *shall give few experimental results.* R.D. Albert in 1959 has measured (p, n) cross-sections in some medium weight nuclei between the energy range 4 to 5.5 MeV, and used the Bjorklund and Fernback optical potential which has the form

$$U(r) = V_\rho(r) + iWq(r) + \frac{\lambda\hbar^2}{4m_0^2c^2}\frac{V}{r}\frac{d\rho}{dr}(\vec{\sigma}.\vec{I}) \qquad ...(8)$$

where $\rho(r) = [1 + e^{(r-R)/a}]^{-1}$, $q(r) = e^{-[(r-R)/b]^2}$, $R = r_0A^{1/3}$ and m_0 = mass of the electron.

The optical potential function represented by equation (8) has also been used by Shore, Wall and Irvine to analyze angular distribution of 7.5 MeV protons scattered elastically by vanadium. They used two sets of parameters corresponding to surface absorption and to volume absorption. Recently, Rosen, Beery, Goldhaber and Auerbach have proposed a spin dependent optical potential by analyzing and measuring 80 separate angular distribution data.

Halbert, Bassel Satchler carried out an analysis of the d-d scattering data at 11 MeV and at 11.8 MeV. They used nuclear potentials of the form

$$V(r) = V\left[1+e^{\left(r-r_0A^{1/3}\right)/a}\right]^{-1} + iW\left[1+e^{\left(r-r_wA^{1/3}\right)/a_w}\right]^{-1} \qquad ...(9)$$

for volume absorption and

$$V(r) = V\left[1+e^{\left(r-r_0A^{1/3}\right)/a}\right]^{-1} - ia_wW_0\frac{d}{dr}\left[1+e^{\left(r-r_wA^{1/3}\right)/a_w}\right]^{-1}$$

for surface absorption.

In each case uniform charge distribution of radius 1.3 $A^{1/3}$ F was assumed for the Coulomb potential. They obtained an excellent fit for the target nuclei Ni, Zr, Ag and Sn and a reasonably good fit for Ca and Ti. The case of the deuteron is complicated because, due to polarization or stripping of the deuteron by Coulomb interaction, non-elastic channels may couple to the elastic channel and may change the optical parameters.

The optical potential may contain spin-orbit terms when the incident particle has a spin other than zero. The strength of the spin-orbit potential is adjusted to fit the measured polarization of the scattered particles. It has been found from experiments that the idea of the optical potential is not limited to the calculation of cross-sections of neutron reactions and incident energies of a few million electron volts but is applicable to the description of reactions of protons and heavier particles in the region of high energies as well.

The basic idea of the shell model is present in the optical model analysis of nuclear reactions. If the imaginary parts of the potential were absent, we should have an ordinary Schrodinger equation and corresponding stationary orbits. The shell model shows that such an approximation has considerable merit.

Statistical Theory of Nuclear Reactions

The theoretical treatment of the interactions between an incident particle and the nucleons in the nucleus requires the use of a nuclear model. In light nuclei the energy levels of the compound nucleus are generally well separated and the theory of the resonance reaction. In the heavy nuclei, the levels are closely spaced. If the excitation energy is sufficiently high, overlapping levels will be excited and the no. of excited levels is too great to be treated individually, and thus requires Fermi statistics. The compound nucleus is pictured then as a hot liquid or solid from which various particles evaporate. From heavy nuclei, neutrons evaporate most easily even at high energies of excitation but proton emission is also common from lighter nuclei.

In statistical thermodynamics, the entropy S(E) of the system is related to ρ(E), the density of states per unit energy interval, according to the Boltzmann formula as

$$S(E) = \log \rho\,(E). \qquad ...(10)$$

More formal derivation of eqn. (10) is also obtained by defining the thermodynamic free energy F(β) of the nucleus at temperature $T = 1/\beta$ as

$$e^{-\beta F}\int \rho(E')e^{-\beta E'}dE'. \qquad ...(11)$$

We can invert the Laplace transform to obtain

$$\rho(E') = \frac{1}{2\pi i}\int e^{\beta(E'-F)}d\beta. \qquad ...(12)$$

This integral can be solved by using *steepest descent method (saddle point method)*. The integrated has a saddle point at $\beta = \beta(E)$ given by $d(\beta F)/d\beta = E$, where E is assumed as independent variable. This gives that

$$\frac{d}{dE}\left[\beta(E-F)\right] = \frac{d}{dE} = \beta = \frac{1}{T}, \qquad ...(13)$$

where $\quad S(E) = \beta(E-F) = \int_0^E \beta(E')dE$

Thus integration of eqn. (119) gives

$$\rho(E) = \frac{e^S}{\sqrt{\left(-\frac{2\pi dE}{d\beta}\right)}}.$$

$\therefore$

$$\frac{d}{dE}\log\rho(E) = \frac{dS}{dE} - \frac{1}{2}\frac{d}{dE}\log\left(-\frac{dE}{d\beta}\right) = \frac{1}{T} - \frac{1}{2}\frac{d}{dE}\log\left(T^2\frac{dE}{dT}\right)$$

As the last term of this eqn. is negligible, hence we get

$$\frac{d}{dE}\log\rho(E) \simeq \frac{d\,S(E)}{dE} = \frac{1}{T}. \qquad ...(14)$$

This is nothing but eqn. Let us now find the dependence of nuclear entropy S and the energy E of the nucleus on the nuclear temperature T. If the nucleons in nucleus are assumed to be as an ideal Fermi gas, we know that the thermal heat capacity $c = dE/dT$ is proportional to T and, therefore, $E \propto T^2$. Hence

$$E = aT^2, \qquad ...(15)$$

where a is a constant. From eqns (15) and (16), we get

$$\frac{d}{dE}\log\rho(E) = \left(\frac{a}{E}\right)^{1/2} \quad \text{or } \rho(E) \sim e^{2(aE)^{1/2}} \qquad ...(16)$$

As approximate result obtained by Ericson is as given below.

$$\rho(E) = \frac{1}{16}\left(\frac{\pi^2}{a}\right)^{1/4} E^{-5/4} e^{2(aE)1/2} \quad ...(17)$$

DIFFERENT STAGES OF A NUCLEAR REACTION

Detailed theories of nuclear reaction were patterned after the two nuclear models. The rigid drop model and the shell model. In the *compound nucleus* theory it was assumed that a nuclear projectile incident on a nucleus would interact strongly with all the nucleons in the nucleus and quickly share its energy with them.

In the *reaction theory* (optical model theory), it was proposed that an incident nucleon would interact with the nucleus via the shell model potential. The success of the optical model suggests that the Bohr theory of compound nucleus is in need of modification. A more general scheme of nuclear reaction this been described by Weisskopf. According to him the nuclear reaction proceeds through three stages : the *independent particle (optical model) stage, compound nucleus stage and the final stage.*

During the first stage, the interaction between the incident wave and nuclear potential may lead to partial reflection of the incident wave, called *shape elastic scattering.* The part of the wave function which enters the nucleus undergoes absorption. This process leads to the second stage. Some of the possible absorption processes are :

1. Ejection nucleons in a collisions with incident particle a direct interaction,
2. Multiple collisions of the incoming particle with several nucleons of the target nucleons,
3. The excitation of some type of collective motion such as surface vibrations of the target nucleus,
4. The formation of compound nucleus, without remembering details of the initial stage of formation.

The third stage of the reaction is the more or less rapid transition to the final state. It is concerned with the way the reaction products are produced, *i.e.,* on the second stage. Either disintegration or de-excitation of the compound nucleus will occur. Actually there is no sharp division between the different possibilities in a compound system. Two extreme cases, direct interaction and the compound nucleus, correspond to two general aspects individual and collective particle movements respectively.

Direct Reactions

A direct reaction is one which proceeds without the formation of a compound nucleus. The time during which the incident particle interacts with the target nucleus is very much shorter than the life of a correspond compound nucleus. Formation of compound nucleus is more likely at low energies whereas the direct reaction mechanism will prevail at higher energies. The term direct reaction is used for a variety of nuclear processes including *inelastic nuclear collisions, stripping* and its inverse, the *pick up reaction.* We shall discuss briefly the stripping reaction and its inverse reaction.

Stripping Reactions

The term stripping is used for a type of direct reaction in which the incoming compound particle splits into two fragments, one of which absorbed by the target nucleus and the other continues more or less undisturbed. In this type of reaction target nucleus captures one or two, sometimes three nucleons from the incident particle without the formation of a compound nucleus as an intermediate stage. The remaining portion of the projectile is usually proton, neutron or deuteron. Stripping reactions of several types, *e.g.,* (d, p), (d, n), (t, p), (t, d) and (α, p) are known to occur at high particle energies with many different nuclei. In pick-up reaction incident particle removes one or two nucleons from the target nucleus. Reactions of this type are (p, α), (p, t), (p, He^3). (d, t) and (d, He^3). The theories of stripping and pick-up reactions are very similar and the angular distributions are also quite similar.

Stripping reaction was first recognized by Oppenheimer and Phillips in 1935 in analyzing the low energy (d, p) reactions. It was observed experimentally that (d, p) reactions were more frequent than (d, n) reactions. They explained the reaction by stating that the deuteron behaves as a resistively loose combination of a neutron and a proton. When it approaches a nucleus, the electrostatic repulsion of the positive charges tends to force the proton away, but the neutron is not affected. If the energy of the incident deuteron exceeds the neutron binding energy (2 MeV), the proton will be repelled and neutron will enter the target nucleus. The (d, n) reactions are of particular interest as they are used as a means for obtaining neutrons of high energy. In the high energy region the (d, p) and (d, n) reactions are equally probable.

At medium energies (10 – 20 MeV), the stripping reactions has been very useful for the determination of the energies as well as the spins and parities of excited states of nuclei. In the (d, p), say, stripping reaction the target nucleus accepts a neutron of orbital angular momentum l_n directly into one of the levels of the final nucleus. The proton proceeds in a direction determined by l_n and with an energy equal to the reaction energy Q for the formation of the level into which its partner was captured plus the kinetic energy of the deuteron. Consider a vector diagram for the (d, p) reaction. The deuteron approaches with momentum $k_d\hbar$ and the proton goes off with momentum $k_p\hbar$ at an angle θ with the direction of incident deuteron. The momentum of captured neutron may be obtained from the cosine rule applied to momentum triangle.

$$k_n^2 = k_d^2 + k_p^2 - 2k_d k_p \cos\theta. \qquad ...(1)$$

A semi-classical consideration of the conservation of angular momentum suffices to give a qualitative understanding of the angular distribution of the protons after the (d, p) reaction. At the point of impact of the deuteron on the surface of the nucleus, there is a splitting into proton and neutron. The neutron subsequently penetrates the nucleus having an impact parameter r and is captured by it. The transfer of momentum $k_n\hbar$ is associated with the transfer of a quantized orbital angular momentum. Hence we have

$$(k_n \times r)\,\hbar = l_n\hbar \text{ or } k_n R \sin\beta = l_n. \qquad ...(2)$$

If $l_n/k_n > R$, the nuclear radius, it is not possible for the d-n interaction to occur. This in order to occur a stripping process, we must have $l_n/k_n \le R$. Hence

$$k_d^2 + k_p^2 - 2k_d k_p \cos\theta \ge \frac{l_n^2}{R^2} \qquad ...(3)$$

or

$$\cos\theta \le \frac{k_d^2 + k_p^2}{2k_d k_p} \frac{l^2}{2k_p k_d R^2}. \qquad ...(4)$$

The first term on the R.H.S. is always ≥ 1, as $a^2 + b^2 \ge 2ab$. The second term increases quadratically with the quantum number l. There is a lower limit for the angle θ that depends on l_n. Equation (4) shows that the preferred angle of scattering θ will increase with increasing l_n. Choosing

$$E_d = 7 \text{ MeV} \qquad (\therefore k_d = 0.82 \text{ f}^{-1}),$$

$$E_p = 13 \text{ MeV} \quad (\therefore k_p = 0.79 \text{ f}^{-1}),$$

R = 6f and applying cosine rule we get θ_{max} = 16°, 29° and 42° corresponding to l = 1, 2, 3 respectively. For each l_n values there are secondary maxima also. In general the differential cross-section is largest for $l_n = 0$ and decreases as the angular momentum transfer increases. The position of a peak or a pattern of peaks in the angular distribution of the emitted protons yields the value of l_n.

The analysis of the angular distribution of these protons provides information about angular momentum of these captured neutrons and on the spin and parity of the final nucleus. Let us consider the initial and final states of the nucleus with angular momenta I_i, I_f, the orbital angular momentum of deuteron with respect to the target nucleus l_d and the deuteron spin s_d. If in (d, p) stripping reaction, the proton escapes at an angle θ with respect to the initial direction of motion of the deuteron and neutron is captured, with orbital angular momentum l_n, then we can write

$$I_f = I_i + I_d + s_d - I_p - s_p = I_i + I_n + s_n. \qquad ...(5)$$

The orbital angular momentum of the captured neutron l_n is thus restricted by the inequality

$$I_f + I_f + \frac{1}{2} \geq l_n \geq \| I_i - I_f \left| -\frac{1}{2} \right| \qquad ...(6)$$

Law of conservation of parity $\Pi_f = \Pi_i(-1)^{ln}$ demands that initial and final states have same parity if l_n is even, and opposite parities if l_n is odd.

The quantum theory of low energy stripping reactions was first developed by Butler, using Born approximation which assumes that the plane or any particle whatever close to the nucleus is not distorted by the presence of nuclear potential. Agreement with experiment is excellent over a wide range of target nuclei and deuteron energies. Improvements have been made which lead to better agreement. They require the use of DWBA (distorted wave Born approximation). There are a number of and it is very likely that a better fit could be obtained.

At very high energies the stripping phenomenon may be described by considering one of the nucleons as absorbed by a nucleus and the other as continuing along its trajectory almost unperturbed with the initial velocity of the deuteron. If proton is captured by the target nucleus, the energy spread and the angular spread of the neutron beam are

$$\Delta E = 1.5\ (E_d B_d)^{1/2}, \text{ and } \Delta^{\theta} = 1.6 \left(\frac{B_d}{E_d}\right)^{1/2} \quad ...(7)$$

where B_d is the binding energy of the deuteron and E_d its kinetic energy.

Stripping Reactions and the shell model

It was pointed out first by Bethe and Butler in 1952 that stripping reactions in some cases could be used to measure the purity of shell model states. In stripping reaction the captured nucleon, say the neutron, occupies one of the available quantum states in the target nucleus and the other, say the proton, carries some information concerning this state. Conversely in pick-up reaction the picked up nucleon is removed from an occupied quantum state and the emerging deuteron conveys some information about this level. Thus we see that the stripping can be used the study unoccupied nuclear states and pick-up to investigate occupied states.

For example let us consider the reaction ${}_6C^{12}$ (d, p) ${}_6C^{13}$. According to the shell model C^{12} has spin 0 and positive parity and C^{13} has spin 1/2 and negative parity. By the use of parity conservation law

$$(-1) = +(1)(-1)^{l_c}.$$

Thus l_c of the captured particle must be odd.

We know that the angular momentum of the captured particle

$$I_c = I_f - I_i = \frac{1}{2} - 0 = \frac{1}{2}.$$

Since $\quad I_c = I_c + s_c \quad$ or $\quad I_c = I_c - s_c$

$$\therefore \quad I_c = \vec{\frac{1}{2}} - \vec{\frac{1}{2}} = 0 \text{ or } 1.$$

The $l_c = 0$ is not acceptable as it violates parity law, hence $l_c = 1$. Thus if the shell model prediction of C^{13} spin and parity is correct, then we should observe a peak around 17° in the differential scattering cross-section corresponding to $l = 1$ capture. This is exactly what is observed experimentally.

In the reaction P^{31} (d, p) P^{32}, the target nucleus has angular momentum $I_i = \frac{1}{2}^+$ (an odd $2s_{1/2}$ proton) and the ground state of P^{32} is expected to be formed by adding a neutron in a $d_{3/2}$ state. The two l-values can combine to $I_f = I_p + I_n = 1^+$ or 2^+. If we apply the model independent selection rule we find that the ground state (1^+ state) can be reached both by an $l = 0$ and $l = 2$ neutron, whereas the shell model state $d_{3/2}$ requires that $l = 2$. Bethe and Butler raised the question : Will a stripping experiment yield an angular distribution consistent with $l = 2$, with $l = 0$ or with a mixture of these?

In the same year, after performing experiment, Parkinson, Beach and King found an angular distribution consistent with $l = 2$ and very little (if any) trace of $l = 0$ contribution. In this particular example the test is very sensitive because $l = 0$ stripping gives a cross-section that is one order of magnitude larger at the maximum than does $l = 2$ stripping under similar circumstances. The result is, therefore, a great tribute to the shell model.

NUCLEAR CHAIN REACTION

The particular processes with which we are concerned are nuclear fission reactions in which fresh neutrons arc supplied by the fission processes. Thus to maintain a chain reaction, the neutrons must be produced at a constant rate and number of neutrons released per fission, v, must be sufficiently greater than 1 to compensate neutron loss. Since v is constant for a given fissionable material and, therefore, beyond human control. The only alternative is to reduce the neutron loss. The achievement of a chain reaction with uranium depends on a favourable balance among following six processes:

(1) Neutron may escape without being captured.

(2) Neutron may be absorbed by other materials present in the assembly without causing fission.

(3) Neutron may be absorbed by U^{235} without causing fission.

(4) Neutron may be absorbed by U^{235} nucleus causing fission.

(5) Neutron may be absorbed by U^{238} without producing fission.

(6) Neutron may be absorbed by U^{238} and may cause a fission when its energy is greater than the threshold energy.

Thus we see that processes (1) and (3) create new neutrons while other remaining processes remove available neutrons from the assembly. *A device for releasing nuclear energy through fission at a controlled rate is called nuclear reactor or chain reacting pile.* In nuclear weapons, uncontrolled nuclear reactions are used as most rapid and largest energy released is desired. The neutron balance in a thermal reactor can be described in terms of a cycle that shows what happens to the neutrons.

Consider this cycle starting with the fission of U^{235} nucleus by a thermal neutron, v fast neutrons are emitted in this process. Some of them, having energy greater than the fission threshold for U^{238}, can cause fissions in U^{238} with a consequent increase in the number of fast neutrons. The total number of fast neutrons from fission is thus increased from v to v$\in$, where ε is greater than one and is called the *fast fission factor*. In certain reactors $\in = 1.03$.

From a finite size reactor core, some fast neutrons may leak out before being slowed down to thermal energies. To account for this, we introduced l_f the leakage factor for fast neutrons and find that the number of fast neutrons remaining in the core long enough for slow down is $v\varepsilon (1 - l_f)$. These neutrons are slowed-down by collisions with moderator atoms, but during the slowing down process some of them may be captured by U^{238} to form U^{239} which decays to Np^{239} and then to Pu^{239}, particularly at energies in the region of strong resonances. We introduce the resonance escape probability p and find that out of $v\in (1 - l_f)$ neutions, a fraction $v\varepsilon (1 - l_f)$ p escapes while the fraction $v\in (1 - l_f)(1 - p)$ is captured and goes to form Pu^{239}.

The neutrons which escape are slowed down to thermal energies. Some of these thermal neutrons may leak out of the core. To account for this we introduce l_{th}, the *leakage factor for thermal neutrons* and find that the number of slow or thermal neutrons remaining in the core is $v\varepsilon (1 - l_f)\, p\, (1 - l_{th})$. A fraction of this remaining number is absorbed by uranium and the remaining other part by other materials. The fraction of thermal neutrons absorbed by the fuel as compared to all thermal neutron absorptions in the assembly is called the *thermal utilization factor f.* If we introduce this term we obtain the number of thermal neutrons that are actually absorbed by the fuel, which is $v\in (1 - l_f)\, p\, (1 - l_{th})f$.

Not all of the neutrons absorbed in the fuel caue of U^{238} nuclei, some are absorbed by U^{238} to form Pu^{239} while others are absorbed m U^{235} to form U^{236}. If σ_a is the cross-section of thermal neutrons absorbed in uranium and σ_f that of neutrons cause fission of U^{235}. Thus the final number of second generation neutrons absorbed in the fuel is $v\in (1 - l_f)\, p\, (1 - l_{th})\, f\sigma_f/\sigma_a$. This quantity is known as *reproduction factor or multiplication factor* and is denoted by k. The quantity $v\,(\sigma_f/\sigma_a)$ represents the number of fast fission neutrons, produced by each thermal neutrons, absorbed by the fuel and is called η. This quantity can be expressed as

$$\eta = v \frac{N_{235}\,\sigma_f(U^{235})}{N_{235}\,\sigma_f(U^{235}) + N_{235}\,\sigma_r(U^{235}) + N_{238}\,\sigma_o(U^{238})} \qquad ...(1)$$

where σ_r (U^{235}) stands for the cross-section for irradiative capture by U^{235}. For natural uranium $N_{235}/N_{238} = 0.00715$. It has been found that v = 2.5. The experimental values for the microscopic cross-section are σ_f (U^{235}) = 549 barns, σ_r (U^{235}) barns, σ_a (U^{238}) = 2.8 barns, hence η = 1.32.

$$\therefore \qquad k\ \eta \in (1 = l_f)\ p\ (1 - l_{th})f \qquad ...(2)$$

If the core is of very large size, the leakage of fast as well as slow neutrons can be disregarded. The multiplication factor is then denoted by kw and is given by

$$k_\infty = \eta\varepsilon pf \qquad ...(3)$$

This relation is known as the *four-factor formula.*

In an infinite system the condition of the self-sustaining chain reaction is that each neutron generation just replaces the previous one *i.e.*, k = l. The system is then said to be critical. Fora finite system, because in each neutron generation some neutrons are loset from the sides of the system, the *critical condition* is that k must be greater than unity, the exact amount by which it has to be greater than unity depending upon the shape of the system and the arrangement of the materials in the system, etc. If k is greater than unity for the infinite system, the number of neutrons will increase steadily. This is known as a *super-critical* state. On the other hand for k less than unity the number of neutrons would decrease to zero and this state is known as *sub-critical* state.

In the special case of a reactor in which the fuel contains U^{236} only and no U^{238}, both the factors e and p are practically unity. The formula for kw thus reduces to

$$k_\infty = \eta f \qquad ...(4)$$

For an assembly of finite dimensions, the effective multiplication factor keff will be less than k_∞ by a factor $(1 - l_f)\ (1 - l_{th})$. Thus

$$k_{eff} = k_\infty\ (1 - l_f)\ (1 - l_{th}) \qquad ...(5)$$

The magnitude of keff determines the speed with which the number of neutrons builds up and the rate at which fission occurs. In a nuclear bomb type of assembly, this build up must take place very rapidly,

($k_{eff} > 1$), whereas in industrial and research reactors this self-multiplication must be slow enough to allow the fission rate to remain always under the control of the operator ($k_{eff} \simeq 1$).

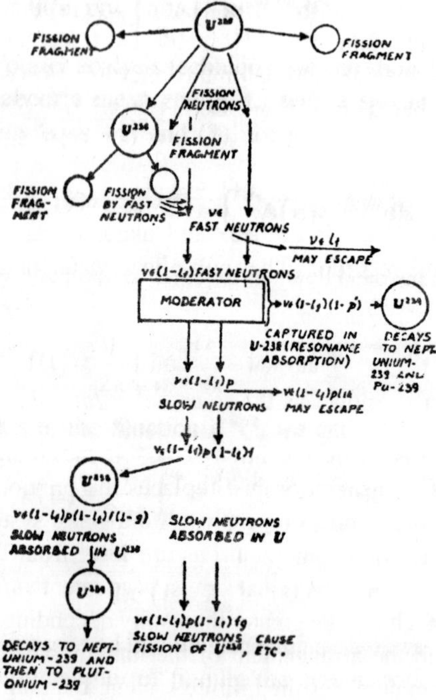

Fig 3 : Neutron cycle.

If $k_{ef} > 1$ for an assembly, we can decrease it by progressively reducing the reactor size, there by increasing the neutron loss through leakage from the assembly. The reactor size of the assembly for which $k_{eff} = 1$ is called the critical size.

THE CRITICAL SIZE OF A REACTOR

The size of a reactor that operates under the condition of exact balance between neutron production and neutron loss is known as critical *size and the reactor itself is said to be critical.* For such a reactor, a relation between its geometric properties and the material properties of the assembly is known as the critical equation. For all types of reactors, both homogeneous and heterogeneous, we start with eqn.

$$\nabla^2 \phi - \frac{3}{\lambda_{tr}\lambda_a}\phi + \frac{3q}{\lambda_{tr}} = 0 \qquad ...(6)$$

where q is the slowing down density which is a function of the space co-ordinates r and the neutron age τ. Thus the Fermi age equation can be written as

$$\nabla^2 q(r,\tau) = \frac{\partial q(r,\tau)}{\partial \tau} \qquad ...(7)$$

The solution of this equation can be written in the form of a product of two functions R (r) and T (τ). We can thus write

$$q(r, \tau) = R(r)\, T(\tau). \qquad ...(8)$$

By taking the second derivatives with respect to the space coordinates and first derivative with respect to τ only and substituting in equation (6), we get

$$T(\tau)\, \nabla^2 R = R(r)\, \partial T/\partial \tau$$

or $$\frac{\nabla^2 R}{R(r)} = \frac{1}{T(\tau)} \frac{\partial T}{\partial \tau} = -B^2 \text{ (say)} \qquad ...(9)$$

where $-B_2$ is a constant. Thus we have two differential equations.

$$\frac{1}{T(\tau)} \frac{\partial T}{\partial \tau} = -B^2 \text{ or } \frac{\partial T}{\partial \tau} + B^2 T(\tau) = 0 \qquad ...(10)$$

and $$\frac{\nabla^2 R}{R(r)} = -B^2 \text{ or } \nabla^2 R + B^2 R(r) = 0 \qquad ...(11)$$

The solution of eqn (10) is

$$T(\tau) = T_0 e^{-B^2 r} \qquad ...(12)$$

where T_0 is the value of T initially when $\tau = 0$.

At the beginning of the slowing down process,

$$q(\tau = 0) = q_0 = R(r)\, T(0) = R(r)\, T_0 \qquad ...(13)$$

$$\therefore \quad q = R(r)\, T_0 e^{B2} = q_{0e}{-B^2 r} \qquad ...(14)$$

We can also express q_0 in terms of the physical properties of the assembly. Number of neutrons per unit volume per sec. available for slowing down is (given by the rate of production of fission neutrons and is equal to the neutron multiplication factor $\epsilon f \eta$ per thermal neutron absorbed. The rate of thermal neutron absorption per unit volume of the reactor is $\phi \Sigma_a$. Hence

Initial slowing down density = rate of fast neutrons available or $$q_0 = \phi(r)\ \Sigma_a \varepsilon f \eta \qquad ...(15)$$

Thus equation (14) becomes

$$q = \phi(r)\ \Sigma_a \varepsilon f \eta\, e^{-B^2 r} \qquad ...(16)$$

If we consider neutron absorption by the moderator, then the corrected eqn. obtained by multiplying the solution by p is

$$q = \phi(r)\ \Sigma_a \varepsilon f \eta p\, e^{-B^2 r}$$

$$= k_\infty \Sigma_a \phi e(r)\, e^{-B^2 r} \qquad ...(17)$$

Substituting this value of q in eqn. (6), we get

$$\nabla^2\phi - \frac{3}{\lambda_{tr}\lambda_a}\phi + \frac{3k_\infty e^{-B^2 r}}{\lambda_{tr}\lambda_a}\phi = 0$$

or using $\frac{1}{3}\lambda_{tr}\lambda_a = L^2$, we get

$$\nabla^2\phi + \left(\frac{k_\infty e^{-Bh^2 r - 1}}{L}\right)\phi = 0 \qquad ...(18)$$

This is a general *diffusion eqn. for an actual reactor.*

Reactor Buckling

Using eqns. (7) and (17), we get

$$k_\infty \Sigma_a e^{-B^2 r}\,(\nabla^2\phi + B^2\phi) = 0,$$

or $$\nabla^2\phi + B^2\phi = 0 \qquad ...(19)$$

Since the thermal neutron flux distribution ϕ (r) depends on the geometry of the assembly, the value of B^2 is similarly determined by the reactor geometry and is known as geometrical buckling, denoted by B_g^2. Combining eqns (18) and (19), we get

$$\left[\frac{k_\infty}{(1 + L^2B^2)}\right] e^{-B^2 r} = 1 \qquad ...(20)$$

This is the critical equation. The left hand side is also denoted by k_{eff}. The numerical value of B^2 depends upon the physical properties of the reactor material and is, therefore, called the *material buckling* (B_m^2) of the reactor. Geometrical buckling B_g^2 decreases as we increase geometrical dimensions of a critical reactor. But this increase results in a k_{eff} greater than unity. As material buckling $B_m{}^2$ does not change with the reactor size. Hence for a super critical reactor ($k_{eff} > 1$), $B_m{}^2$ must be greater than $B_g{}^2$ Similarly, a reduction in size makes the reactor subcritical ($k_{eff} < 1$) for which $B_g{}^2$ is greater than $B_m{}^2$.

Nonleakage Factors

Fraction of fast neutrons that does not leak out of the assembly during slowing down, l_f, is given by the ratio of the actual production rate of thermal neutrons over the maximum possible for a reactor of infinite size. Thus

$$l_f = \frac{k_\infty \Sigma_a \phi e^{-B^2 r}}{k_\infty \Sigma_a \phi} = e^{-B^2 r} \qquad ...(21)$$

Thermal non-leakage factor l_{th} = 1—Thermal leakage factor

$$= 1 - \frac{\text{Thernal diffusion rate}}{\text{thermal diffusion rate + Thermal absorption rate}}$$

$$= 1 - \frac{-\frac{1}{3}\lambda_{tr}\nabla^2\phi}{-\frac{1}{3}\lambda_{tr}\nabla^2\phi + \Sigma_a\phi} = - \frac{\Sigma_a\phi}{-\frac{1}{3}\lambda_{tr}\nabla^2\phi + \Sigma_a\phi}$$

$$= \frac{1}{1 + \frac{1}{3}\lambda_{tr}\lambda_a B} = \frac{1}{1 + B^2L^2} \qquad ...(22)$$

$$\therefore \quad k_{eff} = k_\infty\, l_{th} l_f = k_\infty\, e^{-B^2 r}/(1 + L^2B^2)$$

$$= \frac{k_\infty}{(1 + B^2\tau)(1 + L^2B^2)} = \frac{k_\infty}{1 + B^2(L^2 + \tau)} = \frac{k_\infty}{1 + B^2M^2} \qquad ...(23)$$

Hence $L^2 + \tau$ is replaced by M^2, the migration area. M is known as migration length. Thus for large thermal critical reactor

$$k_{eff} = k_\infty/(1 + B^2M^2) = 1 \qquad ...(24)$$

The shape of nuclear reactors are almost exclusively either a rectangular parallelepiped, a sphere, or a cylinder. The most suitable co-ordinate systems are cartesian for the first, spherical for the second and cylindrical for the last. We shall summarised here the solutions for these three co-ordinate systems.

Rectangular Parallelepiped of Sides a, b, c.

The equation of neutron flux in this reactor in cartesian co-ordinates is given by

$$\frac{\partial^2 \phi}{\partial x^2} + \frac{\partial^2 \phi}{\partial y^2} + \frac{\partial^2 \phi}{\partial z^2} + B^2 \phi = 0 \qquad ...(25)$$

The solution, the neutron flux, will be the function of x, y, z and can be written as the product of three functions as :

$$\phi = X(x)\, Y(y)\, Z(z). \qquad ...(26)$$

$$\therefore \quad \frac{1}{X}\frac{\partial^2 X}{\partial x^2} + \frac{1}{Y}\frac{\partial^2 Y}{\partial y^2} + \frac{1}{Z}\frac{\partial^2 Z}{\partial z^2} + B^2 = 0 \qquad ..(27)$$

Let us assume that

$$\frac{1}{X}\frac{\partial^2 X}{\partial x^2} = -\alpha^2,\ \frac{1}{Y}\frac{\partial^2 Y}{\partial y^2} = -\beta^2 \text{ and } \frac{1}{Z}\frac{\partial^2 Z}{\partial z^2} = \gamma^2, \qquad ...(28)$$

$$\therefore \quad \alpha^2 + \beta^2 + \gamma^2 = B^2 \qquad ...(30)$$

The solution of equation $\left(\frac{1}{X}\right)\frac{\partial^2 X}{\partial x^2} = -\alpha^2$ is supposed to be

$$X = A_1 \cos \alpha x + A_1' \sin \alpha x \qquad ...(30)$$

Because sine function is not a symmetric function, hence the arbitrary constant A_1' has to vanish. Using now the boundry condition $\phi = 0$ for $x = \alpha/2$, we get $A_1 \cos a\, \alpha/2 = 0$ or $\alpha = \pi/a$.

Thus the solution becomes

$$X = A_1 \cos (\pi/a)\, x \qquad ...(31)$$

Similarly we have $\quad Y = A_2 \cos (\pi/b)\, y \qquad ...(32)$

and $\quad Z = A_3 \cos (\pi/c)\, z \qquad ...(33)$

$$\phi = A \cos (\pi/a)x \cos (\pi/b)y \cos (\pi/c)z \qquad ...(34)$$

and $\quad B^2 = (\pi/a)^2 + (\pi/b)^2 + (\pi/c)^2. \qquad ...(35)$

It can be shown that the minimum critical volume of a parallelepiped is a cube (a = b = c) for which $B^2 = 3\pi^2/a^2$.

$$\therefore \quad V_{min} = a^3 = 3^{3/2} + \pi^3/B^2 = 161/B^3 \qquad ...(36)$$

Spherical Reactor of Radios R

Because of spherical symmetry, if we take the origin of the reference system at the centre of the sphere, the solution will be independent of θ and ϕ. The diffusion equation thus can be written as

$$\frac{\partial^2 \phi}{\partial r^2} + \frac{2}{r}\frac{\partial \phi}{\partial r} + B^2\phi = 0 \qquad ...(37)$$

The solution of this equation is

$$\phi = (A_1/r) \sin Br + (A_2/r) \cos Br. \qquad ...(38)$$

With the use of the boundary conditions : ϕ is finite at $r = 0$ and is zero at $r = B$, we get $A_2 = 0$ and $BR = \pi$ or $R = \pi/B$. Thus the solution for the spherical reactor is

$$\phi = (A_1/r) \sin (\pi/R)r. \qquad ...(39)$$

Under the critical condition the minimum volume

$$V_{min} = \frac{4}{2}\pi R^2 = \frac{4}{3}\pi\left(\frac{\pi}{B}\right)^2 = \frac{130}{B^2} \qquad ...(40)$$

Cylindrical Reactor

The diffusion equation in cylindrical co-ordinates with cylindrical symmetry (term containing derivative of angular co-ordinates $\theta = 0$) is

$$\frac{\partial^2 \phi}{\partial r^2} + \frac{1}{r}\frac{\partial^2 \phi}{\partial r} + \frac{\partial^2 \phi}{\partial z^2} + B^2\phi = 0 \qquad ...(41)$$

The solution of this equation can be written as the product of two functions as

$$\phi = R(r)\, Z(z) \qquad ...(42)$$

Thus equation (41) can be written as

$$\frac{1}{R}\frac{\partial^2 R}{\partial r^2} + \frac{1}{rR}\frac{\partial R}{\partial r} + \frac{1}{Z}\frac{\partial^2 Z}{\partial z^2} + B^2 = 0 \qquad ...(43)$$

Let $$\left(\frac{1}{Z}\right)\frac{\partial^2 Z}{\partial z^2} = -\alpha^2 \qquad ...(44)$$

The solution is $Z = A_2 \cos \alpha z + A_2' \sin \alpha z$.

To have cylindrical symmetry A_2' has to vanish because since function is not a symmetric function. We then have

$$Z = A_2 \cos \alpha z \qquad ...(45)$$

By using boundary condition $\phi = 0$, for $z = H/2$, we have

$$A_1 \cos \alpha H/2 = 0$$

$$\therefore \qquad \frac{\alpha H}{2} = \frac{\pi}{2} \text{ or } \alpha = \frac{\pi}{H} \qquad ...(46)$$

Hence $\qquad Z = A_2 \cos (\pi/H/)z. \qquad ...(47)$

Substituting the value of a and hence that of term containing Z in equation (43), we have

$$\frac{d^2R}{dr^2} + \frac{1}{r}\frac{dR}{dr} + \left(B^2 - \frac{\pi^2}{H^2}\right) R = 0 \qquad ...(48)$$

Writing $B^2 = \frac{\pi^2}{H^2} = K^2$ and multiplying by r^3, we have

$$r^2 = \frac{d^2R}{dr^2} + r\frac{dR}{dr} + K^2r^2R = 0$$

Calling $rK = \mu$, we can write finally

$$\mu^2 = \frac{d^2R}{d\mu^2} + \mu\frac{dR}{d\mu} + \mu^2R = 0 \qquad ...(49)$$

It is a linear differential equation of second order in cylindrical co-ordinates and known as the Bessel's eqn. of zero order. Its solution is the Bessel functions of zero order and can be written as

$$R = A_3J_0(\mu) \qquad ...(50)$$

Combining the results we have finally

$$\phi = AJ_0(Kr) \cos (\pi/H)z. \qquad ...(51)$$

From the Bessel function tables of zero order we find that $J_0(Kr)$ reaches the value 0 when $Kr = 2.405$. It corresponds to r = radius R of the cylinder. Hence $K = 2.405/R$.

Substituting this value of K in equation (51), we have

$$\phi = AJ_0\left[\left(\frac{2.405}{R^2}\right)r\right] \cos\left(\frac{\pi}{H}\right)z \qquad ...(52)$$

$$\therefore \qquad B^2 = \frac{}{H^2 + (2.405/R)^2} \qquad ...(53)$$

For the smallest critical volume of the cylinder $V = (\pi R^2H)$ should be minimum. Hence

$$\frac{dV}{dH} = \frac{d}{dH}\left(\pi \times \frac{H^2(2.405)^2}{B^2H^2 - \pi^2} \times H\right) = 0$$

or $$B^2 = \frac{2\pi^2}{H^2} \qquad ...(54)$$

Therefore, $$V_{min} = \frac{\pi(2.405)^2}{2\pi^2}\left(\frac{\sqrt{3}\pi}{B}\right)^3 = \frac{148}{B^3} \qquad ...(55)$$

From the above results it is apparent that for a nuclear reactor using the same materials the most convenient shape is the spherical one as this shape requires the minimum volume. It is also clear from above result that the minimum volume of any of the shapes considered is proportional to the inverse of tie coefficient B^3. The quantity B depends upon the average number of neutrons created per fission, the diffusion length and the slowing down length. As these depend upon reactor material, the quantity B is known as the *material buckling* and denoted by B_m. Hence the minimum volume is defined as the critical condition under which geometrical buckling is equal to the material buckling. This determines the amount of material required for the reactor to become critical.

The Effect of a Reflector

The critical size of a reactor core can be reduced by surrounding it with good moderating material. The latter has the properties of being a good neutron scatterer and does not usually absorb neutrons to any great extent. This material acts as a reflector. Graphite is often used for this purpose. The reactor with a reflector has certain distinct advantantages over the bare reactor which are stated as follows :

(a) Improved Neutron Economy

The reflector knocks neutrons which leave the reactor back into it and thus reduces neutron leakage from the core. The reflector also acts as a moderator for the fast neutrons that have entered it from the core. Since the absence of neutron absorbing material in the reflector reduces neutron loss due to resonance absorption, the moderation of fast neutrons in the reflector will be more efficient that in the core itself.

(b) Possibility of Fuel Saving

For constant core size reacton the improvement in the neutron economy reduces the amount of fuel or the fuel concentration required to achieve criticality. It means that reflector increases fuel saving. We also know that the reflector decreases the leakage or neutrons from the reactor and thus decreases the critical volume. Hence it is possible to reduce the size of the nuclear reactor including the reflector to below the size of a similar barer eactor with the sums fuel concentration.

(c) Improved Reactor Power Utilization

Since the power production rate is proportional to the average neutron flux, the reactor can be operated at a higher total power output for the same maximum neutron flux. The improvement in the power utilization is a consequence of the flattening of flux across the reactor core occuring due to the presence of the reflector. By virtue of this (flux flattening) effect the power production is also move uniform over the core volume.

ANALYSIS OF REACTION CROSS-SECTION

Consider an incident plane wave of neutrons travelling parallel to the z-axis. This plane wave has a spatial part e^{ikz}, where $k = 2\pi/\lambda$, and can be expressed as a super-position of spherical waves with different orbital angular momentum quantum numbers relative to the target nucleus. The expansion is

$$e^{ikz} = e^{ikr\cos\theta} = \sum_{i=0}^{\infty} i^i \sqrt{[4\pi(2l+1)]}\, j_l(kr)\, Y_{l,0}(\theta) \qquad ...(1)$$

This solution involves Spherical Bessel functions j_l (kr) and spherical harmonics $Y_{l,m}(\theta)$. As there is cylindrical symmetry about the z-axis the order m of the spherical harmonics is zero. If r, the distance of the incident particle from the target nucleus, is very large so that $kr \geq l$, then the form of the function j_l (kr) is

$$j_i(kr) = \frac{\sin\left(kr - \frac{1}{2}l\pi\right)}{kr} = i\frac{e^{-i(kr-l\pi/2)} - e^{+(kr-l\pi/2)}}{2kr}.$$

$$\therefore \quad e^{ikz} = \frac{\sqrt{\pi}}{kr}\sum_{l=0} \sqrt{(2l+1)}\, i^{l+1}\left[e^{-i(k_r - l\pi)} - e^{+i(k_r - l\pi/2)}\right] Y_{l,0}(\theta) \qquad ...(2)$$

The first exponential term represents a series of spherical waves coming towards the origin with momentum $k\hbar$ and angular momentum $l\hbar$, the second exponential term represents a series of a outgoing spherical waves with just the same properties. The above expression describes an undisturbed plane wave. If however, a nucleus is located at the origin, the amplitudes and phases of the outgoing spherical waves from the origin are, in general, changed. Let the outgoing part of the partial wave with orbital angular momentum l be changed by a factor η_l. This may be a complex number. This is related to the phase shift by $\eta_l = |\eta_l|\, e^{2i\sigma l}$. Elastic scattering will take place, when $|\eta_l| = 1$. If $|\eta_l| < 1$ both elastic scattering and nuclear reaction will take place. Thus the wave function for the disturbed wave is

$$\phi(r) = \frac{\sqrt{\pi}}{kr}\sum_{l=0}^{\infty}\sqrt{(2l+1)i^{l+1}\left[e^{-i\left(k_{r-l\pi/2}\right)-\eta_l e^{+i\left(k_{r-l\pi/2}\right)}}\right]}\,Y_{l,0}(\theta) \quad ...(3)$$

The elastically scattered wave function ψ_{sc}, which is the difference between the disturbed wave and the incident wave, is given by

$$\psi_{sc} = \psi(r) - e^{ikx}$$

$$= \frac{\sqrt{\pi}}{kr}\sum_{l=0}^{\infty}\sqrt{(2l+1)i^{l+1}}\left(1-\eta_l\right)e^{i(kr-l\pi/2)}\,Y_{l,0}(\theta) \quad ...(4)$$

Since $\psi(r) \propto e^{ikz} + f(\theta)\, e^{ikr}.r$, hence

$$f(\theta) = \frac{\sqrt{\pi}}{k}\sum_{l=0}^{\infty}\sqrt{(2l+1)i^{l+1}}\left(1-h_l\right)e^{-il\pi/2}\,Y_{l,0}(\theta).$$

The scattered cross-section is obtained by dividing the number N_{sc}. of scattered particles per second by the number of incident particles per unit area per second which for the plan wave e^{ikx} is v, the velocity if the particles. Consider a sphere of radius r_0. If r_0 is very much larger than the range of the nuclear forces, than the number of scattered particles per second N_{sc} is equal to the flux of ψ_{sc} this sphere.

$$\therefore N_{sc} = \frac{\hbar}{2iM}\int\left(\frac{\partial\psi_{sc}}{\partial r}\psi^*_{sc} - \frac{\partial\psi^*_{sc}}{\partial r}\psi_{sc}\right)_{r=r_0} r_0^2 \sin\theta\, d\theta\, d\phi, \quad ...(5)$$

where M is the mass of the incident particle. Here integral is extended over the surface of the sphere. Substituting values of ψ_{sc} and ψ^*_{sc},

$$N_{sc} = \frac{\hbar}{2iM}\int\left[f^*(\theta)\frac{e^{-ikr0}}{r_0} f(\theta)\left(ik\frac{ikr_0}{r_0}\frac{1}{r_0}\frac{e^{ikr_0}}{r_0}\right)\right.$$

$$-(\theta)\frac{e^{ikr^0}}{r_0}f^*(\theta)\left(-ik\frac{e^{ikr_0}}{r_0}\frac{1}{r_0}\frac{e^{ikr_0}}{r_0}\right)\Bigg]r_0^2 \sin\theta\, d\theta\, d\phi$$

$$=\frac{h2ik}{2iM}\int f^*(\theta)\, f(\theta).\frac{1}{r_0^2}r_0^2 \sin\theta\, d\theta\, d\phi$$

$$\because \quad \frac{\hbar k}{M}=\left(\frac{h}{2\pi}\right)\left(\frac{2\pi}{\lambda}\right)\left(\frac{1}{M}\right)=\frac{p}{M}=v,$$

$$\therefore \quad N_{sc}=\frac{\hbar k}{M}\int |f(\theta)|^2 \sin\theta\, d\theta\, d\phi = v\int |f(\theta)|^2\, d\Omega \qquad ...(6)$$

Because of the orthogonality and normalization of the $Y_{l,0,}$ we get

$$N_{sc}=v\frac{\pi}{k^2}\sum_{i=0}^{\infty}(2l+1)\,|1-\eta_l|^2 \qquad ...(7)$$

Thus the scattering cross-section is given by

$$\sigma_{sc}=\frac{N_{sc}}{v}=\frac{\pi}{k^2}\sum_{i=0}^{\infty}(2l+1)\,|1-\eta_l|^2 \qquad ...(8)$$

If we define

$$\sigma_{l,\,sc}=\pi\lambda\!\!\!{}^{-}{}^2\ (2l+1)\ |1-\eta_l)^2, \qquad ...(9)$$

then eqn. (8) can be rewritten as

$$\sigma_{sc}=\sum_{i=0}^{\infty}s_{l,\,sc}\,. \qquad ...(10)$$

Thus we see that in the integral cross-sections the contributions of the partial waves for different *l* are simply added.

We now derive an expression for the reaction cross-section σ_{Re}. It is determined by the number N_{Re}, the number of particles which enter a sphere of radius r_0 around the enter without leaving it again through the entrance channel. It is, therefore, equal to the net flux into this sphere as computed from the complete wave-function $\psi(r)$.

$$N_{Re}=-\frac{h}{2iM}\int\left(\psi^*.\frac{\partial\psi}{\partial r}\psi\frac{\partial\psi^*}{\partial r}\right)_{r=r0} r_0^2 \sin\theta\, d\theta\, d\phi \qquad ...(1)$$

The minus sign is used. because inward flux N_{Re}, is positive. Using the orthogonality and normalization of the Y_l, 0, we get

$$N_{Re} = \nu \frac{\pi}{k^2} \sum_{i=0}^{\infty} (2l+1)\left(1-|\eta_l|^2\right) \quad ...(12)$$

$$\therefore \ \sigma_{Re} = \frac{\pi}{k^2} \sum_{i=0}^{\infty} (2l+1)\left(1-|\eta_l|^2\right) = \pi\lambda\!\!\!^{-}{}^2 \sum_{i=0}^{\infty}(2l+1)\left(1-|\eta_l|^2\right). \quad ...(13)$$

Using the definition

$$\sigma_{l,\,Re} = \pi\lambda\!\!\!^{-}{}^2 \ (2l+1)\ (1-|\eta_l|^2). \quad ...(14)$$

We can represent eqn. (46) as $\sigma_{Re} = \sum_{i=0}^{\infty} \sigma_{l,\,Re}$. ...(15)

The total cross-section for a given l is the sum

$$\sigma_{l,\,t} = \sigma_{l,\,sc} + \sigma_{l,\,Re}$$

$$= \pi\lambda\!\!\!^{-}{}^2 (2l+1)\left[|1-\eta_l|^2 \_ \left(1-|\eta_l|^2\right)\right]$$

$$= 2\pi\lambda\!\!\!^{-}{}^2 \ (2l+1)\ [1 - R_e(\eta_l)], \quad ...(16)$$

where $R_e(\eta_l)$ is the real part of the complex number η_l.

Conclusion

1. If $\eta_l = 0$, then $s_{l,\,Re}$ takes on its maximum value $\pi\lambda\!\!\!^{-}{}^2(2l+1)$, however the scattering cross-section takes on the finite value. Scattering and reaction cross-sections are then both of the same size.
2. If $\eta_l = -1$, then $\sigma_{l,\,Re} = 0$ and the scattering cross-section $\sigma_{l,\,sc}$ is maximum having value $4\pi\lambda\!\!\!^{-}{}^2(2l+1)$.
3. If $h_l = 1$, then $\sigma_{l,\,sc} = 0$. It also implies that there are no reactions, as there is no case in which reactions occur without elastic scattering.
4. If $|\eta_l| = 1$, then $\sigma_{l,\,Re} = 0$, but $\sigma_{l,\,sc}$ may be finite for the complex value of η_l.
5. $|\eta_l|^2$ can never be negative, otherwise the outgoing wave will have a high intensity than the incoming one. Thus the condition $|\eta_l|^2 \leq 1$ insures that the cross-section $\sigma_{l,\,Re}$ does not become negative.

If the incident neutron has a high enough energy, the reduced wavelength $\lambda\!\!\!^{-}$ is small compared to the radius R of the nucleus. Suppose the nucleus is completely black to the incident neutrons, *i.e.*, all neutrons which hit the nucleus are absorbed by it and, therefore, lead to a reaction, all the particles with $l \leq R/\lambda\!\!\!^{-}$ strike the nucleus. We can express this as

$$\eta_l = 0 \text{ if } l \leq \frac{R}{\lambda\!\!\!^{-}} \text{ and } \eta_l = 1 \text{ if } l > \frac{R}{\lambda\!\!\!^{-}}. \quad ...(17)$$

Therefore all incident particles with $l \leq \frac{R}{\lambda\!\!\!^{-}}$ react and all those with $l > \frac{R}{\lambda\!\!\!^{-}}$ pass by. Thus eqn. gives

$$\sigma_{Re} = \sum_{i=0}^{R/\lambda\!\!\!^{-}} \pi\lambda\!\!\!^{-2}(2l+1) = \pi\,(R+\lambda\!\!\!^{-})^2. \quad ...(18)$$

When σ_{Re} has its maximum value, $\sigma_{Re} = \sigma_{Re}$, and thus we obtain the apparently paradoxical result that the total cross-section is twice them geometrical cross-section of the nucleus:

$$\sigma_l = \sigma_{Sc} + \sigma_{Re} = 2\pi(R+\lambda\!\!\!^{-})^2 \sim 2\pi R^2. \quad ...(19)$$

The physical meaning of the results involves the diffraction of the incident wave to form a shadow behind the absorbing centre. Since any neutron striking the nucleus reacts, therefore, the nucleus will appear opaque to a fast neutron.

General Aspect of Reactor Design

In spite of numerous possible variations in the design and components of reactor systems, there are number of general features which all reactors possess in common to a greater or lesser extent Let us consider and compare few features in the given order.

Fuel

The material containing the fissile isotope is called the reactor fuel. The fuels that can be used are uranium containing U^{235} in its natural concentration of 0.715% or in an enriched proportion. Th^{232} and U^{235}, which can be converted partly at least into fissile material. The form in which a fuel is used in a reactor depends upon various circumstances. Solid fuel elements are used in most of the reactors. Some reactors have been designed which make use of fluid fuels.

MODERATORS AND REFLECTORS

Materials used to reduce the neutron energy, known an moderators, are graphite, light water, heavy water, beryllium and its oxide and possibly certain organic compounds. As previously noted, the good moderators are usually materials of low mass number having small absorption cross-section and large slowing down power. The ordinary water has the greatest *sdp* but an appreciable neutron absorption cross-section. It is attractive as a moderator because of its low cost. Heavy water is the best available material for neutron moderation. It has a very small absorption cross-section and the highest moderating ratio of all moderators. Carbon in the form of graphite has been employed as moderator or reflector in a number of reactors. It has a very small absorption cross-section and the second highest moderating ratio of all moderators. Both beryllium and its oxide are good moderators having small absorption cross-section and the best sdp of all metals. Metallic beryllium is much more expensive than other moderators. Hydrogeneous organic compounds have proved to be suitable as moderators.

Above mentioned materials arc not suitable for use as the reflector for the reactor in which fast neutrons are used. To avoid this situation the reflector must be a material of high mass number.

Reactor Coolants

The materials employed, to remove the heat that is being generated in the reactor core as a result of fissioing taking place, are known as *coolants*. These materials are circulated through the core tor the purpose of abstracting the heat and transferring it to the outside of the core. An ideal coolant should have as little effect on the neutron as possible, ideally it should not absorb nor moderate the neutrons, should not react chemically with the other materials which it contacts at the system, should not break up under irradiation, should acquire intense long lived radioactivity during its passages through the reactor, should have a low vapour pressure at the operating temperature of the reactor, should be reasonably easy to handle, should be able to remove large amounts of heat for a small expenditure of pumping power and should not be costly. Actually no one coolant meets all these requirements, a final choice for any particular reactor is a compromise based on these requirements. The materials proposed for coolants are ordinary water, heavy water, liquid metals such as sodium or sodium potassium alloy or mercury, organic liquids and gases. Each type has its advantages and disadvantages.

At ordinary temperatures both heavy and light water are good coolants. For reactors operating at high temperatures, high pressures are required to prevent boiling. Liquid metals have been proposed for use at higher temperatures. The best of these is sodium. It does not require pressurization even at very high temperatures. On the other hand, it is very reactive with water and oxygen and becomes radioactive due to neutron capture when passed through reactor. As a compromise between water and liquid sodium, certain organic compounds e.g. polyphenyls, were suggested. However the heat removal properties of these hydrocarbons are inferior to those of water or liquid sodium. In several of the earliest reactors air was used as a coolant. It is not satisfactory at high temperatures because it reacts chemically with so many materials. Carbon dioxide is also employed in large power reactors. Since it is used as coolant in graphite-moderated reactors. The chemical reaction with graphite is a problem at temperatures greater than 500°C. By the addition of some carbon-monoxide or methane it can be used upto 600°C. Actually helium is a good coolant but it is very costly.

CONTROL MATERIALS

In thermal reactors control is achieved by means of a neutron absorbing materials. The material should not become radioactive as a result of neutron capture. Became of its availability and ease of fabrication cadmium was used in early reactors. It can be utilized at low temperature work only because it has low melting point. An alloy of silver with 15 per cent of indium and 5 percent of cadmium has been also employed in some reactors. This alloy has a higher melting point than Cd and large cross-section for neutron capture also over a wider energy range. The most common neutron absorber used for reactor control is boron. This element has very high melting point and large cross-sections for neutron absorption-Boron is often used incorporation with stainless steel, aluminium or carbon.

The control elements are commonly located in the core in the form of either rods or plates, but in some reactors it is more convenient to have the control elements in the reflector close to the core. In a thermal reactor the control rods are moved in to decrease the fission rate or neutron flux and out to increase it. In fast reactors materials cannot be used because the cross-sections are small. In this case control is achieved by adjusting neutron escape by moving part of the reflector or some of the fuel elements.

REACTOR SHIELDING

All nuclear reactors, except those operating at very low powers, are sources of intense neutron and γ-radiation and therefore, represent hazard to persons in the immediate vicinity of the reactor. Provisions for their health protection are made by surrounding the reactor core with a radiation shield. This shield is generally known as the *biological shield* because its primary function is of health protection. It consists generally a layer of concrete, about 6 to 8 ft. thick and is capable of absorbing both γ-rays and neutrons. In reactors operating at high powers, that part of the shield which is in immediate contact with the core heats up considerably and requires special cooling facilities in order to prevent it from cracking or suffering other heat damage. Shield required for this purpose is known as thermal shield. It is fairly close to the core and consists of a few inches of iron or steel.

SOLVED EXAMPLES

Example 1:

When F^{19} is bombarded with protons a (p, n) reaction with subsequent α-emission occurs. Calculate the excitation energy of the compound nucleus that corresponds to the resonance with a proton energy of 4.99 MeV.

$${}_1H^1 + {}_9F^{19} \rightarrow ({}_{10}Ne^{10})^* \rightarrow {}_{10}Ne^{19} + {}_0n^1.$$

The convertible energy of the protons or the energy of relative motion is

$$E = \frac{E_0}{1+\frac{m}{M}} = \frac{4.99}{1+\frac{1}{19}} = 4.745 \text{ MeV.}$$

This binding energy contribution of the absorbed proton in the compound nucleus is

$$\Delta E = [M({}_9F^{19}) + M({}_1H^1) - M({}_{10}Ne^{28})]mu$$

$$= [18.998405 + 1.007276 - 19.992440]mu$$

$$= 0.013241 \text{ mu} = 12.332 \text{ MeV}$$

∴ Excitation energy of the compound nucleus

$$= 4.745 + 12.332 = 17.077 \text{ MeV}$$

Example 2:

A thin sheet of Co^{59}, 0.04 cm, thick, is irradiated with a neutron beam of flux density 10^{12} neutrons per cm^2 per sec for a period of 3 hr. If the cross-section for neutron capture by Co^{59} is 30 barns, calculate the number of nuclei of the isotope Co^{50} produced at the end of the irradiation period per cm^2 and the initial β-activity of the sample. Given: half-life of Co^{60} 5.2 years and density of Co^{59} 8.9 gm/cm^3.

Solution:

$$\text{Number of reactions/second/m}^2 = N_0\sigma It = N\left(\frac{\rho}{M}\right)\sigma It$$

$$= 6.02252 \times 10^{26} \times \left(8.9\times\frac{10^3}{59}\right) \times 30 \times 10^{-28} \times 10^{16} \times 0.04\times10^{-2}$$

$$= 0.092 \times 10^{15}.$$

$\therefore$ No. of Co^{60} nuclei produced = No. of transmutation in 3 hours

$$= 1.092 \times 10^{15} \times 3 \times 60 \times 60 = 118 \times 10^{17}.$$

$$\text{Initial activity per m}^2 = \lambda \times 118 \times 10^{17} = \frac{0.693\times118\times10^{17}}{5.2\times365\times24\times60\times60}$$

$$= 5 \times 10^6 \text{ decays/sec} = 134 \text{ micro curies.}$$

Example 3:

Calculate k_∞ for an enriched uranium-graphite moderated assembly which contains 400 molecules of moderator per molecule of uranium and a U^{238}/U^{235} ratio of 75. Given: For graphite, $\sigma_a = 0.0032$ barn, $\sigma_s = 4.8$ barns. $\xi = 0.758$; For uranium, $\sigma_f(235) = 590$, $\sigma_a(235) = 698$, $\sigma_a(238) = 2.75$ and $\nu = 2.46$.

Solution:

Let us first calculate absorption cross section $\sigma_a(U)$ and number of fission neutrons per neutron absorbed η_u for the enriched fuel.

$$\sigma_a(U) = [N(235)\,\sigma_a(235) + N(238)\,\sigma_a(238)]/[N(235) + N(238)]$$

$$= \frac{\sigma_a(235) + \sigma_a(238)\,N(238)/N(235)}{1 + N(238)/N(235)} = \frac{698 + 2.75 \times 75}{1 + 75}$$

$$= 11.89 \text{ barns.}$$

$$\eta_u = [N(235)\ \sigma_f(235)\ v]/[N(235)\ \sigma_a(235) + N(238)\ \sigma_a(238)]$$

$$= \frac{v\sigma_f(235)}{\sigma_a(235) + \sigma_a(238)\ N(238)/N(235)} = \frac{2.46 \times 590}{698 + 2.75 \times 75} = 1.60$$

Calculate f and p and then k_∞ as we did, in the last example.

This value of p is considerably greater then the values obtained for natural uranium. Since f is evaluated with respect to U^{235}, the matching value of η to be used now is 2.08. Hence

$$k_\infty = \epsilon\eta\ pf = 1 \times 2.08 \times 0.957 \times 0.754 = 1.5.$$

Example 4:

Calculate the critical volume of a spherical thermal reactor that uses U^{235} as fuel and heavy water as moderator in molecule ratio of 1 : 10^5.

Solution:

$$\text{Thermal utilization factor } f = \frac{\sigma_a(U)}{\sigma_a(U) + \sigma_a(M)\dfrac{N_m}{N_u}}$$

$$= \frac{698}{698 + 0.00092 \times 10^5} = 0.8948$$

$$\therefore k_\infty\ \eta f = 2.08 \times 0.8948 = 1.819$$

Thermal diffusion length for the mixture

$$L^2 = L_m{}^2\,(1 - f) = 1.70^2$$

$$(1 - 0.8948) = 0.3038\ m^2$$

$$\therefore \quad k_{eff} = \frac{k_\infty e^{-B^2}}{1 + L^2B^2} = \frac{1.819\ e^{-0.125B^2}}{1 + 0.3038\ B^2} = 1$$

or $B^2 = 2.192\ m^2$

$\therefore$ Critical volume $V_C = 120/53 = 37\ m^3$.

Example 5:

What would be the length of she side of a cubical reactor having geometrical buckling 65. If the reactor is spherical, calculate its critical radius.

Solution:

Geometrical buckling $B^2 = \frac{\pi^2}{a^2} + \frac{\pi^2}{b^2} + \frac{\pi^2}{c^2}$

For a cubical reactor a = b = c

$\therefore\ B^2 = \frac{3\pi^2}{a^2}$ or $a^2 = \frac{3\pi^2}{B^2} = 0.46154$

$\therefore$ a = 0.674 m = 67.4 cm

Critical radius of a spherical reactor

$R = \pi/B = 3.141/8 = 0.39$ m = 39 cm.

Example 6:

Calculate the Q-value for the formation of P^{30} in the ground state in the reaction Si^{29} (d, n) P^{30} from the following cycle of nuclear reactions:

(i) $P^{31} + \gamma \rightarrow P^{30} + n - 12.37$ MeV

(ii) $P^{31} + p \rightarrow Si^{28} + He^4 + 1.909$ MeV

(iii) $Si^{28} + d \rightarrow Si^{29} + p + 6.246$ MeV

(iv) $2d \rightarrow He^4 + 23.834$ MeV

Solution:

Reaction (i) can be written as

$P^{30} + n \rightarrow P^{31} + g + 12.37$ MeV. ...(ii)

Formation of P^{30} is given by the reaction

$Si^{29} + d \rightarrow P^{30} + n + Q.$

Combination of these equations with reactions (ii), (iii) and (iv) gives

$4d = 2He^4 + Q + 44.359$ MeV

$Q = 2(2d - He^4) - 44.359$

$= 47.668 - 44.359 = 3.309$ MeV.

Example 7:

A tritium gas target is bombarded with a beam of monoenergetic protons of kinetic energy 3 MeV. What is the K.E. of the neutrons emitted at 30° to the incident beam? Atomic masses are : H^1 1.007276 mu, n^1 = 1.008665 mu, ${}_1H^3$ = 3.016056 mu, and He^2 = 3.016030 mu.

Solution:

Nuclear reaction in question can be written as

$$_1p^1 + {}_1H^3 \rightarrow {}_2He^3 + {}_0n^1 + Q,$$

where $Q = m(_1p^1) + m(_1H^3) - m(_2He^3) - m(_0n^1)$

$= 1.007276 + 3.016050 - 3.016030 - 1.008665$

$= -\ 0.001369 \text{ mu} = -\ 1.2745 \text{ MeV}.$

The K.E. associated with the emitted neutron is given by eqn. (11) as

$$\sqrt{E_n} = u \pm \left(u^2 + v\right)^{1/2},$$

$$\text{where } u = \frac{\left(m_p m_n E_p\right)^{1/2}}{mH_e + m_n} \cos\theta = \frac{(1.007276 \times 1.008665 \times 3)^{1/2}}{3.016030 + 1.008665} \times \frac{\sqrt{3}}{2}$$

$= 0.3753$

$$\text{and } v = \frac{m_{He} Q + E_p\left(m_{He} - m_p\right)}{m_{He} + m_n}$$

$$= \frac{-3.016030 \times 1.2745 + 3 \times 2.008754}{3.016030 + 1.008665} = 0.5424$$

$\therefore\ \sqrt{E_n} = 0.3753 \pm 0.8266 = 1.2019$

or $E_n = 1.444$ MeV.

Example 8:

The total cross-section of nickel for 1 MeV neutrons is 3.5 barns. What is the fractional attenuation of a beam of such neutrons on passing through a sheet of nickel 0.01 cm in thickness? Given, density of nickel 8.9 gm/cm^3.

Solution:

Macroscopic cross-section $\Sigma = N_0\ \sigma = \left(\frac{\sigma}{M}\right) N\sigma$

$$= \left(\frac{8.9 \times 10^3}{58}\right) \times 6.02252 \times 10^{26} \times 3.5 \times 10^{-28}$$

$= 32.37\ m^{-1}$

We know the relation $\sigma I_0 e^{-\Sigma x}$ or $2.3026 \log_{10} \frac{I_0}{I} = \Sigma x$

$$\therefore \quad \frac{I_0}{I} = \text{Anti log} \frac{32.37 \times 0.01 \times 10^{-2}}{2.3026} = 0.0014.$$

Example 9:

Estimate the relative probabilities of (n, n) and (n, γ) in indium, known to have a neutron resonance at 1.44 eV with a Γ of 0.1 eV and cross-section of 28000 barns.

Solution:

At resonance ($E = E_0$) Breit-Wigner formula simplifies to

$$\sigma(n, \gamma) = \frac{\lambda^2 \Gamma_n \Gamma_\gamma}{4\Gamma^1}.$$

The de Brogile wavelength $\lambda = \frac{h^1}{\sqrt{(2mE)}} = 2.4 \times 10^{-11}$ m.

$$\Gamma_n \Gamma_\gamma = \frac{4\sigma\Gamma^2}{\lambda^2} = \frac{4 \times 2.8 \times 10^{-24} \times (0.1)^2}{\left(2.4 \times 10^{-11}\right)^2}$$

$$= 1.5 \times 10^{-4}$$

Since $\qquad \Gamma_\gamma \simeq \Gamma = 0.1$, hence $\frac{\Gamma_n}{\Gamma_\gamma} = 0.015.$

Example 10:

The neutron capture reaction of Au^{197} at neutron energies upto a few hundred eV is characterized by a number of resonances. The most prominent is at 4.906 eV and has $\Gamma_\gamma = 0.124$ eV and $\Gamma_n = 0.007\ E^{1/2}$ eV. The compound nucleus formed in this resonance absorption has spin 2. Calculate the peak cross-section.

Solution:

$\Gamma_n = 0.007\ E^{1/2} = 0.007 \times (4.906)^{1/2} = 0.0155$ eV.

We know the relation for resonance cross-section

$$\sigma(n,\gamma)=\frac{\lambda^2}{\pi}\frac{2I_c+1}{2(2I+1)}\frac{\Gamma_\gamma\Gamma_n}{\left(\Gamma_\gamma+\Gamma_n\right)^2}=\frac{h^2}{2\pi mE}\frac{2I_c+1}{2(2I+1)}\frac{\Gamma_\gamma\Gamma_n}{\left(\Gamma_\gamma+\Gamma_n\right)}$$

$$=\frac{\left(6.625\times10^{-34}\right)^2}{2\times3.14\times1.67\times10^{-27}\times4.906\times1.6\times10^{-15}}\times\frac{2(2+I)}{2(3+1)}$$

$$\times\frac{0.124\times0.0155}{(0.1395)^2}$$

$$= 3.446\times10^4 \text{ barns.}$$

Example 11:

Calculate k_∞ for a homogereous, natural uranium-heavy water moderated assembly which contain 50 molecules of moderator per molecule of uranium. Assume natural uranium to contain one part U^{235} to 139 parts of U^{238} and use the following constants : For uranium, $\sigma_a(U)$ = 7.68 barns, $\sigma_s(U)$ barns; and for D_2O, σ_a = 0.00092 barn, σ_s = 10.6 barns; and ξ = 0.570.

Solution:

Thermel utilization factor $f=\dfrac{N_u\sigma(U)}{N_u\sigma_a(U)+N_u\sigma_a(M)}$

$$=\frac{1}{1+\dfrac{N_m\,\sigma_a(M)}{N_u\sigma_a(U)}}=\frac{1}{1+\dfrac{0.00092}{7.68}}=0.994$$

Number of U^{238} atoms only per unit volume $N_0 = N_u(139/140)$

$$\therefore\quad \frac{\Sigma_s}{N_0}=\frac{N_m}{N_0}\,\sigma_s(M)=N_m\cdot\frac{140}{139N_u}\times\sigma_s(M)$$

$$=\frac{N_m}{N_u}\times\frac{140}{139}\times\sigma_s(M)=50\times\frac{140}{139}\times10.6=530\text{ barns}$$

$$\therefore\quad \int(\sigma_s)_{eff}\frac{dE}{E}=3.85\left(\frac{\Sigma_s}{N_0}\right)^{0.415}=60.2\text{ barns}$$

$\therefore$ Resonance escape probability $P = e^{-60.2/530\times0.57} = 0.820$

Hence $k_\infty = \epsilon\eta pf = 1\times1.34\times0.820\times0.994 = 1.09$

This assembly is, therefore, not capable of sustaining a chain reaction.

Example 12:

Estimate the relative loss due to leakage for a critical thermal reactor employing U^{235} and graphite in on atom ratio of 1: 705. Given : For uranium $\sigma_a = 0.003$ barn, $L_m = 54$ cm, $\tau_0 = 364$ cm^2. For uranium $\sigma_a = 698$ barns and $\gamma_1 = 2.08$. Also given $p = 1$ and $\in = l$.

Solution:

Since $p = 1$ and $\in = 1$, hence $k_\infty = \eta f$.

$$\text{Thermal utilization factor } f = \frac{N_{0(235)}\,\sigma_{a(235)}}{N_{0(235)}\,\sigma_{a(235)} + N_{0(c)}\,\sigma_{a(c)}}$$

$$= \frac{1}{1 + \dfrac{N_{0(c)}\,\sigma_{a(c)}}{N_{0(235)}\,\sigma_{a(235)}}} = \frac{1}{1 + \dfrac{10^5 \times 0.003}{698}}$$

$$= 0.6995.$$

$\therefore \quad k_\infty = 2.08 \times 0.6995 = 1.454.$

Thermal diffusion length for the mixture $L^2 = L_m^{\ 2}\,(I - f)$

$$= (0.54)^2\,(1 - 0.6995) = 8.76 \times 10^{-2}\ \text{m}^2$$

We know that for critical reactor

$$k_{eff} = \frac{k_\infty\, e^{-B^2\tau}}{1 + L^2B^2} = \frac{1.454\, e^{-364\times10^{-4}B^2}}{1 + 8.76\times10^{-2}\ B^2} = 1$$

$\therefore \quad B^2 = 3.27$. Substituting this value of B^2, we get fast non-leakage probability $l_f = e^{-B^2\tau}$

$\therefore$ Fast leakage probability $= 1 - l_f = 1.112 = 11.2\%$.

Thermal non-leakage probability $l_{th} = 1/(1 + L^2B^2) = 0.778$.

$\therefore$ Thermal leakage probability $1 - l_{th} = 1 - 0.778 = 0.222$

Since only a fraction of 0.888 of the original neutrons have reached thermal energies. Hence the fraction of original number lost by thermal leakage, is given by

$$0.222 \times 0.888 = 0.197 = 19.7\%$$

$\therefore$ Total leakage factor $= 11.2\% + 19.7\% = 30.9\%$.

It can also be obtained by other method.

Total non-leakage factor $l_f\, l_{th} = \frac{k_{eff}}{k_\infty} = \frac{1}{1.464} = 0.688$

$\therefore$ Total leakage factor = 1 – 0.688 = 0.312 = 31.2%.

Example 13:

The world's first water boiler reactor, the LOPO nuclear reactor, used an enriched uranium sulphate solution made up as follows:

Element	*weight*	*σ_a(barn)*	*σ_s(barn)*
U^{235}	*580 gm*	*698*	*10*
U^{238}	*3378 gm*	*2.75*	*8.3*
S	*534 gm*	*0.49*	*LI*
0	*14065 gm*	*0.0002*	*4.2*
H	*1573 gm*	*0.33*	*20*

Calculate k_∞ for this reactor.

Solution:

Thermal utilization factor

$$f = \frac{1}{1 + \frac{N_{(238)}\sigma_{a(238)}}{N_{(235)}\sigma_{a(235)}} + \frac{N_{(s)}\sigma_{a(s)}}{N_{(235)}\sigma_{a(235)}} + \frac{N_{(0)}\sigma_{a(0)}}{N_{(235)}\sigma_{a(235)}} + \frac{N_{(H)}\sigma_{(H)}}{N_{(235)}\sigma_{a(235)}}}$$

$$f = \frac{1}{1 + 5.75 \times 3.94 \times 10^{-3} \times 6.76 \times 7.15 \times 10^{-4}}$$

$$+ 356 \times 2.86 \times 10^{-5} + 637 \times 4.73 \times 10^{-4}$$

$$= 0.754$$

$$\frac{\Sigma_s}{N_0} = \frac{N_{(0)}}{N_{(238)}}\sigma_{s(C)} + \frac{N_{(H)}}{N_{(238)}}\sigma_{s(H)} = \frac{879.25}{14.19} \times 4.2 + \frac{1573}{14.19} \times 20 = 2475 \text{ barns}$$

$$\therefore \int_E^{E_0} (\sigma_a)_{eff} \frac{dE}{E} = 3.85 \left(\frac{\Sigma_s}{N_0}\right)^{0.415} = 99 \text{ barns}$$

and $$\frac{N_0}{\xi\Sigma_s} = \frac{N_0}{\xi_{(0)}\sigma_{s(0)} N_{(0)} + \xi_{(H)}\sigma_{(H)} N_{(H)}}$$

$$= \frac{14.19}{0.120 \times 4.2 \times 879 + 1 \times 120 \times 1573} = \frac{1}{2249}$$

$$\therefore \quad p = e^{-99/2249} = 0.957$$

EXERCISES

1. A critical reactor of cubical shape uses U^{235} and ordinary water in a molar ratio of 1:400. Calculate the critical mass of U^{235} for this reactor.
2. An indium foil of 2 cm^2 cross-section and 10^{-3} cm, thickness is exposed to a spread beam of neutrons of uniform energy. Calculate the number of neutron capture that will occur during a 3 min exposure of the foil. Given; I = 5 × 10 neutrons/cm^2 sec. and σ_u = 190 barns.
3. The B^{10} (α, p) C^{13} reaction shows among other a reasonance for an excitation energy of the compound nucleus of 13.23 MeV. Calculate the mean life of the nucleus for this excitation if level width is 130 keV.
4. Calculate the threshold energy required to initiate and the reaction P^{31} (n, p) Si^{31}, given M_p = 1.00814, M_n = 1.00898, M_P = 30,98356 & M_{Si} = 30.98515.

   Calculate also the maximum energy of β-decay of Si^{31} to P^{31}.
5. Assuming a resonance cross-section 5.5 × 10^4 barns for Cd^{113} for Cd^{113} at a resonance energy of 0.176 eV. Calculate the relative probability of a neutron emission as compare to a resonance capture if Γ = 0.113 eV and $\Gamma_\gamma << \Gamma_n$.
6. Calculate and compare the collision and absorption mean free paths for neutrons in graphite. Given : σ_s = 4.8 barns, σ_a = 3.2 × 10^{-3} barns and ρ = 2.25 gm/cm^3.
7. Calculate the thermal utilization factor for an aqueous fuel solution containing 10% by weight, of U^{235}.
8. Calculate k_∞ for a homogeneous, natural uranium-graphite-moderated assembly with contains 300 molecules of graphite per molecule of uranium. Assume natural uranium to contain one part of U^{235} to 139 parts of U^{235}. Given; For natural Uranium, σ_a = 7.68 barns, σ_s = 8.3 barns; For graphite, σ_a = 0.0032 barn, σ_s = 4.8 barn, and ξ = 0.158.
9. Calculate k_∞ lot an enriched uranium-graphite-moderated assembly, using 400 molecules of graphite to 1 molecule of uranium and a U^{238}/U^{235} ratio of 70.
10. A nuclear reactor has an effective k of 1.03. How many generations will be required to double the number of neutrons?

What time will be required for doubling, assuming an initial energy of 1.5 MeV and a mean free path for fission of 8.0 cm?

11. Liquid sodium used as a reactor coolant has a cross-section for thermal activation of 600 mb. The circulating sodium spends about 1/4 of its time in the reactor core, where the thermal flux is 8.6×10^{12} n cm^{-2} sec^{-1}. What is the specific activity of the coolant at equilibrium? How long will be required for the coolant to reach 3/4 of equilibrium activity?
12. A power reactor operating at a power level of 10^5 watts goes prompt critical due a failure of the control rod system. Assume a multiplying factor of 1.09, and that the negative temperature coefficient stopped the reaction in 5 m sec. Estimate the thermal energy generated during the runaway.
13. In the B^{10} (α, p) C^{13} reaction with 4.77 MeV α-particles the two most energetic proton groups which were observed to be emitted at the angle of 90° with the direction of the incident α-particle beam had energies of 6.84 MeV and 3.98 MeV respectively. Give the information about the energy levels of the residual nucleus.
14. The Q-valve of the reaction $Ra^{224} \rightarrow Em^{222} + He^4$ is 4.88 MeV. The radium nucleus is originally at rest. Calculate the K.E. of each disintegration product.
15. Calculate p for a homogeneous, natural uranium-BeO assembly with molecular ratio of 1 : 200.

2

INTERACTIONS OF THE PARTICLES

THE GRAVITATIONAL INTERACTION

The first force that any of us discover is gravity. It holds the moon and earth together, keeping the planets in their solar orbits and binds stars to form our galaxy. Newton gave a formula $F = Gm_1m_2/r^2$ for the interaction between two masses. The gravitational effect does not depend on the colour, size, charge, velocities, spin and angular orientation but depends on the magnitude of the inertia. The gravitational force between two nucleons separated by a nucleon diameter is

$$F = G\frac{m_1m_2}{r^2} = 6.7\times10^{-11}\frac{\left(1.7\times10^{-27}\right)^2}{\left(10^{-15}\right)^2} \approx 2\times10^{-34} \text{ newtons} \quad ...(1)$$

and the gravitational attraction is only about 2×10^{-49} joule. Hence we sec that it plays no role in particle reactions. In the ninteenth century the forces were thought to be propagated by fields, space warped for particular effects. In the twentieth century these fields are explained in terms of agents or messengers which actually propgate the effect. Gravitation can thus be explained in terms of the interactions of *gravitons*. Their mass must be zero and therefore, their velocity must be that of light. *As the gravitational, field is extremely weak, the gravitons cannot be detected in laboratory.*

ELECTROMAGNETIC INTERACTIONS

All of the ordinary chemical and biological effects are due to the interaction of electric charges and the fields they produce. The term electromagnetism is because the electricity and magnetism are both part of the same phenomenon. The appropriate law for the interaction of point charges bears the name of Coulomb ($F = q_1q_2/4\pi\epsilon_0r^2$). For two protons,

10^{-15} metre apart, the repulsion force will be $9 \times 10^9 \times (1.6 \times 10^{-19})^2/(10^{-25})^2 \approx 30$ newtons. It is about 10^{35} times greater than the gravitational attraction caused by the mass. The energy released by the complete separation of these protons would be 3×10^{-14} joules.

If the particles are not at rest but are moving, the field will not only be an electric field but would be new one depending on the velocity and magnitude of the charge. When the charge is accelerated, the energy is radiated out in the form of an electric and megnetic pulses. This energy comes from the agent which accelerates the charge. The pulse is called a *photon* and travels with the velocity of light. If the source charge is accelerating in an oscillating fashion, the propagated signal will consist of successive waves of electric and magnetic fields or the *radio-photons*. Thus we see that the photons are emitted and reabsorbed by a charge. *Interaction between two charged particles consists of an exchange of these photons*. The strength of the electromagnetic interact n is given by the dimensionless fine structure constant $\alpha(= e^2/\epsilon_0 \hbar_c = 1/137)$, and is due to photon exchanges.

The electromagnetic interaction is charge dependent. In terms of isobaric spin, the interaction depends on T_s and is governed by the isospin rule $\Delta T = 0, \pm 1$. All Other quantities such as charge, baryon number, lepton number, hypercharge, parity, strangeness number are conserved.

The capture of photon can effect the production of mesons or hyperons by an electromagnetic interaction

$$\pi^o + p$$
$$\gamma + p$$
$$\Lambda^o + K^+.$$

An example of a irradiative capture reaction is

$$\pi^- + p \rightarrow n + \gamma.$$

The neutral particles such as

$$\pi^o \rightarrow \gamma + \gamma, \ \Sigma^o \rightarrow \Lambda^o + \gamma,$$
$$\eta^o \rightarrow \pi^+ + \pi^- + \pi^o \qquad \eta^o \rightarrow \gamma + \gamma,$$

decay electromagnetically since these processes involve no change of strangeness The decay processes such as $\Sigma^+ \rightarrow p^+ + \gamma$ are forbidden because the change $\Delta S = 1$ is required. The paradox that the decay of netural particle is by electromagnetic nteractior is resolved by introducing as an intermediate step in the overall reaction. The virtual production of a *nucleon-anti-nucleon (or electron-positron)* pair. Thus we have

$$\pi^o \xrightarrow{\text{strong}} \underset{\text{virtual}}{\left(N+\overline{N}\right)} \xrightarrow{\text{electromagnetic}} \gamma + \gamma.$$

The process of mutual annihilation of particles and anti-particles is an example of electromagnetic interaction

STRONG INTERACTION

The strong nuclear interaction is independent of the electric charge. The force is same between p-p and n-n. For this purpose the proton and neutron are, one but in different electric charge states. Strong interactions involve mesons and baryons. The range is very much shorter than that of gravitational or electromagnetic interaction. Strong interaction energy falls off rapidly when the distance between two particles increases. Yukawa in 1935 predicted the existence of heavy quanta, which played the same role in nuclear forces (or strong forces) as photons in electromagnetic ones. From estimates of the range of nuclear forces. Yukawa predicted that the new particles, called mesons, should have a mass of the order of 200 to 300 electron masses. In the chapter of nuclear forces, we outlined an elementary theory of pion-nucleon interaction and introduced the concept of a nucleon charge g analogous to the electric charge e. The strength of the nuclear interaction is represented by the magnitude of the dimensional coupling constant $\gamma^2/4\pi\hbar c$ ($\approx$ 14). It is about a thousand times the electromagnetic coupling constant α.

Strong interactions between elementary particles are responsible for the total cross sections as a function of energy. The strong interaction is a short range force ($\approx 10^{-15}$ m), conserves baryon number B, charge Q, hypercharge Y, parity π, isospin T and its component T_s. It is responsible for kaon production, however the decay of mesons, nucleons and hyperons proceeds by an electromagnetic or weak interactions.

WEAK INTERACTION

The weak interaction is responsible for the decay of strange and non-strange particles and for non-leptonic decay of strange particles. The numerical constant, which is characterstic of the weak interactions, is obtained from Fermi's theory of decay. Its value is $g_F = 1.41 \times 10^{-62}$ Jm^2. In analogy with the expression for the other interactions, the dimensionless weak interaction coupling constant is of magnitude

$$\left[\frac{gF^2}{\left(\hbar_c\right)^2}\right]\left[\frac{m_{rc}}{\hbar}\right]^4 \simeq 5 \times 10^{-24}.$$

Consider the reactions which do not involve a change of strangeness and yet which must be due to weak interaction. The neutron decay is the proto-type of all the β-decays :

$$n \rightarrow p + e^- + \overline{\nu_e}.$$

The nature of such an eqn. is that the reaction can go in either direction so long as energy is conserved find that any participant can be replaced on opposite side by its anti-particle, i.e. $p \rightarrow n + e^+ + \nu_e$. Another variation of this four fermion interaction is the proton capture of an anti-neutrino $\left(\overline{\nu_e} + p \rightarrow \nu + e^+\right)$. Another example of the four-fermion interaction is the muon decay $\left(\overline{\mu} \rightarrow e^- + \nu_\mu + \overline{\nu_e}\right)$. There is also a coupling among muons, nucleons and neutrinos, thus

$$\mu^- + p \rightarrow n + \nu_\mu.$$

There are however other types of weak interactions which require coupling between other pairs of fermions.

$$\Lambda^o\ (S = -1) - \pi^-\ (S = 0) + p\ (S = 0).$$

There is no neutrino involved here, strangeness changes by + 1 and only two fermions are involved. For four fermions one can assume this decay as through a virtual stage $\left(\Lambda^o \rightarrow \overline{p} + n^o + p^+\right)$. In the first vitual step, four fermions are involved. In the next stage, the strong nuclear forces come into play.

There are restrictions whether one transition is forbidden or allowed : *In a weak interaction involving change in strangeness of baryons or mesons, the change in strangeness must be equal to the change in charge.* The change in isospin T and its component T_z may be nonzero.

The strangeness and isospin are not meaningful for leptons, and are useful when hadrons are involved in the weak interactions. The lepton number is conserved in these interactions. The parity is not conserved, but CP and CPT are conserved.

Similar to the graviton for gravity, photon for elecromagnetism and mesons for the strong nuclear force there is one agent for the weak interaction.

It would also be a boson of mass above 800 MeV and is named as the *intermediate charged vector Boson* and given the symbol W. Its half life against decay into electron neutrino or muon-neutrino would be less than 10^{-17} sec.

CONSERVATION OF ANGULAR MOMENTUM

The conservation of angular momentum includes both types (orbital and spin) of angular momentum together. The first is given by the motion of the object as a whole about any chosen external axis of rotation. The second is the intrinsic angular momentum of each object about an axis through its own centre of mass. Strongly interacting fermions have a half integer spin (s = 1/2 for Ξ, Σ, Λ, n and p; s = 3/2 for Ω), strongly interacting bosons (η, K, π) have s = 0, weakly interacting fermions (leptons μ, e, ν_e, ν_μ) have s = 1/2, massless bosons (electromagnetic interacting γ-rays) have s = 1 and gravitions have s = 2.

Conservation of Energy

Conservation of energy on other hand seems more complicated with elementary particles because a large fraction of the total energy is oftenly interchanged between rest energy associated with mass and kinetic or potential energy. The sum of these three, the total energy is always Conservation of Charge

The most familiar of the conservation laws is the conservation of electric charge. The charge is conserved in all processes and no exceptions are known. We note that all elementary charges are 0, or –1; multiple charges are not found.

CONSERVATION OF BARYON NUMBER

The number of baryons minus the number of anti-baryons is conserved. In other words the net baryon number in any process always remains unchanged. All normal baryons such as p^+, n^o, Λ^o, Σ^+, Σ^-, Σ^-, Ξ^-, Ξ^o and Ω^- have a baryon number of +1, the corresponding anti-particles known as anti-baryons have a baryon number of – 1. All the mesons have a baryon number of zero. For example, the reaction $\Lambda^o \rightarrow \pi^+ + \pi^-$ is *allowed* because the baryon Λ^o is replaced by the baryon p^+, keeping the total number of baryons constant. The reaction $\Lambda^o \rightarrow + p^- + \pi^+$ is *forbidden* because one baryon is replaced by one anti-baryon changing the baryon number by – 2.

CONSERVATION OF LEPTON NUMBER

The number of leptons minus the number of anti-leptons is conserved. In other words *the net leptons number in any process always remains conserved.* The ordinary electron, negative muon and neutrinos all have a lepton number + 1, the corresponding anti-particles known as antileptons have a lepton number of –1. The reaction $\mu^+ \rightarrow e^+ + n + \overline{\nu}$ is allowed because a lepton number of –1 is replaced by a lepton number (–1) + (+1) + (–1) = – 1.

The reaction $n \rightarrow p + e^- + \overline{\nu}$ is allowed by both conservation of baryons and conservation of leptons. No exception to the rules of conservations of baryons and leptons has been found in the many elementary particle reactions so far studied one by one. Conservation of leptons has a significance for strong interactions. There are other conservation laws which are not applicable to weak interactions. The property that is conserved in strong interactions only is known as isospin. Other properties which are not conserved for all the three interactions, but are conserved in one or two interactions only, are hypercharge, *strangeness, purity*, invariance under charge conjugation and invariance under CP conjugation.

Conservation of Isospin

According to the ordinary idea of isotopic spin, each nuclear particle posseses a certain total isotopic spin T and each possible projection of this isotopic spin along a certain axis T_3, appears to us as a different charge state of the corresponding particle. In the case of nucleons, T = 1/2 and the 2T + 1 = 2, possible values of T_3 are + 1/2 (for the proton state) and – 1/2 (for the neutron state). For the pions, T = 1 and so there are 2T + 1 = 3 charge states. The triplet consists of π^+, π^0 and π^- particles and the values of T_3 are +1.0, –1, respectively.

B is +1 for the proton, neutron and hyperons and is zero for the pions. Inserting the value of T_3 and B, one obtains :

$$Q_p = +\frac{1}{2}+\frac{1}{2} = +1,\ Q_n = -\frac{1}{2}+\frac{1}{2} = 0 \qquad ...(1)$$

$$Q\pi^+ = +1+0 = 1,\ Q\pi^o = 0+0 = 0,\ Q\pi^- = -1+0 = -1.$$

Analytically, if B is the baryon number, T the isotopic spin quantum number, T_3 the component of T and Q the charge in units of the electron charge, the relation between these quantities is

$$Q = T_3 + \frac{B}{2}. \qquad ...(2)$$

Isospin numbers are associated with *hadrons* (particles that can exhibit strong interactions) but not with leptons. The isospin component

T_3 is conserved in both strong and electromagnetic interactions but not in weak interaction.

Conservation of Hypercharge

A quantity called *hypercharge* (the twice the average charge of the members of the group), is also conserved in strong and electromagnetic interactions. For example, for the triplet π^+, π^0, π^-, average charge is zero and hence all these three mesons have a hypercharge of zero. The hypercharge of the pair of the particles K^+ hnd K^0 is +1 and that of the pair of anti-particles K^- and $\overline{K}^0$ is – 1. Thus the alternative definition is that it is twice the difference between the actual charge Q and the isospin component T_3 of a particle. Thus hypercharge

$$Y = 2(Q - T_3). \qquad ...(3)$$

Conservation of Strangeness

The concept of strangeness has found wide application in particle physics. It is an additional quantum number which describes the interactions of elementary particles. It has been chosen in such a manner that it becomes zero for all the well known particles (non-strange particles). Rochester and Butler, in 1947, at the University of Manchester arranged a magnetic cloud chamber and after about a year of operation reported certain new types of forked or V shaped tracks in cloud chamber photographs of cosmic rays. These new V-particles were also investigated intensively by cloud chamber groups from various institutions for several years. By 1953, at an International Conference on Elementary particles at Bagneres a decision was taken to name those particles of mass greater than π-mesons, but smaller than protons, *K-particles*, while those particles whose masses were greater than protons were to be termed hyperons.

One of the most common V-particles (Λ^o) was neutral and decayed ($\Lambda^o \rightarrow p + \pi^-$) in time 2.5×10^{-10} seconds. The question arised if the Λ^o could interact strongly, why did not its decay go via strong interactions with a life time of ~ 10^{-23} sec instead of the observed 10^{-10} sec? To explain this Pais suggested the *hypothesis of associated production. The V-par tides can interact strongly and, therefore, are produced only in pairs, once separated each number can decay into ordinary particles only through the weak interaction. A typical example is*

$$\pi^- + p^+ \rightarrow \Lambda^o + K^o$$
$$\rightarrow p^+ + \pi^- \rightarrow \pi^+ + \pi^-.$$

Both the associated creation of the strange particles and their individual stability against immediately decay were the features that earned them the title *strange.* Both features can be explained by insisting that the total *strangeness* must remain constant in fast particle reaction.

If the baryons are arranged in columns according to their electric charge (plus under plus, minus under minus, neutral under neutral). The electric charge centres do not thus occur in the same vertical line. The electric charge centre of the nucleon is at + 1/2, halfway between p and n. The charge centre of Λ^o is at 0. The triplet sigma is centred at 0, but the doublet xi is centred at – 1/2. The Ω^- singlet is at charge – i. If we take the charge centre of the nucleon doublet arbitrarily to be reference origin, then we have

For Λ^o ... $\Delta Q = Q_\Lambda - Q_N = -1/2$; for Σ-hyperons ... $\Delta Q = Q_\Sigma - Q_N = -1/2$: for Ξ-doubtlet...$\Delta Q = -1$ for Ω^- ... $\Delta Q = -3/2$.

By defining strangeness quantum number as

$$S = 2\Delta Q, \qquad ...(4)$$

We obtain S = 0 for the nucleons and non-zero for hyperons (S = – 1, for Λ^o and Σ's – 2 for Ξ's and – 3 for Ω^-). Once a hyperon or K-meson is produced and gets beyond the influence of the collision, it can decay. As the decay products have strangeness zero hence the strangeness remains conserved in fast nuclear progresses. Once produced and separated from each other, they must wait for some weaker interaction to allow them to decay. Strangeness conservation governs only the strong interactions and not the weak. Even the weak interactions have some respect for strangeness. It is found experimentally that weak decays change strangeness as little as possible.

$\Delta S = \pm 1$. No example with $\Delta S = \pm 2$ has been seen.

The examples are :

$$\Sigma^+ \rightarrow \Lambda^o + e^+ + \nu_e, \qquad \Delta S = 0$$

$$\Sigma^- \rightarrow n + e^- + \overline{\nu}_e, \qquad \Delta S = 1$$

$$\Lambda^o \rightarrow p + e^- + \overline{\nu}_e, \qquad \Delta S = 1.$$

* M. Gell Mann in U.S.A. and T. Nakano and K. Nishijima in Japan independently suggested this strangeness quantum number. They suggested a scheme, known as Gell-Mann and Nishijima scheme after their names, for this quantum number. In this relation (2) is replaced by

$$Q = T_3 + \frac{1}{2} Y = T_3 + \frac{1}{2}(B + S), \quad ...(5)$$

where $Y = S + B$ ...(6)

Since B is conserved always, the strangeness like the hypercharge is conserved in strong and electromagnetic interactions.

To cover a whole range of strongly interacting particles, the formalism has also been extended to mesons. Since the hyperons all have negative strangeness, they can be produced only in association with K-mesons of positive strangeness. On the assumption that strangeness is a conserved quantity in a reaction such as

$$p^+ + p^+ \rightarrow p^+ + \Lambda^o + K^+$$

$$S: \quad 0 + 0 \rightarrow 0 + (-1) + S_k,$$

strangeness $S_k = +1$. Similar argument gives strangeness S = l for neutral kaon. From the associated production reaction

$$\pi^- + p \rightarrow \Lambda^o + K^o$$

$$S: \quad S_\pi + 0 = -1 + 1,$$

it follows that $S_\pi = 0$. It also applies to π^+, π^o and η^o mesons. Since for mesons the baryon number B = 0, hence strangeness S = hypercharge Y.

Thus we see that strangeness is not an independent new quantity, but is related to a combination of Q, T_2 and B, each of which is regulated by conservation laws.

Let us now see what kinds of particle type can be formed by various choices of T, B and S. *If S = 0, there are three possibilities,*

(a) *B = 0, T = 0 yields Q = 0 (natural meson), η^o meson.*

(b) *B = 0, T = 1 yields Q = + 1, 0, – 1 (pions).*

(c) *B = 1, T = 1/2 yields Q = + 1, 0 (nucleons).*

If S = 1, the multiple charge can be avoided only when

(a) *B = 1, T = 0 and Q = + l (Baryon singlet).*

(b) *B = 0, T = 1/2 and Q = +l, 0 (K^+, K^o).*

If S = – S, the multiple charges can be avoided only when

(a) *B = 0, T = 1/2 and Q = 0, – 1 ($\overline{K}^o$ and K^-).*

(b) B = l, T = 0 and Q = 0, (singlet baryon Λ^o).

(c) B = l, T = l and Q = +l, 0, – 1 (Σ+, Σ^o and Σ^-).

If S = – 2, the multiple charges can be avoided only when

(a) B = 1, T = 1/2 and Q = 0, – 1 (hyperons Ξ^o, Ξ^-).

If S = – 3, the multiple charges can be avoided only when

(1) B = 1, T = 0 and Q = – 1 (Ω-hyperon).

In the case of hadrons, the strangeness must be conserved ($\Delta S = 0$) for fast reactions. The decays of kaons and hyperons are very slow because they involve a breakdown of strangeness conservation. The decays $\Xi^- \rightarrow n + \pi^-$ and $\Xi^o \rightarrow p + \pi^-$ which involve $\Delta S = 2$ would be expected to be exceptionally slow, and the decays $\Xi^- \rightarrow \Lambda^o \pi^-$ and $\Xi^o \rightarrow \Lambda^o + \pi^o$ with $\Delta S = 1$ to be slow. Since charge and baryon number are always conserved hence for weak decay processes, eqn (5) gives

$$|\Delta S| = 2\,|\Delta T_2|$$

or
$$|\Delta T_3| = \frac{1}{2}. \qquad ...(7)$$

As T_3 is the component of total isospin along a particular direction, hence the general form of eqn (7) for weak decays is

$$|\Delta T| = \frac{1}{2}. \qquad ...(8)$$

Let us apply conservation laws to high energy pion-nucleon collisions which often give large quantities of kaons. Examples of possible equations are

$$\pi^+ + n^o \rightarrow \Lambda^o + K^+$$
$$\rightarrow K^o + K^+.$$

Using the conservation of baryons and strangeness, we see that the second reaction violates baryon conservation and, therefore, cannot occur. Similarly equation $\pi^- + p^+ \rightarrow \Lambda^o + K^o$ is possible whereas $\pi^- + p^+ \rightarrow \Lambda^o + \pi^o$ is not possible by strangeness violation.

Let us consider equation $p^- + p^+ \rightarrow 2\pi^+ 2\pi^- + \pi^o$. Applying various conservation laws and remembering that the pairs of pions are ejected with opposite isospins, we have

$$Q = -1 + 1 \rightarrow 2 - 2 + 0, \qquad \therefore \delta Q = 0$$

$$B = -1 + 1 \rightarrow 0 + 0 + 0, \qquad \therefore \delta B = 0$$

$$T = \frac{1}{2} + \frac{1}{2} \rightarrow 0 + 0 + 1, \qquad \therefore \delta T = 0$$

$$Y = -1 + \rightarrow 0 + 0 + 0, \qquad \therefore \delta Y = 0$$

$$S = 0 + 0 \rightarrow 0 + 0 + 0, \qquad \therefore \delta S = 0.$$

Charge Conjugation

Charge conjugation is defined as the interchange of particles and anti-particles. It does not simply mean a change over the opposite electric charge or magnetic moment, the sign of other charge quantum numbers [hypercharge Y, baryon number B, lepton numbers (l_e, l_μ)] is also reversed without changing mass M and spins. Thus a unitary operator, also known as charge conjugation operator C, satisfies the following relations :

$$CQC^{-1} = -Q,\ CYC^{-1} = Y,\ CBC^{-1} = -B,\ Cl_eC^{-1} = -l_e$$

and $Cl_\mu C^{-1} = l_\mu$.

Some elementary particles e.g., γ, n^o – mesons and the positronium atom ($e^- + e^{+'}$) are transformed into themselves by charge conjugation. They are their own anti-particles. These are known as *selfconjugate or true neutral particles*. The neutron (B = 1, Y = 1) and K^o – mesons (Y = 1, B = 0) are not invariant under C.

A system is said to possess charge conjugation symmetry or to be invariant under charge conjugation if the system (or the process) is such that it is impossible to know that it has undergone charge conjugation. For example, the operation C converts the negative pion decay $\left(\pi^- \to \mu^- + \overline{\nu_\mu}\right)$ into the positive pion decay $\left(\pi^+ \to \mu^+ + \nu_\mu\right)$, since the π^+ is the anti-particle of π^-. Until about two decades ago it was believed that the entire universe is invariant under C. In the end of 1956, experiments revealed that weak interactions violated it. It turned out that the μ^+ and μ^- decay electrons have angular distributions of opposite asymmetry : that the π^+ and π^- decay muoni have opposite polarizations and that while a free neutrino is left handed an anti-neutrino is right handed. Charge conjugation applied to a free moving neutrino then results in a process which does not exist in nature.

The charge conjugate of the Dirac equation, which corresponds to the wave functions of positrons, has the form

$$C = \text{i.e.}^{i\phi}\begin{pmatrix} 0 & \sigma_y \\ \sigma_y & 0 \end{pmatrix} = e^{i\phi}\, i\, \alpha_y.$$

The phase ϕ is arbitrary. For zero phase ...(9)

$$C = i\alpha_y, \qquad ...(10)$$

where α_y is a square matrix and σ_y is the Pauli spin matrix having value

$$\sigma_y = \begin{vmatrix} 0 & -i \\ i & 0 \end{vmatrix} \quad ...(11)$$

For a single photon state

$$C\,|\gamma> = -\,|\gamma>. \quad ...(12)$$

A state vector for n-photons

$$C\,|n\gamma> = (-1)^n\,|n\gamma>. \quad ...(13)$$

Thus for n-quanta the eigenvalue of C is $(-1)^n$. Since π^o mesons decay through electro magnetic interaction into two photons $\pi^o \rightarrow 2\gamma$, it follows that the π^o-meson is in eignstate of C with eignvalue +1, *i.e.*

$$C\ \pi^o > = |\pi^o>$$

On the other hand $|\pi^o>$ and $|\pi^->$ and not eignstates of C as

$$C\,|\pi^+> = -\,|\pi^-> \text{ and } C|\pi^-> = -\,|\pi^+>.$$

As the triplet spin state is symmetrical and the singlet spin state anti-symmetrical, hence to exchange an electron with a positron we must induce factor $(-1)^{s+1}$ as well as a factor $(-1)^l$. Thus we have

$$(-1)^{l+s+1}\ C = -1 \text{ or } C = (-1)^{l+s}.$$

This relation gives that the singlet ground state (l = 0) decay into two photons (C = 1) and the triplet ground state decays into three photons.

SPACE-INVERSION INVARIANCE (PARITY)

The parity principle says that there is a symmetry between the world and its mirror image. This may be defined as reflection of every point in space through the origin of a co-ordinate system $x \rightarrow -x$, $y \rightarrow -y$ and $z \rightarrow -z$. If a system or process is such that its mirror image is *impossible to obtain in nature*, the system of process is said to violate the law of parity conservation.

Human body is a good example of mirror symmetry. The body of a car is symmetric except for the position of the steering wheel. If we were looking at the mirror image of a normal car, it seems to violate the symmetry but it is not the case as it is also possible to design a car with the steering wheel on the other side. The mirror view of a printed pure looks wrong. But there is nothing impossible about it. A printer could design inverted type and produced a page. One can read the page from right to left. The type of printing is not unnatural but is unconventional and unfamiliar.

All phenomena involving strong and electromagnetic interactions alone do conserve parity. In these cases the systems can be classified by the eigen values of the parity operator P. For a single particle Schrodinger wave function ψ, the result of the parity operation is

$$P\ |\psi(x)> = e^{i\alpha}\ |\psi(+x)\ \Delta, \qquad ...(1)$$

As α is an arbitrary real phase, hence can be set equal to zero.

$$P|\psi(x)> = |\psi(-x)> \qquad ...(2)$$

$$\text{and } P^2|\psi(x)> = |\psi(x)>. \qquad ...(3)$$

It shows eignvalues of P as + 1 or – 1.

The parity of the photon depends upon the mode of transition, it is due to the change of the sign of electromagnetic current j under the parity operation. The nucleons and electrons are assigned *positive* or *even* intrinsic parity. The pions have *negative or odd* parity as they involve in strong interactions with nucleons. K-mesons and η^o-meson have negative parity. $\Lambda^o - \Xi^- -$, $\Sigma - \Omega$-hyperons have positive intrinsic parity. All anti-particles of spin 1/2 are of opposite parity to the corresponding particle, while the bosons and their anti-particles have the same parity.

The conservation of parity requires that the Hamiltonian of a free system commute with the parity operator.

$$(PH - HP) = 0. \qquad ...(4)$$

The transition probability must be scalar it may contain pseudoscalar operator (I.p). The conservation of parity prevents the mixing of even and odd operators in the amplitude. Thus for the non-conservation of pairty in β-decay, the transition probability must contain both scalar and pseudoscalar terms; or the number of electrons emitted parallel and anti-parallel to the spin of the source should be different.

The weak decay of the K-mesons, which was difficult to reconcile with parity conservation and known as the τ-θ puzzle, was explained by Lee and Yang. They suggested that *the weak interaction was not invariant to space reflection.* In 1956, Wu and others, using polarized Co^{60} nuclei, found that the direction of emission of electrons in the transformation to Ni^{60} was preferentially opposite to the spin direction.

The value of the pseudoscalar I. p, where I is the nuclear spin and p the electron momentum, was measured and found to be different from 0.

COMBINED INVERSION (CP)

Landu (1956) advanced a hypothesis to the effect that any physical interaction must be invariant under simultaneous reversal of position coordinates and change over from particles to anti-particles. For example a neutrino has a definite helicity and its parity conjugate has opposite helicity. The charge conjugate of the neutrino also has opposite helicity. *Thus under the combined operation* PC (*or* CP) *the neutrino changes to anti-neutrino.* The combined operation also known as *combined parity* (charge and space) *is conserved in most of the known physical processes.* Let us consider the decay of the positive pion,

$$\pi^+ \rightarrow \mu L^+ + \nu_\mu L.$$

Here subscript L indicates that neutrino and +ve muon fly apart with left handed spin. As the C-inversion changes particles into anti-particles and vice-versa, whereas the P-inversion converts left handed motion to right handed motion. Hence

C-inversion : $\pi^- \rightarrow \mu L^- + \nu_\mu L$ impossible process

P-inversion : $\pi^+ \rightarrow \mu R^+ + \nu_\mu R$ Impossible process

CP-inversion : $\pi^- \rightarrow \mu R^- + \nu_\mu R$ possible process

Let us consider the case of the β-decay of polarized nuclei (*e.g.* Co^{60}). The interpretation of the parity non-conservation, charge non-conservation and conservation under the combined operation. In this figure B shows the direction of a magnetic field due to current loop, used for polarizing the nuclei. It represents the nuclear spin and thus known as *polarization vector.*

The upper diagrams represent the result of the reflection of the process shown in the lower diagrams of Fig. 1. Fig. 1(a) shows that the space reflection creates a different system, as it changes the direction of decay arrows but not the direction of B. Fig. 1(b) represents the charge conjugation only. This process leaves the direction of the decay unchanged, although electrons are replaced by positrons and the polarization direction is reversed. Fig. 1(c) shows the combined effect of CP operation. This shows that under CP reflection, we obtain the process of decay of the anti-nucleus. From the above results we conclude that the reflection type of symmetry can be obtained by the combined operation of C and P only in weak interactions. Number of other examples show that the weak interactions do not grossly violate the combined CP-invariance.

Unfortunately the decay of K-meson is not invariant under the combined operation CP.

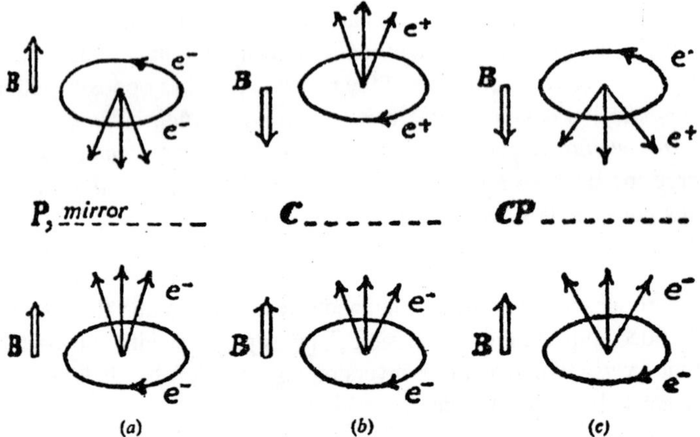

Fig. 1 : (a) Parity mirror, (b) charge conjugation mirror and (c) CP-mirror.

Time Reversal

The time reversal operator is defined as that operator which reverses the direction of time, or the direction of all motions. Under this operation displacement, acceleration and electric fields remain invariant but momenta, angular momenta and magnetic fields invert their signs. If the time reversed process is impossible to occur in nature we can say that the process violates time reversal symmetry. In order to imagine a process under time reversal, it is convenient to imagine that a film of the process is being non-backwards.

Time reversal invariance finds its simplest application in the world of particles, where it appears to govern the strong and electromagnetic interactions and possibly also the weak. It also shows that a particle possessing time reversal symmetry cannot have electric and magnetic dipole moments simultaneously. The time reversal process is the creation of an electron-positron pair by the collision of two photons.

Time reversal invariance is satisfied in quantum mechanics if the Hamiltonian H is time independent and real. In this case $\psi^*(x, -t)$ is the time reversal wave function of $\psi(x, t)$. Thus time reversal operation T changes ψ as

$$T\ \psi(x, t) = \psi^*(x, -t). \qquad ...(1)$$

The motion of a particle is an external fixed magnetic field is not invariant under inversion of time. The relativistic treatment of time reversal shows that the inversion of time axis inverts the sign of the electrostatic potential. The (π^o-) mesic field, like the magnetostatic potential, is odd under time reversal in order to ensure that the interaction is time reversible.

Combined Inversion of CPT

The strong and the electromagnetic interactions are invariant under the separate operations of C, P, and T. The weak interaction does not conserve parity and also is not invariant under charge conjugation. All the interactions are invariant under the combined *strong reflection* operation CPT, irrespective of the order of the operations. No example of a violation of the CPT theorem is known. It follows that if T invariance is satisfied for all interactions, then these interactions will also be invariant under the combined operation of CP. The existence of CP violating interactions means that the analysis is not quite accurate.

The quantities conserved have the quantum numbers, which behave in two different ways, when one considers a system formed by the combination of two other systems. The quantum numbers, such as *angular momentum, isospin, strangeness, baryon number, lepton number, electric charge* are called *additive.* On the other hand, quantum numbers, such as *parity, invariance under charge conjugation, invariance under time reversal* are called *multiplicative.*

Let us now discuss the production and properties of elementary particles.

Electrons and Positrons

The electron is the first and best known elementary particle. It was discovered in 1897 by *J J. Thomson.* The mass of the electron at rest is 9.1091×10^{-31} kg. All other elementary particles have larger mass than that of the electron. Thus it is convenient to use the *mass of the electron as a wait for a mass.* The electron has a negative electric charge of magnitude 1.6021×10^{-19} coulomb and a spin 1/2. Its spin can still have either one or the other of the two possible conditions, usually known as *spin up or spin down.* According to the Dirac quantum mechanical theory of the electron the magnetic dipole moment associated with the electron spin should be exactly one Both magneton ($eh/4\pi m_e$). Another prediction of this theory is that the $2S_{1/2}$ and $2P_{1/2}$ atomic levels of an one electron atom should be degenerative. Both these predictions are wrong, the magnetic moment is about 0.1 per cent greater than the Dirac value and the leaves have a small energy difference.

In *1930, C.D. Anderson* in the United States and *P.M.S. Blackett* in England observed a particle corresponding to the electron, but having positive charge. This particle was called positive electron or *positron.* Just previously *P.A.M. Dirac*, in England, has predicted the existence of a positron in his relativistic theory of the free electron. Dirac suggested that electron positron pair cannot be created by bombarding particles or photons of energy less than the threshold energy $2mc^2$ (1.02 MeV). At the time of passage through ordinary matter, the positron may join an electron to form a positron-electron system, called *positronium*, which lasts for a measurable time before combining to produce *annihilation* radiation. Because of this annihilation property, the *positron is said to be the anti-particle to the electron.*

The lowest Bohr orbit of positronium is one for which n = 1 and l = 0 (S-State). This state has fine structure owing to the spins of the particles. The atom is in a ^1S state when the two spins are oppositely directed and in a ^3S State when the spins are parallel. Theoretical calculations by J. Pirenne (1944) and J.A. Wheeler (1946) showed that the lifetime of the singlet should be of the order of 1.3×10^{-10}. The annihilation radiation emitted by the combining of a positron-electron pair in the singlet state should consist of two gamma ray photons emitted simultaneously in opposite directions. Theoretical calculations by Ore and Powell (1949) showed that the lifetime of the triplet state should be about 1.5×10^{-7} sec. The annihilation should consist of three gamma ray photons travelling away from their origin at mutual angles of 120. We can explain the production of two in one case and of three γ-ray photons in the other as the result of the conservation of spin in the annihilation process. In the singlet state net spin in zero, two photons (spins +1 and –1) are expected in opposite directions in order to satisfy the law of conservation of momentum. In the triplet state (spin +1 or –1), three photons (+1, +1, –1 or –1, –1, +1) are expected. M. Deutsch (1951) measured experimentally the time between the formation of positrons and their subsequent annihilation and found the value 1.5×10^{-7} sec. This is the experimental evidence for the existence of the triplet state (*ortho*) of positronium.

Anti-Protons and Protons

Since early in the present century, the proton has been recognized as an essential component of matter. It is readily available for study and is the nucleus of the common hydrogen atom *(singly charged positive hydrogen atom H^+)*.

The name proton, was suggested by Rutherford. Since the weight of the hydrogen atom is 1837 m, hence the weight of the proton is 1836 m_e. The charge of the proton is positive and exactly equal in magnitude to that of the electron. It is a fermion of intrinsic spin half. It is useful to regard the proton, as a particular charge state $T_3 = +1/2$ of a single particle, the nucleon, of total isospin T = 1/2. Proton has a magnetic moment of 2.79275 nuclear magneton ($\mu_n = \mu.eh/4\pi m_p$).

Like the electron, the proton has a counter part of opposite charge with which it undergoes annihilation. Just as positron-electron pairs are formed by the interaction with matter of rays or particles having sufficient energy (1.02 MeV), so it should be possible to produce proton-anti-proton pairs in a similar manner, but it would require about 2000 times energy.

Until 1955, the required energy was available only in high energy particles in cosmic radiations and the evidence of the antiprotons could be obtained from cosmic ray events in photographic emulsions and cloud chambers. With the completion of the 6 BeV bevatron at the University of California it was possible to perform an unambiguous experiment for the production and detection of anti-protons. The first convincing evidence of the anti-proton came from the experiment performed by Chamberlain, Segre Weigand and Yspsilantis in 1955.

The schematic diagram of their detecting apparatus, which is as a mass spectrograph, is shown in Fig. 2. Protons of 6 BeV energy are incident on a copper target T. Anti-protons are produced by collisions between the incoming protons and protons in the target nuclei ($p+p \rightarrow 3p+p^-$). Negatively charged particles were selected by a mass spectrograph consisting of a series of deflecting magnets, focusing magnets, scintillation counters and Cerenkov counters. This selected beam of particles was expected to about 50,000 negative pions to every anti-proton. These particles all have the same momentum (1.19 BeV/c) and thus travel in the same trajectory in a magnetic field. Because of its larger mass the proton had a velocity of only 0.78c while the pion had a velocity of 0.99c.

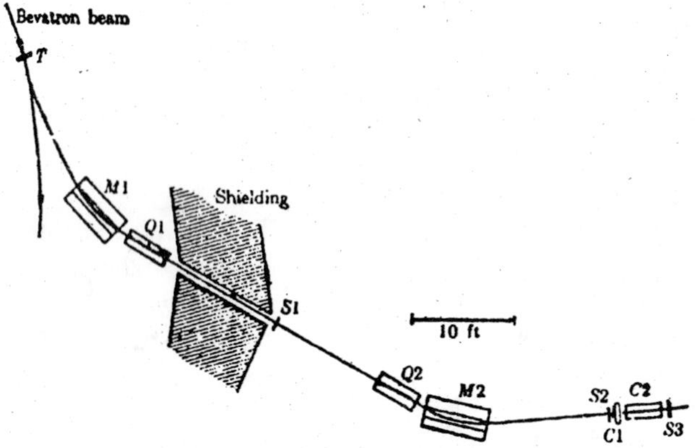

Fig. 2 : Experimental arrangement for the detection of anti-proton.

Following arrangements were made for the identification of the anti-protons. The distance between plastic scintillation counters S_1 and S_2 was 40 ft.; the time of flight for anti-protons was 51 n sec. and for pions 40 n sec. Thus we see that the pulse emitted from counter S_2 occurs 51 n sec later than that from S_1. To obtain a coincidence the signal from S_1 was delayed enough that the two signals arrived at the coincidence circuit at exactly the same time. Sometimes there would be two different mesons passing through the two counters in such a way that a pulse from the second counter would be produced just 51 n sec. after a pulse from the first.

This gave accidental coincidences appearing like anti-protons. To eliminate this difficulty the experimenters introduced the use of *Cerenkov counters*. The Cerenkov counter C_1, of chemical composition $C_8F_{16}O$ and refractive index 1.28, was designed to respond to particles with velocities greater than 0.79 c. Any pulse from C_1 cannot be due to an anti-proton having velocity 0.78c. The second Cerenkov counter C_2, of fused quartz and refractive index 1.46, was designed to respond only to particle whose velocities were between 0.75c and 0.78 c. It was converted in anticoincidence with C_1, so that the pulses in C, were those produced by anti-protons only and in C_1 by negative pions only. The third scintillation counter S_a certifies that an anti-proton has not been scattered through a large angle in C. The distance between S_2 and S_3 was only a few inches.

In the experiment performed by Hill, Johanson and Gardner, the beam of 6.2 BeV protons were incident on a photographic emulsion. The proton produced a star (star A in Fig. 2) in a collision with a nucleus. One of the particles produced in this event was an anti-proton, which proceeded in the emulsion and found to give another star, whose several prongs were corresponding to particles appeared in this process. Measurements showed that the total energy released was greater than expected from the mass of the anti-proton alone, so that another particle, either a proton or neutron must have been annihilated simultaneously. The annihilation of an anti-proton with a proton or neutron differs from that of a positron and electron in that several charged and neutral n-mesons together with protons and occasionally (if sufficient kinetic energy is supplied) deaterons and K-mesons are usually produced, rather than two or three quanta. The mass of the anti-proton has not been measured very accurately but it is presumably exactly equal to that of the proton. *Its electric charge is equal in magnitude, but negative in sign*, confirmed by the fact that protons annihilated with anti-protons. The energy released in proton antiproton annihilation is 1836.12 times the energy released in electron-proton annihilation.

Elementary Particles

In 1932, when Chadwick identified the neutron and Heisenberg suggested that atomic nuclei consisted of neutrons and protons, it seemed as if p, n and e^- were sufficient to account for the structure of matter. Besides these there was the photon, *the intermediary* or *field particle for electromagnetic forces*, such as exist between the nucleus and electrons in the atom. If anti-matter exists it would then be made up of anti-electrons, i.e. positions, anti-protons and neutrons. Thus we see that seven particles could explain both matter and anti-matter. In 1935, Yukawa postulated the existence of another particle, with a mass m $\simeq$ 200 m_e *as the field particle for the strong nuclear forces*. Recently the extensive studies made partly on high energycosmic ray particles and even more, with the help of high energy accelerators have revealed the existence of numerous new nuclear particles. Apart from a dozen or so the particles have very short lifetimes, very much less than 10^{-6} sec. They cannot therefore be regarded as normal constituents of matter. They arc characterised by the parameters, miss, spin, electric charge and magnetic moment They have been described by such adjectives, as *fundamental, strange* and *elementary*, but none of these is quite appropriate. The word fundamental implies that the particles are the basic building blocks of matter, but unstability of most of the particles indicates that the great majority are certainly not. It is true that their behaviour was strange in the early 1950, but it is much less now. For the want of better one the term elementary particles is now commonly used. These particles are elementary in much the same sense as are the chemical elements.

TRANSURANIC ELEMENTS

The capture of a neutron by uranium does not necessarily lead to the fission of the compound nucleus. Uranium-238, for example, has a small capture cross-section for low energy neutrons. The importance of the capture of neutrons by ${}_{92}U^{238}$ is that it leads to the formation of new elements known as *iransvranic elements* as they lie beyond uranium in the penodic system. In 1934, Fenni suggested the possibility that the bombardment of uranium with neutrons might result in the production of elements of atomic number greater than 92. If neutron is captured by U^{238}, the following reaction should occur.

$${}_{92}U^{238} + {}_{0}n^{1} \rightarrow ({}_{92}U^{239}) \rightarrow {}_{92}U^{239} + \gamma$$

If the ${}_{92}U^{239}$ were than to decay by electron emission, the result would be a nuclide with $Z = 93$, an isotope of an unknown element. It was also considered that the new element might decay by electron emission to form an isotope of another new element of atomic number 94. Between 1934 and 1939 many attempts were made to make and identify transuranic elements.

A—Neptunium (Z = 93)

McMillan and Abelson in 1940 discovered the first transuranic element by bombarding uranium with large energy neutrons. The capture of neutrons led to the formation of a new isotope of uranium with mass number 239, which decays by β-emission with a half life of 23 min resulting in the formation of a new element. This was later named *neptunium* (Np), by the discoverers, after the *name of the planet Neptune lying beyond Uranus in the solar system*. The element thus discovered was found itself to be P-active and to decay with a 2.3 day period. The following are the nuclear reactions :

$${}_{92}U^{238} + {}_{0}n^{1} \rightarrow {}_{92}U^{239} + \gamma$$

$${}_{92}U^{239} \rightarrow {}_{93}Np^{239} + {}_{-1}e^{0} \qquad (T = 23 \text{ min.})$$

Mc-Millan and Abelson performed chemical experiments with the minute quantities *(tracer amounts)* of the element formed in the above process and showed that its oxidation states differ from those of uranium.

In this way the existence of new element was established by chemical experiments. By microchemical tests the chemical properties of Np were ascertained and its position in the periodic table confirmed as following uranium:

Soon after the discovery of the 2.3 day isotope of neptunium, Seaborg, McMillan, Kennedy and Wahl in Berkeley at the end of 1940 identified another isotope by the bombardment of uranium oxide with fast deuterons. The product is β-active, decaying with a half life of 2.1 days.

$$_{92}U^{238} + {}_{2}H^{2} \rightarrow {}_{93}Np^{238} + 2\ {}_{0}n^{1}$$

The two isotopes of Np mentioned so far are too unstable to carry out chemical analyses on them. The most stable isotope was discovered by Wahl and Seaborg in 1942 by the action of fast neutrons on uranium with mass number 238. The following are the processes:

$$_{92}U^{238} + {}_{0}n^{1} \rightarrow {}_{92}U^{237} + 2\ {}_{0}n^{1}$$

$$_{92}U^{237} \rightarrow {}_{93}U^{237} + {}_{-1}e^{0}\ (\beta^{-}) \qquad (6.8 \text{ days})$$

$$_{93}Np^{237} \rightarrow {}_{91}U^{233} + {}_{2}He^{4}\ (\alpha) \qquad (2.2 \times 10^{6} \text{ days})$$

Besides the above mentioned three isotopes of neptunium, eight others, with mass number ranging from 231 through 241, have now been produced by other reactions. Some of the isotopes and the reactions by which they are made are:

Np^{231} ... U^{233} (d, 4n), U^{235} (d, 6n), U^{238} (d, 9n)

Np^{234} ... Pa^{231} (α, n), U^{235} (d, 3n), U^{235} (α, p4n)

Np^{235} ... U^{235} (α, p3n), U^{235} (d, 2n)

Np^{236} ... U^{235} (d, n), U^{235} (α, p2n), U^{238} (α, 4n), Np^{237} (d, t)

Np^{238} ... U^{238} (d, 2n), U^{238} (α, p3n)

Np^{239} ... U^{238} (d, n), U^{238} (α, p2n)

Neptunium is a silvery metal, not affected by air. Its measured density is about 19.5 gm/cc at normal temperature hut varies with the temperature. The melting point of metallic neptunium is 640°C.

Unlike rhenium, neptunium does not precipitate with hydrogen sulphide in acid solution, it is not reduced 10 the metallic state by Zn, neither does it have an oxide volatile at red heat. In a reducing (SO_2) solution neptunium could be precipitated with cerium as fluoride, CeF_3, or with thorium as thorium, Th $(IO_3)_4$. After oxidation with an acid bromate solution the neptunium activity could be separated with sodium uranyl acetate. Subsequent work has shown that neptunium resembles uranium in the respect that both exist in III, IV, V and VI oxidation states. The main difference is that neptunium VI is more easily reduced to the IV and III states than is uranium VI. The most stable solid compounds of neptunium are those of the oxidation state IV but for the later the uranyl IV compounds are the most stable.

If the dioxide of neptunium is heated in a mixture of hydrogen and hydrogen fluoride at 500°C, the trifluoride is formed. Thus

$$2NpO_2 + H_2 + 6\ HF \rightarrow 4Np\ F_3 + 4H_2O$$

Upon heating the above product in a current of oxygen and hydrogen fluoride, also at 500°C, we have

$$4NpF_3 + O_2 + 4HF \rightarrow 4Np\ F_4 + 2H_20t$$

If the trifluoride is heated in fluorine, the hexafluoride NpF_6, is formed. It is a solid that vaporises at low temperatures like uranium hexafluoride. Similar to uranium tetrachloride, neptunium tetrachloride is formed by heating the dioxide of neptunium in carbon tetrachloride vapour according to the equation

$$NpO_2 + 2CCl_4 \rightarrow Np\ Cl_4 + 2COCl_2.$$

Further $NpCl_4$ is reduced to the trichloride by means of hydrogen at 450°C similar to uranium. Two bromides and an iodide of neptunium have also been made by a method similar to uranium. The identity of all the halides was established by X-ray diffraction studies which showed them to have the same crystal structure as the corresponding uranium compounds. Neptunium trifluoride reduces to the elemental state by heating at 1200°C in the presence of barium vapour.

B—Plutonium (Z = 94)

When the β-active isotopes of neptunium decay, the daughter element should have the atomic number 94. This element was discovered by Seaborg, McMillan, Wahl and Kennedy in the winter of 1940-41. This element was named by its discoverers as *plutonium, symbol Pu*, after *Pluto, the planet beyond Neptune*. They bombarded uranium with deuterons and showed the following reactions.

$$_{92}U^{238} + {_1H^2} \rightarrow (_{93}U^{240}) \rightarrow {_{92}Np^{238}} + 2\ {_nN^1}$$

$$_{92}Np^{238} \rightarrow {_{94}Pu^{238}} + {_{-1}e^0} \qquad (T = 2.1\ \text{days})$$

$${}_{94}Pu^{238} \rightarrow {}_{92}U^{234} + {}_{2}He^{4} \qquad (T = 86.4 \text{ years})$$

The important fissible species plutonium 239, which McMillan and Abelson had sought as the decay product of Np^{239}, was characterized in March 1941 by Kennedy Seaborg, Segre and Wahl.

$${}_{93}Np^{238} \rightarrow {}_{94}Pu^{239} + {}_{-1}e^{0} \qquad (T = 2.3 \text{ days})$$

$${}_{94}Pu^{239} \rightarrow {}_{92}Pu^{235} + {}_{2}He^{4} \qquad (T = 24000 \text{ years})$$

Many new isotopes of plutonium are formed by cyclotron α-bombardment with uranium isotope as well as by the β-decay of the corresponding neptunium isobars. Known isotopes have all mass numbers from 232 to 246. In addition a useful method for the production of plutonium isotopes with mass numbers exceeding 239 is to expose plutonium 239 to neutrons in a nuclear reactor, such as:

${}_{91}Pu^{232}$ ... ${}_{92}U^{235}$ (α, 7n)

${}_{94}Pu^{234}$ ... ${}_{92}U^{233}$ (α, 3n)

${}_{94}Pu^{237}$ ... ${}_{92}U^{235}$ (α, 2n), ${}_{92}U^{238}$ (α, 5n)

${}_{94}Pu^{240}$ ... ${}_{92}U^{238}$ (α, 2n)

${}_{94}Pu^{241}$ ... ${}_{92}U^{238}$ (α, n).

All the even numbered isotopes except Pu^{242} as well as the familiar 239-isotopes arc alpha-emitters. The isotopes having mass numbers 241, 243, 245 and 246 are beta-emitters.

Plutonium metal is *silvery white* but it tarnishes rapidly due to reaction with oxygen in the air. It also reacts readily with water and hydrogen gas. Its melting point is 640°C. Upto melting state is exists in six different solid modifications. Except four there are two forms existing between 319° and 451°C and between 451° and 470°C respectively which have the very unusual property of contracting when heated. All the forms have heat and electrical conductivities which are exceptionally low for a metal. The density of plutonium at normal temperatures is 19.7 gms/c.c which decreases with the temperature upto about 16 gm/c.c. The Chemistry of plutonium is in many respects very similar to that of neptunium. Like neptunium, plutonium resembles uranium in many of its chemical properties, the main difference lies in the stability of the IV state in plutonium whereas with uranium it is the VI state which is the most stable. The roost stable oxide is PuO_2. If it is heated to about 1500°C in a vacuum it forms the lower oxide Pu_2O_3. In addition there is a mixed oxidation state varying in composition from Pu_2O_3 to roughly Pu_4O_7. The halides PuF_3, $PuBr_3$ and PuI_3 may be prepared by the methods as described for the neptunium compounds. The PuF_4 is stable, although it can be reduced to the trifluoride. Plutonium formed a solid hexafluoride, PuF_6, which sublimes readily, which is less stable than uranium hexafluoride.

Pu-III can be precipitated from solution as the hydroxide $Pu(OH)_3$ and the fluoride PuF_3. Aqueous solutions of the +ve Pu-III ion are oxidized in air to the Pu-IV state. As Pu-IV is most stable we know many solid compounds of this category. The fluoride PuF_4, the iodate $Pu(IO_3)_4$ and the hydroxide $Pu(OH)_4$ are insoluble and can be precipitated from aqueous solution. The chloride, nitrate, perchlorate and sulphate are soluble in water. Pu-IV solutions can be reduced to the Pu-III state by means of warm SO_2 by hydroxylamine, by iodide ion or by zinc amalgam. The conversion of Pu-IV to the VI state requires the use of fairly strong oxidizing agents, as acid solutions of permanganate or dichromate, ceric ions or pursulphate in the presence of silver ions. A solution of Pu-VI in dilute hydrochloric acid can be partially reduced electrolytically to the Pu-V state.

The first microgram samples of the metal were prepared by H.L. Baumbach and P.L. Kirk by reducing plutonium tetrafluoride with barium vapour at 1400°C. At present plutonium is made by heating the tetrafluoride with calcium in closed vessel. The Pu^{239} is manufactured in large amounts in nuclear reactors and can be used for the production of nuclear power or atomic bombs. This plays an important part in an atomic bomb project because it is fissionable with both slow and fast neutrons.

C—Americium (Z = 95)

The third transuranic element was discovered in 1944 by Seaborg James and Morgan, by bombarding uranium 238 with α-particles from a 40 MeV cyclotron. It has been named *americium*, symbol *Am*, by analogy with its rare earth homologue europium and is formed from one of the unstable isotopes of plutonium according to the reactions.

$$_{92}U^{238} + {}_2He^4 \rightarrow {}_{94}Pu^{241} + {}_n n^1$$

$$_{94}Pu^{241} \rightarrow {}_{95}Am^{241} + {}_{-1}e^0 \qquad (T = 13 \text{ years})$$

This isotope of Am is an α-emitter having half life 500 years.

$$_{95}Am^{241} \rightarrow {}_{93}Np^{237} + {}_2He^4 \qquad (T = 500 \text{ years})$$

More substantial quantities of Am211 were produced by exposure of Pu^{239} 10 thermal neutrons in a nuclear reactor. Two stages of neutron capture resulted in the formation of Pu^{241} as [Pu^{239} (n, γ) Pu^{240}; Pu^{240} (n , γ) Pu^{241} Am^{241} has also been obtained directly by the (α, 2n) reaction with Pu^{239}. It is an a-emiter with a half life of 433 years.

Isotopes in the range 237-246 have been formed by a series of reactions with reactor neutrons and using high energy particles from accelerators. Among them isotope (241) has been used extensively in the study of the chemistry of americium. Several of the isotopes of lower mass number have been obtained by bombardment of Pu^{239} with protons or deuterons, or of Np^{237} by α-particles. Decay of the β-emitting isotopes of Pu yields isotopes of americium with the same numbers.

It is a silvery metal with a density of 13.7 gm./c.c. at ordinary temperatures. Its melting point is 1340°C. During the winter of 1945-46, B.B. Cunningham prepared some pure compounds of americium, such as AmF_3 and Am_2O_3. Since that time a number of other solid compounds have been prepared including $AmCl_3$. $AmBr_3$, AmI_3 and AmO_2. The III oxidation state of americium is most stable but there is evidence for the existence of both higher and lower states. Aw-III can be oxidized to the VI state by using very strong oxidizing agent like ammonium persulphate. The V state of americium is obtained most readily by reduction of the VI state in solution by means of chloride or bromide ions. It can also be formed by oxidation of the III state with hypochlorite. The relative stability of the Am-W state is shown by the fact that the method used for the preparation of the tetrafluoride and tetra-chloride of neptunium yield only the corresponding trihalides of americium. Tribromide and tri-iodide of americium are made by heating the oxide together with aluminium in the vapour of the appropriate halogen similar to those of neptunium and plutonium. The action of barium vapour on AmF_3 at temperatures over 1000°C gives elemental americium.

D—Curium (Z = 96)

G.T. Seaborg, James and Ghiorso in 1944 identified the 4th transuranic element of atomic number 96, which has been named *curium*, symbol *Cm*, in honor of Marie and Pierre Curie for their pioneer work on the radioelements. This was first obtained by bombarding Pu^{239} with cyclotron accelerated a particles and found to be an α-emitter of half life 162.5 d.

$$_{94}Pu^{239} + {}_{2}He^{4} \rightarrow {}_{96}Cm^{242} + {}_{0}n^{1}$$

The same nuclide was later found to be formed in the β-decays of the 16 hr. isomeric (metastable) state of Am^{242}. Thus

$$_{95}Am^{242} \rightarrow {}_{96}Cm^{242} + {}_{-1}e^{0}$$

Several different isotopes of curium are formed when Pu^{239} is bombarded with a-particles of 30 to 40 MeV. Such as:

$$_{94}Pu^{239} + {}_{2}He^{4} \rightarrow {}_{96}Cm^{241} + 2\ {}_{0}n^{1}$$

$$_{94}Pu^{239} + {}_{2}He^{4} \rightarrow {}_{96}Cm^{240} + 3\ {}_{0}n^{1}$$

These are followed by the radioactive disintegration of Cm as.

$$_{96}Cm^{241} \rightarrow {}_{94}Pu^{237} + {}_{2}He^{4}$$

$$_{96}Cm^{241} + {}_{-}e^{0} \rightarrow {}_{95}Am^{241} \qquad (T = 35 \text{ days})$$

$$_{96}Cm^{240} \rightarrow {}_{94}Pu^{236} + {}_{2}He^{4} \qquad (T = 26.8 \text{ day})$$

There are thirteen isotopes known, ranging from $_{96}Cm^{238}$ to $_{96}Cm^{250}$. The lower isotopes, namely curium-238, 239. 240 and 241 result from the same bombardment process, *i.e.*, Pu^{239} by α-particles, as yields $_{96}Cm^{242}$. The higher isotopes are obtained either as a result of radioactive decay or by several stages of neutron capture, starting with either U^{238}, Pu^{239} or a curium isotopes of lower mass number.

Curium, in the form the oxide Cm_2O_3 was first obtained as a pure compound in 1947. The (111) oxidation state of curium is extremely stable. There is no evidence for any other state in solution. However the solid tetrafluoride, CmF_4 a curium (IV) compound, is known. From the theoretical point of view curium's solutions are colourless. It exhibits strong absorption of ultraviolet radiation $_{96}Cm^{240}$ disintegrates by spontaneous fission with a half life of 7.9 × 105 years.

E—Berkelium (Z = 97)

The fifth transuranic element was discovered by Thompson, Ghiorso and Seaborg at the end of 1949 by bombarding americium with 35 MeV alpha particles. It has been named berkelium, symbol Bk, in honor of the city of Berkeley, where the discovery of this and of other related elements was made. The nuclear reaction being

$$_{95}Am^{241} + {}_{2}He^{4} \rightarrow {}_{97}Bk^{243} + 2\ {}_{0}n^{1}$$

This isotope disintegrates mostly by electron capture and has a life time of 4.6 hours. Thus

$$_{97}Bk^{243} + {}_{-1}e^{0} \rightarrow {}_{96}Cm^{243}$$

$$_{97}Bk^{243} \rightarrow {}_{95}Am^{239} + {}_{2}He^{4} \qquad (T = 4.6 \text{ hours})$$

At present there are eight isotopes of berkelium available, but only in extremely small quantities. Their masses range from 243 to 250. Most of them, have longer half-lives than that of Bk-243. Berkelium-247 is an alpha emitter with half-life of 1380 years. It can be obtained in traces by bombardment of curium isotopes with alpha particles. The berkelium-249, a β-particle emitter with a half-life of 314 days, is available in fair quantities as a result of multiple neutron capture and occasional β-decay In a reactor starting with ${}_{96}Pu^{239}$ or ${}_{96}Am^{243}$ or other heavy nuclides.

F—Californium (Z = 98)

Thompson, Ghiorso, Seaborg and Street in March 1950 produced a new element by bombarding curium-242 with 35 MeV alpha particles. They named it californium, symbol Cf, for the state and university of California. The reaction being

$${}_{96}Cm^{242} + {}_{2}He^{4} \rightarrow {}_{98}Cf^{244} + 2\ {}_{0}n^{1}$$

followed by

$${}_{98}Cf^{244} \rightarrow {}_{96}Cm^{240} + {}_{2}He^{4} \qquad (T = 25 \text{ min.})$$

We also occur reaction of (α, n) type instead of (α. 2n) and the product is californium–245, half life 44 minutes.

$${}_{98}Cm^{242} + {}_{2}He^{4} \rightarrow {}_{98}Cf^{245} + {}_{0}n^{1}$$

$${}_{98}Cf^{245} \rightarrow {}_{96}Cm^{241} + {}_{2}He^{4} \qquad (T = 44 \text{ min.})$$

Eleven californium isotopes in the mass number range from 244 to 254 have been identified. In addition to the use of alpha particle reaction? and successive neutrons capture processes, some of the isotopes of californium have been obtained by bombardment of uranium-238 with cyclotron ions heavier than α-particles such as ${}_{6}C^{12}$ and ${}_{7}N^{14}$ ions according to the following reactions:

$${}_{92}U^{238} + {}_{6}C^{12} \rightarrow {}_{98}Cf^{244} + 6\ {}_{0}n^{1}$$

$${}_{92}U^{238} + {}_{6}C^{12} \rightarrow {}_{98}Cf^{246} + 4\ {}_{0}n^{1}$$

$${}_{92}U^{238} + {}_{7}N^{14} \rightarrow {}_{98}Cf^{248} + 3\ {}_{0}n^{1} + {}_{1}p^{1}$$

The californium–254 is formed by a series of neutron captures and beta decays. Its half life is 60 days. It decays almost entirely by spontaneous fission, rather than by alpha emission. The chemistry of californium is not quite so well established since there is comparatively little material available.

G—Einsteinium (Z = 99) and Fermium (Z = 100)

The new element of atomic number 100 was identified at Berkeley by the ionexchange separation technique. Also element of atomic number 99 was observed both at Berkeley and at the Argonne National Laboratory. By the agreement among the many workers in these laboratories; element 99 was called einsteinium, symbol Es and element 100 was given the name *fermium*, symbol *Fm in honor of Einstein and Fermi respectively*. The work was carried out during 1953 but was announced in June 1955.

An isotope of lower mass number was reported as resulting from the bombardment of uranium with nitrogen ions. The reaction being

$$_{92}U^{238} + _{7}N^{14} \rightarrow _{99}Es^{247} + 5\ _{0}n^{1}$$

Except Es–247, eleven isotopes of einsteinium with mass numbers from 245 to 256 are known. Starting with plutonium–239, long exposure in the materials testing reactor led to the formation of einsteinium–253. The longest lived isotope being Es–254 with half life 270 days. It may be obtainable in weighable amount which can be used for further investigations and production of elements of even higher atomic number.

When accelerated nitrogen nuclei collides with the uranium a complex of mass number 252 forms. This is over energised and loses particles to produce first Es and then by β-decay one with charge 100, *i.e.*, Fm. Fermium has ten isotopes within the range 248–257. The lightest isotopes are produced by heavy ion bom-bardment as given by the reactions.

$$_{94}Pu^{240} + _{6}C^{12} \rightarrow _{100}Fm^{248} + 4\ _{0}n^{1}$$

$$_{92}U^{238} + _{8}O^{16} \rightarrow _{100}Fm^{249} + 5\ _{0}n^{1}$$

$$_{92}U^{238} + _{6}O^{16} \rightarrow _{100}Fm^{250} + 4\ _{0}n^{1}$$

The intermediate isotopes arc produced by the action of alpha-particles on various isotopes of californium. The heaviest isotopes are generally obtained by multiple neutron capture reactions. The longest lived isotope being Fm–253 with half life 4.5 days.

H—Mendelevium (Z = 101)

In 1955, Ghiorso, Harvey, Choppin, Thompson and Seaborg discovered a new element of atomic number 101. *It was named mendelevium, in honour of Mendelyeev*. An invisible amount of Es^{253}, consisting of about 10^9 atoms, deposited on a Au foil, was bombarded with 48 MeV α-particles. The product nuclei were collected and the new element was isolated by the ion-exchange method.

$${}_{99}Es^{253} + {}_{2}He^{4} \rightarrow {}_{101}Md^{255} + 2\ {}_{0}n^{1}$$

This decays by orbital electron capture to ${}_{100}Fm^{255}$ with a half life of 30 minutes. The half life of mendelevium–256 is 90 minutes. A third isotope mass number 257 and about 3 hours half life has also been identified. Other isotopes, ranging from 250 to 254, have also been reported.

I—Nobelium (Z = 102)

The existence of the new clement of atomic number 102 was first sought in 1957 by a team of British Swedish and United State scientists working on the Swedish cycle, tron. *It was called nobelium, symbol No, in honor of Alfred Nobel and the Nobel Institute of Physics in Sweden* where the work was done. In this experiment curium–244 was bombarded with accelerated carbon–13 ions at energies of 65 to 100 MeV. The nuclide produced in this manner was said to have a mass number of 251 or 253 and to emit a-particles of 8.5 MeV energy with a half life of about 10 minutes. The existence of nobelium was finally proved in April 1958 by carbon-12 ions bombardment in the later experiments at Berkely and California. This isotope of nobelium was found to be alpha particle emitter. The following reactions were observed:

$${}_{96}Cm^{246} + {}_{6}C^{12} \rightarrow {}_{102}No^{254} + 4\ {}_{0}n^{1}$$

$${}_{102}No^{254} \rightarrow {}_{100}Fm^{250} + {}_{2}He^{4} \qquad (T = 3 \text{ sec.})$$

$${}_{100}Fm^{250} \rightarrow {}_{98}Cf^{246} + {}_{2}He^{4} \qquad (T = 30 \text{ sec.})$$

In 1965 Russian scientists reported that this isotope of nobelium made by the reactions

$${}_{92}U^{238} + {}_{10}Ne^{22} \rightarrow {}_{102}No^{254} + 6\ {}_{0}n^{1}$$

and $${}_{95}Am^{243} + {}_{7}N^{15} \rightarrow {}_{102}No^{254} + 4\ {}_{0}n^{1}$$

had a much longer half-life probably about 50 seconds. Isotope of nobelium with mass number 252 can be prepared by the reaction

$${}_{95}Cm^{244} + {}_{6}C^{12} \rightarrow {}_{102}No^{252} + 4\ {}_{0}n^{1}$$

It has a half life of about 2.5 seconds. In addition to these isotopes all the others in the mass range from 251 through 257 have been obtained by the bombardment of curium isotopes with accelerated carbon ions.

J—Lawrencium (Z = 103)

The discovery of the new element of atomic number 103 was reported in 1961 by the American research workers at Berkely. *It was given a name lawrencium, symbol Lw, in honour of E.O. Lawrence*, inventor of the cyclotron, in recognition of the contributions to nuclear science made by him. In their experiment lawrencium is formed by bombarding a mixture of californium isotopes with mass numbers 249, 250, 251 and 252 with a beam of ${}_5B^{11}$ and ${}_5B^{11}$ ions. This lawrencium has mass number 257 and is an a-emitter of 8.6 MeV. The half life was found to be about 8 sec.

In U.S.S.R. the isotope ${}_{103}Lw^{256}$ was produced by the reaction

$${}_{95}Am^{243} + {}_8O^{18} \rightarrow {}_{103}Lw^{256} + 5\ {}_0n^1$$

This isotope of lawrencium has a half life of 45 seconds. By the use of the double recoil technique the product of two stages of decay (α-particle emission and electron capture) was collected and identified chemically.

K—Kurchatovium (Z = 104)

In August 1965, Soviet research workers at the laboratory for Nuclear research, Dubna, U.S.S.R., claimed-to have synthesized element 104 by bombarding plutonium-242 with accelerated neon-22 ions. This element was called kurchatorium in honor of I.V. Kurchatov, the first director of the Russian atomic energy project

$${}_{94}Pu^{242} + {}_{10}Ne^{22} \rightarrow {}_{104}Ku^{260} + 4\ {}_0n^1$$

Ghiorso and Sikkeland, in the review published in 1967 of the spontaneous fission half lives of elements of even values of Z from 92 through 102, suggested that the half life of the ${}_{104}Ku^{260}$ should be considerably less than 0.3 sec.

L—Hahnium (Z = 105)

Soviet scientists working with Georgii Flerov at the Joint Institute for Nuclear Research at Dubna have claimed that they have synthesised element 105. In 1967 it was announced that they have produced few atoms of two isotopes (mass number 261, 262) of element 105 by bombarding americium-243 with an isotope of neon-22. The half "life was said to be 0.1 sec. However their results have not been conclusive.

The successful synthesis of this new element has been achieved by Dr. Albert Ghiorso arid his coworkers at the Lawrence Radiation Laboratory of the University of California. Berkely by bombarding Cf^{249} with N^{15} in the Heavy Ion Linear Accelerator of the L.R.L. This synthesis of element 105 has been announced at the American Physical Society meeting held in April 1970. Dr. Ghiorso named the *new element Hahnium symbol Ha, after the name of German Chemist, Professor Otto Hahn*, who had been one of the pioneers of nuclear fission studies in 1930. Dr. Ghiorso and his coworkers used a target made of 60 μ gm of Cf^{249} to produce hahnium atoms, Ha^{260} is produced when Cf^{249} is bombaraed with N^{15}.

$$_{98}Cf^{249} + {}_{7}N^{15} \rightarrow {}_{105}Ha^{260} + 4\ {}_{0}n^{1}$$

Hahnium–260 has a half life of 1.3 seconds and decays to form Lw^{250} and a-particles, Lawrencium in turn decays, into Md^{252} and α-particles.

From theoretical considerations we know that the elements with atomic numbers in the region of 102 to 104 are exceptionally unstable and that a group of elements starting with 105 will have longer half lives.

This idea is now confirmed by the recent discovery of elements of Z-values 116, 124 and 126. Prof. Pal Dirac, R.V. Jentri and T.A. Kahil American scientists, the discoverers, have suggested that these elements have a life-time of the order of million years.

Suggested outer electron configuration for transuranic elements

Np $5f^4\ 6d\ 7s^2$; or	$5f^5\ 7s^2$;	Pu $5f^6\ 7s^2$;	Am $5f^7\ 7s^2$;
Cm $5f^7\ 6d\ 7s^2$;	Bk $5f^8\ 7s^2$;	O^- $5f^9\ 7s^2$;	Cf $5f^{10}\ 7s^2$;
Es $5f^{11}\ 7s^2$;	Fm $5f^{12}\ 7s^2$;	Md $5f^{13}\ 7s^2$;	No $5f^{14}\ 7s^2$;
Lw $5f^{14}\ 7s^2$;	Ku $5f^{14}\ 6d^2\ 7s^2$;	Ha $5f^{14}\ 6d^3\ 7s^2$;	

ELEMENTS IN THE PERIODIC TABLE

Until 1939 four elements were missing from the periodic table. These were the elements of atomic numbers 43, 61, 85 and 87.

A—Technetium

Element 43 was made first in 1937 by bombarding molybdenum of atomic number 42 with deuterons. Neutrons are emitted and the element 43 left behind. Its discoverers Perrier and Segre have called it *technetium*, symbol Tc. This name was derived from the Greek Word technetos, meaning artificial. Longest lived isotope of technetium is obtained in the (d, 2n) reaction with molybdenum-97. G. T. Seaborg and E. Serge in 1939 discovered a isotope of technetium with mass number 99 by the interaction of molybdenum-98 with slow neutrons.

There are 16 isotopes known, ranging from $_{43}Tc^{92}$ to $_{43}Tc^{108}$ with the exception of $_{43}Tc^{106}$. Some are fission products and others have been obtained by the bombardment of isotopes of niobium, molybdenum or ruthenium with accelerated charged particles.

Pure technetium metal has been prepared by passing H^2 gas at 1100°C over the sulphide obtained by precipitation with hydrogen sulphide from hydrochloric acid solution. It has the same crystal structure as rhenium and the adjacent elements osmium and ruthenium.

B—Promethium

From time to time, claims were made by scientists both in America and in Europe to have discovered a missing element 61. But a definite separation and identification of this element was achieved by Marinsky and Glendenin, working in C.D. Coryell's group in connection with the fission product identification programme of the war time atomic energy project. This element has been named *promethium, symbol Pm*, by its discoverers. It is β-active with a 3-7 year period and has mass number 147.

There are 14 isotopes, all radioactive, known to us, ranging from $_{61}Pm^{141}$ to $_{61}Pm^{152}$. An isotope $_{61}Pm^{149}$ with a 53 hour period has been found among the fission products, where it probably results from the negative beta-decay of 1.8 hour neodymium-149. Which is also formed when slow neutrons interact with neodymium-148. An isotope-148 with a 5.4 day period has not been detected as a fission product, but is produced in the (p, n) reaction with neodymium-148. Apart from those present among the fission products, the isotopes arc generally obtained by β^-decay of Nd isotopes, by β^+decay or orbital electron capture of samarium isotopes or by bombardment of these two elements with high energy particle. An isotope with mass number 145 might be expected to be of a relatively long life, *i.e.*, 18 years.

C—Francium

In 1939, Mile M,. Perey in France showed that actinium-227 exhibits a branched disintegration. Besides undergoing β–decay it also simultaneously emits an a-particle. The resultant nuclide has mass number 223 and atomic number 87. The discoverer proposed its name francium, symbol Fr, after her native country. There are 18 other isotopes ranging from 205 through 224 with the exception of an element 216. Few isotopes of Fr have been produced by the thorium bombardment with high energy protons. Its chemical properties are typical of an alkali metal element.

D—Astatine

In 1940 D. R. Corson, K. R. Mackenzie and E. Segre in United States discovered a missing element-85. They found that when bismuth was irradiated with alpha particles (32 MeV) a new element with atomic number 85 was formed. The reaction is as follows.

$$_{83}Bi^{209} + {_2}He^4 \rightarrow {_{82}}85^{211} + 2\ {_0}n^1$$

By orbital-electron capture the above new element decays into polonium-211. Discoverers of this missing element-85 have proposed the name *astatine, symbol At, from the Greek astatos*, meaning unstable, since it is the only member of the halogen group having no stable isotope. Karlik and Bernert of Austria claimed that isotopes (218, 216 and 215) of astatine result from branched disintegrations of radium, thorium and actinium accompanied by the β-particles emission. Some experiments suggest that the production of the 216-isotope in this manner is doubtful.

In addition to these two naturally occuring isotopes, 18 others with mass numbers from 200 through 219, have been reported. Several have been obtained by the bombardment of Bi with α-particles of various energies and some from the interaction of gold with accelerated carbon-12 particles. Element astatine shows some similarities to and some differences from that of the halogens.

CLASSIFICATION OF PARTICLES

The particles are separated into two general groups, called *bosons* and *fermions*. These two groups have different types of spin and their behaviour is controlled respectively by a different kind of statistic (i.e. the Bose statistic or the Fermi statistic, hence the names). Bosons are particles with intrinsic angular momentum equal to an integral multiple of $\hbar$. Fermions are all those particles in which the spin is half integral. The most important difference between the two classes of particles is that *there is no conservation law controlling the total number of bosons in the Universe, whereas the total number of fermions is strictly conserved.*

Boson is a term, which not only includes material particles but also includes those quanta and photons which arise from interactions. Thus in the case of the simple electromagnetic field the bosons are merely the light photons or the X-ray photons. The *photon* has a mass of zero and a spin of unity and is consequently described as a massless boson. A massless boson, called a *graviton* with a probable spin of two units has been postulated as a field particle for gravity. These bosons, created by the electromagnetic field, are essentially of one kind, while the bosons formed in the strong interaction are of two distinct kinds. First there are those which are known as *pions or π-mesons* (π^+, π^- and π^o). The second group of bosons are much heavier than that of pions, and are known is *kaons or K-mesons* (K^+, K^- and K^o).

The fermions fall in two main classes, according to whether they are lighter than mesons, or heavier. Those in the lighter group are often called *leptons (after the Greek word meaning light in weight*), whilst those in the heavier group are called *baryons (after the Greek lord for heavy)*. The leptons are the *electrons, muons* and *neutrinos* and *their anti-particles*. These arc all with masses less than the pions and with spin half. Leptons interact weakly with other particles. The total number of leptons minus the total number of anti-leptons remains unchanged in all reactions and decay processes involving leptons and anti-leptons. The baryons consist of the two nucleons with their *anil-particles* (n^o, $\overline{n^o}$; p^+, p^-) and the *hyperons*.

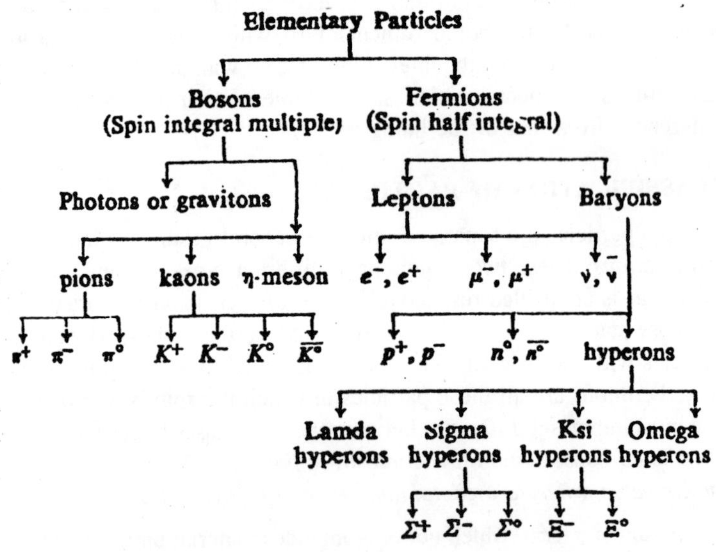

Fig. 3 : Classification of Particles.

Hyperons are the extremely unstable somewhat heavier particles and can be divided into four sub-groups, Λ^o-particle (a neutral particle of mass about 180 m_e), the Σ-particlcs (Σ^-, Σ^o and Σ^+ with masses in the range 2320 to 2340 m_e), the Ξ-particles (Ξ^- and Ξ^o with masses near 2580 m_e) and the Ω^--particle (of mass about 3284 m_e). There is no reason to doubt the existence of the anti-particles of these fermions. The total number of baryons minus the total number of anti-baryons is absolutely conserved in all interactions.

The kaons and pions together with the baryons are placed into a group of strongly interacting particles, called *hadrons*.

Neutrons and Anti-Neutrons

Chadwick in 1932 discovered the neutron through a careful study of the kinematics of α-particle induced reactions with light nuclei. The production and properties have already been given in chapter.

In the series of experiments-which led to the detection of the anti-proton, the anti-neutron was also found. A convincing experimental evidence for the existence of anti-neutrons was announced in September 1956 by B. Cork, G.R. Lambertson and others. When a proton and anti-proton meet, mutual annihilation generally takes place. But if these two particles come fairly close to each other, although not doss enough for annihilation to occur, it is possible for an electrical charge to be transferred from the positive proton to the negative, or vice-versa. As a result, both particles would become electrically neutral, the proton being converted into a neutron and the anti-proton into an anti-neutron.

$$\overline{p} + p \rightarrow \overline{n} + n.$$

Evidence for this charge transfer process was obtained by allowing anti-protons, obtained from the Berkeley Bevatron, to enter a vessel of liquid hydrogen. About three out of every thousand antiprotons interacted With the protons in the hydrogen to form anti-neutrons. The latter were identified by the energy produced in the mutual annihilation of an anti-neutron and a neutron. The release of energy was accompanied by a flash of light in a detector. The recording system was designed to respond only to those associated with the correct amount of energy to be expected from neutron-anti-neutron annihilation. Since neither neutrons nor anti-neutrons have a net electric charge, hence the distinction of a particle from its anti-particle by its charge cannot apply to the neutron and anti-neutron. We do expect, however, that whatever be the internal charge distribution of the neutron, the anti-neutron will have the opposite internal charge distribution.

MASS OF THE MUONS

Early determinations of the mass of the muons in cosmic rays were made by utilizing the tracks in a cloud chamber in a magnetic field or in a photographic emulsion. The quantities measured were momentum and energy. From these one finds the mass. Barkas and co-workers (1956) measured the ratio of the pion-muon mass and the pion-proton mass, by measuring momenta and ranges in photographic emulsion. These measurements gave the masses with an accuracy of one part in a thousand. A complexly different way of obtaining the muon mass is offered by the study of relative transitions in μ-mesonic atoms. In the last few quantum jumps from one orbit to another, muonic X-rays are radiated. For high Z-absorbers, the ^1S muon orbits penerate the nucleus and the finite size of the nucleus strongly affects the energies of the radiated X-rays. For low Z elements, the M and L muonic X-rays have energies approximately the same as given by Bohr theory. In 1960, Devons and coworkers have determined the muon mass by measuring the mass absorption co-efficient of the $^3D_{5/2} \rightarrow {}^2P_{3/2}$ muon transition of phosphorus in lead.

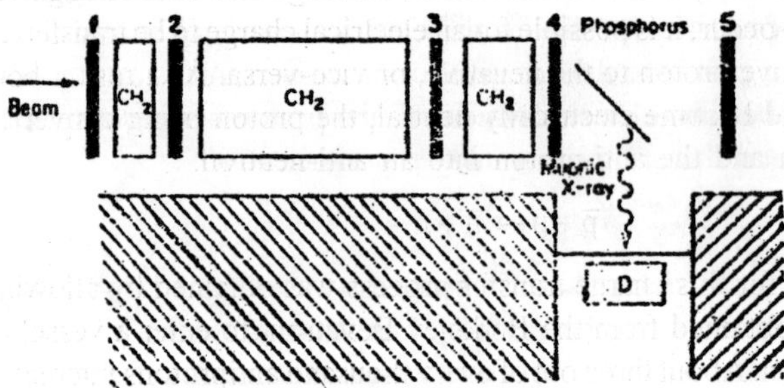

Fig. 4 : Devons experiment for the determination or negative muon rest mass.

A negative muon brought to rest in the phosphorus is recorded by a coincidence between pulses from counters 1, 2, 3, 4 and an anti-coincidence in counter 5. A muonic X-rays photon is recorded by a Nal detector D placed behind a lead absorber. For phosphorus (Z = 15), the strongest line $D_{5/2} \rightarrow P_{3/2}$ has an energy

$$E({}^3D_{5/2} \rightarrow {}^2P_{3/2}) = 88{,}000 \pm 10 \text{ eV}.$$

According to the Bohr theory

$$E = hc\,R\left(\frac{1}{2^2}-\frac{1}{3^2}\right)Z^2(1.001163)\frac{m_\mu}{m_e} \quad ...(1)$$

This relation gives $m_\mu = 206.78\ m_e$. In this method we have assumed that the charge of the muon is exactly equal to that of the electron.

Recent accurate measurements of the gyromagnetic ratio and of the magnetic moment of the muon provide m_μ $(206.77 \pm 0.01)\ m_e$.

Muon

Production of Muons By bombarding graphite with accelerated α-particles of energy 450 MeV it was found that mesons could be created artificially. The charge of muons is equal in magnitude to the charge of the electron. There are no neutral muons. They are usually obtained as a decay product of the pion according to the reactions:

$$\pi^+ \to \mu^+ + \nu_\mu$$

$$\pi^- \to \mu^- + \nu_\mu.$$

The neutrino appearing in this reaction seems to be different from the neutrino of β-decay. One occasionally gets muons at 1 BeV which are interpreted as the product of the reactions.

$$\nu_\mu + n \to \mu^- + p$$

$$\bar{\nu}_\mu + \to \mu^+ + n,$$

However, we know reactions

$$n_e + n \to e^- + p$$

$$\bar{\nu}_e + p \to e^+ + n.$$

Under the conditions of the bombardment one would also expect to produce electrons, *provided the two kinds of neutrinos are identical*. In fact, no electrons have been found, resulting, that *there are two different kinds of neutrinos.*

Although, most common source of muons is the pions decay but they also appear as branchings in other decays such as

$$\Sigma^\pm \to \mu^\pm + \nu + n$$

$$\Delta^o \to \mu^- + \nu + p$$

$$K \to \mu + \nu, \text{ etc.}$$

Muons can also be generated in pairs by photo production, similar to electron pair production.

MUON DECAY AND MEAN LIFE

The decay of muon has been observed in cloud chamber photographs and in nuclear photographic emulsions. In 1949, Leighton, Anderson and Seriff measured the energies of the electrons and positrons produced in the decay of cosmic ray γ-mesons at sea level. They placed Geiger counter controlled cloud chamber in a magnetic field of 7200 gausses and found a range of values from 9 to 55 MeV for the decay electrons. This wide range of energies for the decay electron spectrum suggests that this decay process is accompanied by more than one invisible neutral particle to conserve energy and momentum. The form of the spectrum agrees with the predictions of reactions —

$$\mu^- \rightarrow e^- + \nu_\mu + \bar{\nu}_e$$

$$\mu^+ \rightarrow e^+ + \bar{\nu}_\mu + \nu_e.$$

Of the two neutrinos, one is believed to be of the kind found in beta decay and the other of the kind found in pion decay. Here the choice of $\nu_e + \bar{\nu}_\mu$ is preferred because it associates the e^+ with the ν_e as in beta decay. For electrons having the maximum possible energy, the neutrino pair has zero angular momentum in that direction and the electrons must have the same angular momentum as the decaying muon. The decay of μ^- would require that if e^- is defined as a particle, μ^- must likewise be a particle; e^+ and μ^+ are then anti-particles.

The mean life of the free muons was originally informed from their disappearance in cosmic rays. The most recent work uses the signal from a muon production process $\pi^+ \rightarrow \mu^+$ in a scintillator to start a timing circuit and the subsequent decay electron pulse $\mu^+ \rightarrow e^+$ and to spot it. The lifetimes of μ^+ and μ^- have been compared and are found equal, also to 0.1 per cent as predicted theoertically.

$$\tau_\mu = (2.200 \pm .002) \times 10^{-6} \text{ sec.} \qquad ...(20)$$

From the lifetime and energy release, a coupling constant g can be calculated as in the case of β-decay. The value found is $g_\mu = 1.435 \times 10^{-62}$ Jm^3 which is extremely close to $g\beta = 1.410 \times 10^{-62}$ gm^3. This result emphasises the fact that both β – and μ-decays are the examples of weak interaction.

INTERACTION WITH MATTER

Negative muon captured into Bohr orbit surrounding nuclei may undergo nuclear capture

$$\mu^- + p \rightarrow n + \nu,$$

instead of radioactive decay, analogous to orbital electron capture. The probability of muon capture by a nucleus increases with the atomic number Z of the nucleus. The probabilities of capture and free decay becoming equal when the .atomic number Z is about 12. Consequently, in matter having Z less than 12, most of the negative muons will decay with the normal muon life time, but for Z greater than this value they are captured before they can decay.

The passage of muons through matter is analogous to that of electrons, except that the decay processes decrease with Z. Cosmic rays studies show that fast muons can produce nuclear disintegration stars, which can be interpreted as an electromagnetic process. Similar to the electron scattering, muon beams of energy upto several hundred MeV have been used for nuclear scattering experiments. Slow μ^- from muoniç atoms, slow μ^+ may from muonium. The μ^- may also bind two atoms to form muonic molecule. Pair production of muons (μ^+, μ^-) by high energy bremsstrahlung has also been observed.

Spin and Magnetic Moment of Muons

The experimental measurement of the spin and of the magnetic moment of the muon is intimately connected with the discovery of the non-conservation of parity in muon decay. Assume that a pion comes to rest and decays, emitting a neutrino and a muon in opposite directions. Since the pion has no angular momentum, hence the muon has an angular momentum equal and opposite to that of the neutrino. If the neutrino is polarized with the spin anti-parallel to the direction of motion, so is the muon. This has been verified by direct experiment. If the muons are brought to rest in a target medium, which does not seriously depolarize the muons, we have the emission of a positron (for μ^+ mesons) and two particles of zero rest mass from a polarized source. The positrons tend to travel parallel to the spin direction. Now if we apply magnetic, field to the muon at rest, the field produces a rotation of the direction of polarization with an angular velocity (Larmor frequency)

$$\omega = g\left(\frac{eB}{2m_\mu}\right)2\pi\,\mu_0, \qquad ...(1)$$

where g (= μ/I) is the gyromagnetic ratio, i.e. the magnetic moment in natural $\mu_0 e\hbar/2m_\mu$ divided by angular momentum in units of $\hbar$ By using these phenomena Friedman and Telegdi performed experiments and found g = 2 × (1.0020 ± 0.0005), very close to the g value of the electron. This result is a strong argument for a spin 1/2 of the muon.

The latest and most precise experiment is due to Bailey, Parley, *et al.* This compares the Larmor frequency of the muon in a given magnetic field with the orbital frequency of the particle in the same field. These two are equal, if g = 2. For a finite value of g – 2, a beam of longitudinally polarized muons in a circular orbit will gradually build up misalignment between the spin direction and the direction of motion. The use is made of a strong ring, filled with fast onions produced by pion decay in the same orbit. The muons decay in the flight and the detector circuits are set to respond only the forward going electrons having the highest energy. Since the angular distribution of emission from muon decay is not isotropic, the observed intensity of electron emission is modulated by the muon precession due to the anomalous moment. Counting rate as a function of time after the injection of pious into the ring gives

$$a = 1/2\,(g-2) = (116616 \pm 31) \times 10^{-8} \quad ...(2)$$

for μ^+ and μ^- both.

SYMMETRIES OF THE PARTICLES

Mendelyeev was able to predict correctly the atomic weights and other properties of then unknown elements by means of his periodic system. Similarly, the classification of the elementary particles has also met with success in predicting new particles. The method developed for the arrangement is based on a branch of advanced mathematics, known as *group theory*. It has been found that, in general, *the conservation law represents an invariance which corresponds to an appropriate symmetry operation.* The set of operators that represents the symmetry constitutes the group from which the theory gets its name. The irreducible representations of a group consist of a number of states, quantities or objects to which the symmetry operations are applicable. Thus the appropriate group operation can transform any one of these states into another in the same representation. The fundamental representation is the one containing the smallest number of states for the particular group.

If the system is invariant with respect to displacement in space linear momentum is conserved. Angular momentum is conserved if invariance is w.r.t. angular displacement and energy is conserved if it is w.r.t. time.

The simplest unitary group U (1) contains transformations which add a phase factor only to particle wave functions. The invariance under such transformations gives conservations of charge Q, baryon B, lepton L and hypercharge Y.

UNITARY SYMMETRY [SU (2) SYMMETRY]

We know that proton and neutron are identical as far as the nuclear force is concerned, but differ in their electromagnetic interactions. Thus, it is possible to imagine a group of symmetry operators which could transform a neutron into a proton (or proton into a neutron) in the absence of an electromagnetic field. The proton and neutron would then form the fundamental representations of the group. The existence of such a symmetry implies that something remains constant under the strong interaction. This is known as *isospin* and is 1/2 for proton as well as for neutron. The component of the isospin, T_3 is + 1/2 for the proton and – 1/2 for the neutron. The operators of the symmetry group thus change the co-ordinates of isospin in such a way as to reverse the sign of T_3 It can also be expressed as : *the strong interactions are assumed to be invariant under rotations in the isotopic spin space.*

The particular symmetry group applicable to isospin conservation is a form of *unitary symmetry known as a* U (2), which can be expressed by a set of 2 × 2 matrices. This group may be reduced to a *special unitary group S U* (2), which is also written as SU_2. It is special because a restriction reduces by unity the number of operators in the group. *The two dimensions* refer to the two basic states which make up the fundamental representatation in this case. The restriction of special reduces the number of operators 2 × 2 = 4 to three. The group is then said to have three generators.

By the use of the algebra of the SU (2) group it can be shown that all irreducible representations of the symmetry group consist of a multiplet of 2T + 1 states. All the members of the multiplet have the same isospin T and are essentially identical except for charge. If the symmetry was exact, i.e. isospin is strictly conserved, the components of a multiplet would differ in charge and T_3. The *SU (2) symmetry is violated by the electromagnetic interaction for which conservation of isospin is not applicable.* The nucleon states |p >, | n > have anti-nucleon states |p >, |n >. Omitting | > brackets for clarity and separating the trace from traceless part, the combination of nucleon with an anti-nucleon may be represented as

$$\begin{pmatrix} p \\ n \end{pmatrix} \times (\bar{p}\ \bar{n}) \rightarrow \frac{\bar{p}p + n\bar{n}}{\sqrt{2}} \begin{pmatrix} 1 & 0 \\ 0 & 1 \end{pmatrix} + \begin{pmatrix} (p\bar{p} - n\bar{n})\frac{1}{2} & (p\bar{n}) \\ n\bar{p} & (p\bar{p} - n\bar{n})\frac{1}{2} \end{pmatrix}$$

The first term of r.h.s. represents singlet (η meson, T = 0, I=0⁻) and the second term represents the triplet array (pions T = 1, I = 0⁻). The second term can be written as

$$\begin{pmatrix} \frac{\pi^o}{\sqrt{2}} & \pi^+ \\ \pi^- & \frac{\pi^o}{\sqrt{2}} \end{pmatrix}$$

EIGHTFOLD WAY [SU (3) SYMMETRY]

The SU(3) theory is a generalization of the theory of isospin. This stands for *special unitary group in three dimensions*. The term, *three dimensions* refers to the three basic states which make up the fundamental representation in this case. In a three dimension unitary group there are, in general 3 × 3 = 9 operators, but the restriction of "special" reduces the number in eight. The group is then said to have *eight generators*. Gell-Mann has referred to the resulting group of symmetry operators as the *eightfold way, named for Buddha's Eightfold Path to Nirvana, comprising eight right actions*. Three of the generators apply to *three components of isospin, as in SU (2)* and a *fourth is associated with hypercharge. The remaining four also involve hypercharge in a different way*.

Application of the group algebra showed that the SU (3) symmetry should give rise to six supermultiplets, containing 1, 8, 8, 10, $\overline{10}$ and 27 members. The $\overline{10}$ multiplet is equivalent to the 10 but with hypercharges of opposite signs. In each of these supermultiplets the parity and intrinsic spin of members are the same, while the hypercharge and the isotopic spin are not same. Among above mentioned groups, 8 and 10 member groups are of particular interest.

In the case for B = 0 we may form particle anti-particle states to fill a 3 × 3 array :

$$\begin{pmatrix} p \\ n \\ \Lambda \end{pmatrix} \times (\bar{p}\ \bar{n}\ \bar{\Lambda}) \rightarrow \begin{cases} \frac{1}{3}(2p\,\bar{p} - n\,\bar{n} - \Lambda\,\bar{\Lambda}) \\ n\,\bar{p} \\ \Lambda\,\bar{p} \end{cases}$$

$$\begin{Bmatrix} & p\,\bar{n} & p\,\bar{\Lambda} \\ \frac{1}{3}\left(p\,\bar{p}+2n\bar{p}+\Lambda\,\bar{\Lambda}\right) & & n\,\bar{\Lambda} \\ & \Lambda\,\bar{n} & \frac{1}{3}\left(-p\,\bar{p}-n\,\bar{n}+2\Lambda\,\bar{\Lambda}\right) \end{Bmatrix}$$

It can be identified with known spin zero mesons

$$\begin{Bmatrix} \left(\frac{\pi^{o}}{\sqrt{2}}\right)+\left(\frac{\eta}{\sqrt{6}}\right) & \pi^{+} & K^{+} \\ \pi^{-} & \left(\frac{\pi^{o}}{\sqrt{2}}\right)+\left(\frac{\eta}{\sqrt{6}}\right) & K^{o} \\ K^{-} & \bar{K}^{o} & -\frac{2\eta}{\sqrt{6}} \end{Bmatrix}$$

The neutral η meson is now written as

$$\eta=\frac{\left(p\bar{p}+n\bar{n}+2\Lambda\,\bar{\Lambda}\right)}{\sqrt{6}}.$$

There is in addition the symmetrical neutral combination or singlet

$$\eta'=\frac{p\bar{p}+n\,n+\Lambda\,\bar{\Lambda}}{\sqrt{3}}$$

Since these mesons are formed from fermion particle-antiparticle pairs, hence have odd parity. These eight particles with B = 0, and I^{p} = 0^{-} should be arranged as :

one triplet with	Y = 0,	T = 1	
one doublet with	Y = – 1,	T = 1/2	
one doublet with	Y = –1,	T = 1/2	8 members
one singlet with	Y = 0,	T = 0	

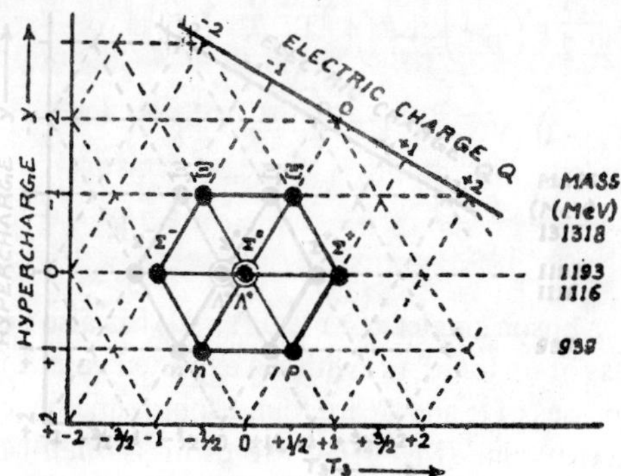

Fig. 5 : Baryon Octet (J^{P} = $1/2^{+}$).

When this was first postulated there were seven ground state mesons. The missing meson was expected to have Q = 0, Y = 0, T = 0 and was predicted both by Gell-Mann and by Ne'eman. This eighth meson was named η-meson. Octects of metastable mesons (zero intrinsic spin) is shown in Fig. 5.

In 1962, Gell-Mann and Ne'eman pointed out that the baryon also formed an octet array. For B = 1, the octet representation of SU (3) gives an octet array

$$\begin{Bmatrix} \left(\frac{\Sigma^o}{\sqrt{2}}\right)+\left(\frac{\Lambda}{\sqrt{6}}\right) & \Sigma^+ & p \\ \Sigma^- & \left(\frac{-\Sigma^o}{\sqrt{2}}\right)+\left(\frac{\Lambda}{\sqrt{6}}\right) & n \\ \Xi^- & \Xi^o & -\frac{2\Lambda}{\sqrt{6}} \end{Bmatrix}$$

These are eight in number (p, n, Λ, Σ^+, Σ^-, Σ^o Ξ^-, Ξ^o). For these $I^p = \frac{1}{2}^+$, B = 1, Y = 1 and T = 1/2 or 0. The eightfold way arrange- ment of this baryon octet is shown in Fig. 5.

For the octet mesons, Gell-Mann suggested a relationship among the average masses of the components, $\eta(M_0)$, π (M_1) and K $(M_{1/2})$, as

$$3M_0^2 + M_1^2 = 4M_{1/2}^2 \qquad \text{...(1)}$$

Another interesting case is that of the boson octet with $I^p = 1^-$. The particles are :

$K^{*o}\left(T_3 = -\frac{1}{2}, Y = 1\right)$, $K^{*+}\left(T_3 = \frac{1}{2}, Y = 1\right)$, $\rho^-(T_3 = -1, Y = 0)$,

ϕ^o & ρ^o $T_3 = 0$, Y = 0), $K^{*-}\left(T_3 = -\frac{1}{3}, Y = -1\right)$,

$\overline{K}^{*o}\left(T_3 = \frac{1}{2}, Y = -1\right)$.

For $I^P = 1^-$, a boson singlet ω^o (T = 0, Y = 0) has also been suggested. Neither the mass of ϕ^o nor ω^o fits with the expected value of M_0 in above, relation. The masses of ω and ϕ^0 are found approximately equally distant from the expected value (M_0 = 930 Me of mass. Gell-Mann and later Sakurai explained this by assuming ω^0 and ϕ^0 approximately the following mixtures of the octet and singlet states

$$\text{Real } \omega = \sqrt{\left(\frac{1}{3}\right)}\phi - \sqrt{\left(\frac{2}{3}\right)}\omega$$

$$\text{Real } \omega = \sqrt{\left(\frac{2}{3}\right)}\phi + \sqrt{\left(\frac{1}{3}\right)}\omega. \qquad ...(2)$$

Without really knowing which particle belongs in the octet state and which in the singlet, the ϕ^o has been arbitrarily placed in the octet.

Gell Mann derived a relationship among the average masses of the components of the four baryon multiplets in the intrinsic spin half octet, given as

$$2(M_N + M_\Xi) = 3\ M_\Lambda + M_\Xi, \qquad ...(3)$$

where M_N is the average nucleon mass and the average masses of the hyperons are indicated by the respective subscripts.

The relations (1) and (2) are special cases of the general formula, given by Okubo,

$$M\ (T, Y) = M_0\ [1 + aY + b\ \{T(T + 1) - 1/4Y^2\}], \qquad ...(4)$$

where a and b are constants for a particular multiplet.

The baryon octet for intrinsic spin 3/2 consists of one doblet ($Y = +1$; resonance states N^{*0} with $T_3 = -1/2$ and N^{**} with $T_3 = 1/2$); one doublet ($Y = -1$; resonance states Ξ^{*-} with $T_3 = -1/2$; and Ξ^{*0} with $T_3 = 1/2$); one triplet ($Y = 0$; resonance states Y^{*-} with $T_3 = -1$, y^{*0} with $T_3 = 0$ and Y^{*+} with $T_3 = +1$.) and one singlet ($Y = 0$, $T_3 = 0$ and member is $Y^{*(0)}$).

A more remarkable prediction was the case of the ten-fold baryon group with intrinsic spin 3/2. Theory predicted that there should be a group of ten members : $Y = 1$, $T = 3/2$, quartet (nucleon resonance denoted by N md exists in four charge states), $Y = 0$, $T = 1$, triplet (hyperon resonance Y_1^* which is containing excited Σ particles), $Y = -1$, $T=1/2$, doublet (hyperon resonance which is equivalent to excited cascade particles Ξ), $Y = -2$, $T = 0$, singlet (a particle with charge -1).

In 1962, when the proposal was made to incorporate the N, Y_1, and Ξ resonances in a decuplet, the tenth particle was unknown. The existence of such a particle was predicted by Gell Mann and was named omega (Ω). The baryon decuplet for intrinsic spin 3/2. Before the Ω^- had been discovered, the values of a and b (eqn. 51) could be found in order to fit the Y = 1, 0 and – 1. Thus the mass of the Y = – 2 member was predicted as 1676 MeV, while the measured value is 1675 MeV. Although all the other members of the baryon decuplet for intrinsic spin 3/2 are resonant states, decaying by the strong interaction, Gell-Mann noted that the Ω^- particle should decay by the weak interaction. Possible decay modes are

$$\Omega^- \rightarrow \Xi^- + \pi^o$$

$$\rightarrow \Xi^o + \pi^-$$

$$\rightarrow K^- + \Lambda^o$$

It is dear that in these modes baryon number is conserved while isospin, hypercharge and strangeness are not conserved. These conservations only hold for strong interactions, hence the decay of Ω^- is by weak interaction.

In a recent development attempts have been made to combine ordinary spin with T and Y. The new group is described as SU (6).

STRUCTURE OF THE NUCLEONS

According to meson theory, the proton as well as the neutron, should be surrounded by a meson cloud and consists of a distribution charge of finite extent. *Electrons are used to probe the electromagnetic structure of the nucleon just as α-particles used to probe atomic structure.* Owing to the small size of the proton, electrons of small wave length (thus of high energy) are required for the experimental investigation of the charge distribution in the nucleons. After the completion of the linear accelerator at Stanford University, Hofstadter and others were able to determine nuclear radius and hence to study the structure of nucleons.

The first observations of a finite size of the proton were made by Hofstadter and McAllister (1955) by examining the scattering at 100 MeV electrons and found a very small effect of finite size. The experiments were further carried out at an electron energy of 236 MeV and found an rms radius of the proton to be equal to 0.78 ± 0.20 fermi. The numerical results were arrived at by assuming that

$$F_1 = F_2 (\equiv F_P)$$

as functions of q and effectively using the shape independent approximation

$$F = 1 - \left(\frac{q^2\alpha^2}{6}\right) + ...,$$

where 'α' is the rms radius of the charge or magnetic moment distribution. It is related more generally to a density distribution through the Fourier transform

$$F = \frac{4\pi}{q}\int_0^{\infty} \rho(r)\exp(i\vec{q}.\vec{r})d\tau,$$

which applies in the non-relativistic limit in which ρ (r) is the static density distribution and is a function of radius. At large angles, where tan θ/2 ≥ 1 and at high energies, it becomes possible to distinguish moment scattering. Hence it becomes possible to find F_1 and F_2 separately. Chambers and Hofstadter (1956) carried out a detailed series of experiments on the proton extending to an energy of 550 MeV with scattering angles lying between 30° and 135° in the lab. frame. They found a very large deviations from point scattering. They ratio F_1/F_2 may be found independently of any model by comparing two cross sections at two different energies and different angles, such that the momentum transfer q is the same at both cross section determinations. $F_1/F_2 = 1.1 \pm 0.2$ and $q^2 = 4 \times 10^{30} m^{-2}$ found from the above experiments.

Using the fact that magnetic moment scattering falls off more slowly at large angles than charge scattering, high energy data may be analyzed to find $F_1 + kF_2$. Since k = 1.79, the data can yield F_2, if F_1 is known even approximately. At small scattering, magnetic scattering is essentially zero and hence F_1 may be investigated alone. This procedure shows that for an exponential model, the charge r.m.s. radius must be close to 0.8 fermi when the magnetic r.m.s. radius is allowed to vary from 0.60 fermi to 1.0 fermi.

From the scattering at 900 MeV and at 1.3 BeV, we find that for high four momentum transfers the charge and anomalous moment from factors differ from one another in the sense that corresponds to a wider distribution for the magnetic moment than for the charge. The equations of theoretical curves show that the proton root mean radii are given as

$$r = r_{pc} = 0.88 \text{ fermi and } r_2 = r_{n\mu} = 0.95 \text{ fermi.}$$

Since neutron targets are not available, the information on the form factors of the neutron is obtained from the scattering of fast electrons by deuterons. Data upto 1.3 BeV have been obtained. The root mean square radii obtained from these results are :

$r_1 = r_{nc} = 0.00$ fermi and $f_2 = r_{n\mu} = 0.87$ fermi.

Charge distributions in the proton and in neutron obtained from above calculations are shown in Fig. 6.

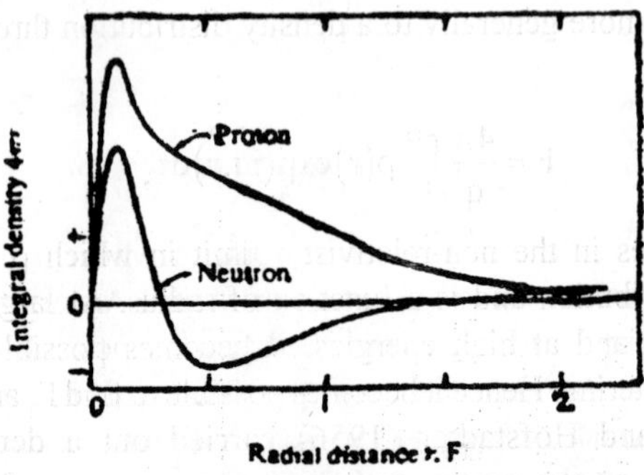

Fig. 6 : The proton and neutron charge density distribution.

The proton has a finite extent as if there were a finite probability of finding it decomposed into a neutron and positive meson cloud. The structure of the neutron can be interpreted as due to the decomposition into a proton surrounded by a negative muon cloud, but apparently produces a positive charge density at larger distances.

PIONS

The result of cosmic ray investigations led to the conclusion that pions are produced both in nucleon-nucleon interaction and in the interaction of high energy photons with nucleons. It is clear from this conclusion that pions can be produced in laboratory by the use of high energy particles obtained by means of an accelerator. The amount of energy which the accelerated particle must have depends on the mass of particle and the nature of the target but it will be always greater then the rest mass of the pion.

Artificial Production

The pions were first produced artificially by Gardner and Lattes in 1948, using a beam of 380 MeV α-particles from the Berkeley Synchrocyclotron. This beam was allowed to fall on the edge of a carbon target and the mesons emitted were detected in nuclear emulsions. For laboratory production of π-mesons, the most obvious target materials are hydrogen and deuterium or substances rich in these isotopes. At the present time, strong beams of charged pions for experimental purposes, with energies covering a wide range, are produced by bombarding solid targets, *e.g.*, Be, with accelerated protons.

The simplest nucleon-nucleon interaction involving protons as projectile and target and producing pions is evidently.

$$p + p \rightarrow \pi^+ + d$$

or $$p + p \rightarrow (n + p).$$

The interaction between a high energy proton and a deuteron may be considered as an interaction between the proton and one of the nucleons which constitute the deuteron. The possible reactions are.

$$p + p \text{ (in d)} \rightarrow \pi^+ + P + n$$

$$p + p \text{ (in d)} \rightarrow \pi^+ + d$$

$$p + n\text{(ind)} \rightarrow \pi^+ + n + n$$

$$p + n\text{(ind)} \rightarrow \pi^- + p + p.$$

The production of pions by gamma radiation incident on nucleons has been studied extensively in the energy range from threshold to 450 MeV and less thoroughly upto 1 BeV.

$$\gamma + p \rightarrow \pi^+ + h$$

$$\gamma + n \rightarrow \pi^- + p.$$

Besides charged pions, neutral pion has been produced by bombarding hydrogen and deuterium with high energy photons.

$$\gamma + p \rightarrow \pi^\circ + p, \gamma + d \rightarrow \pi^\circ + (n + p), \gamma + d \rightarrow \pi^\circ + d$$

and by proton-proton interaction as

$$p + p \rightarrow \pi^\circ + p + p.$$

The source of gamma rays is generally the bremsstrahlung of electrons accelerated by machines. The-production of neutral pions by the interaction of high energy photons with matter was first established by Steinberger, Panofsky and Steller in 1950. In their experimental arrangement a narrow beam of 330 MeV X-rays from the synchrotron was incident on a target (beryllium or hydrogen) and the products were detected by two gamma ray counters making an angle f with each other in a plane at an angle θ with the X-ray beam. Each counter consisted of three crystal scintillation counters with a thin piece of lead between the first two crystals to convert the gamma ray photons into electron-positron pairs. The measured values of the coincidence rates for different values of ϕ and θ agreed with the calculated values based on the assumption that a π^0 meson was disintegrated into two gamma ray photons.

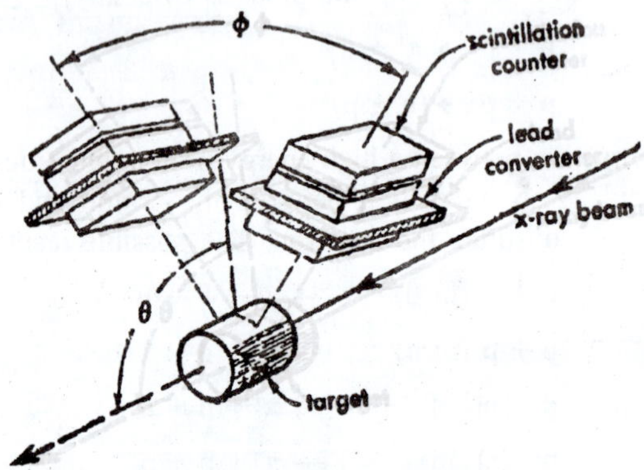

Fig. 7 : Experimental arrangement for the production of π^0-meson.

Pions are commonly formed in the decay of kaons, hyperons and resonant states. Since there is no requirement for the conservation of pions, there is no limitation on their formation even in strong interaction.

MASS OF PIONS

The mass of the charged pion can be determined by methods similar to those used for muons. The rest mass of the negative pion can be found by comparing the observed pionic X-ray energies for the 4f-3d orbit transitions in media of low atomic number ($Z < 20$) with the theoretical expression which contains m as an unknown parameter. Apart from the reduced mass, the only important correction to the Bohr formula is the effect of vacuum polarization (~ 0.2 to 0.3). The rest mass of the negative pion thus obtained is

$$m_\pi = (273.2 \pm 0.1)m_e = 139.577 \pm 0.013 \text{ MeV}.$$

The positive pions have the same mass and can be determined by the same method. The mass of the neutral pion is found by a quite different method. When very low energy negative pions are slowed down in hydrogen they are captured into orbit. These cascade down the optical levels and are ultimately captured by the proton. The reactions observed are:

$$\pi^- + p \rightarrow n + \gamma$$

$$\pi^- + p \rightarrow n + \pi^o$$

$$\pi^o \rightarrow \gamma + \gamma.$$

The first reaction gives rise to a monochromatic gamma ray whose lab energy is peaked around 130 MeV. The second reaction is particularly important for the determination of the π^o-mass. Assume that the negative pion is captured at rest, the π^o and n acquire a relative kinetic energy equal to given by energy mass relation. This energy can be measured by the observation of the gamma spectrum of the decaying π^o. In the co-ordinate system in which π^o is at-rest both gamma rays have equal and opposite momentum. In the laboratory system they appear at an angle to each other and their frequency shows the Doppler shift due to the motion of the π^o. The observed photon energies in the lab system range from

$$E_{\gamma\,(min)} = \frac{1}{2}\, m\pi^o c^2\,(1-\beta)\,(1-\beta^2)^{-1/2} \qquad ...(1)$$

to
$$E_{\gamma\,(max)} = \frac{1}{2}\, m\pi^o c^2\,(1+\beta)\,(1+\beta^2)^{-1/2}. \qquad ...(2)$$

These limits are obtained from relativistic Doppler formula for a π^o decaying at the instant; speed is βc. It is easy to write the rest mass energy difference as

$$(m_{\pi}^- - m_{\pi}^o)\,c^2 = (m_n - m_p)\,c^2 + \frac{\alpha^2 c^2 m_{\pi^-}}{2} + \frac{\frac{1}{2}\left(m_p - m_{\pi^-}\right)\Delta E_{\gamma^2}}{\left[m_{\pi} - \left(m_p - m_p\right)\right] m_n c^2} + ...$$

Here the second term on the right is the pion binding energy in a [1]S Bohr orbit and the third term is the kinetic energy of the π^o. In the second term a is the fine structure constant. Using known values of $\Delta E_{\gamma}(= E_{\gamma,\,max} - E_{\gamma,\,min} = p_{\pi^o}c)$ α, m_n, m_p and $m\pi^-$ we have

$$\left(m_{\pi^-} - m_{\pi^o}\right)c^2 = 4.6 \text{ MeV}$$

or
$$m_{\pi^o c^2} - m_{\pi^- c^2} \quad - 4.6 \text{ MeV}$$

$$= 139.577 \pm 0.013 - 4.6 = 134.977 \pm 0.013 \text{ MeV}.$$

Decays

Unless they have sufficiently high energies (~ 150 MeV), positive pions do not interact to any great extent with nuclei and decay at rest according to the reaction.

$$\pi^o \rightarrow \mu^+ + \nu_\mu.$$

Since positive muon is an anti-particle, the ν_μ must be a neutrino to conserve the muonic lepton number. This decay is due so the weak interaction and the parity is not conserved. The onion is emitted with a unique energy of 4.17 MeV and is longitudinally polarized. This energy is consistent with zero mass for the muon associated neutrino. Pions may also decay according to the reactions

$$\pi^+ \rightarrow e^+ + \nu_e \ (1.24 \times 10^{-2}\%)$$

$$\pi^+ \rightarrow \mu^+ + \nu_\mu + \gamma \ (1.24 \times 10^{-2}\%)$$

$$\pi^+ \rightarrow \pi^o \ e^+ + \nu_e \ (1.5 \times 10^{-6}\%)$$

The first equation requires that the helicity of the positron shall be the same as that of the neutrino. The third equation is analogous to a $0^+ \rightarrow 0^+$ Fermi transition in nuclear β-decay.

Similar decay processes are possible with negative pions.

$$\pi^- \rightarrow \mu^- + \overline{\nu_\mu}, \pi^- \rightarrow e^- + \overline{\nu_e} \text{ and } \pi^- \rightarrow +e^- + \overline{\nu_e}$$

Negative pion decay ($\pi^- \rightarrow \mu^-$) is not observed at rest because of the Coulomb attraction with the nuclei and strong nuclear absorption.

The decay modes of the π^o are

$$\pi^o \rightarrow \gamma + \gamma \qquad (98.8\%)$$

$$\pi^o \rightarrow \gamma + e^+ + e^- \qquad (1.2\%)$$

$$\pi^o \rightarrow e^+ + e^- + e^+ + e^- \qquad (0.003\%)$$

It is evident from the nature of the products that π^o-mesons decay by the electromagnetic interaction while charged pions decay by strong and the weak interactions both.

Mean Life

One of the earliest determinations of the mean lifetime of charged pions was made by J.R. Richardson (1948), using the π-mesons produced by the bombardment of graphite by 380 MeV α-particles. These pions travelled in helical paths in the vertical magnetic field of the cyclotron. Those which rose upward were allowed to travel through 180° before striking a photographic plate, those which spiraled downward were allowed to travel 540° before striking a photographic plate. The number of mesons which decayed while traversing a complete cycle can be determined by the ratio of the number of mesons in the two plates. The time required to traverse a circle in a cyclotron is independent of the speed, hence the average life time of the decaying mesons could be determined. The first result for the mean life of charged pion obtained was 1.1×10^{-8} sec.

A more recent experiment of the decay in flight of positive pions was performed by Durbin, Loar and Havens in 1952. The meson beam produced in the Nevis cyclotron by the bombardment of a beryllium target with 385 MeV protons was deflected by a magnet and directed through three stilbene scintillation counters operated in coincidence. A much larger scintillation counter, C_4, 8" dia and containing xylene, was set at a distance L which could be varied from a few inches to about 8 ft. The coincidence counters C_1, C_2 and C_3 defined the beam of the π-mesons. Some of these mesons decayed in flight between C_3 and C_4, the remaining ones being detected by C_4. After suitable corrections for beam impurities, angular divergence, accidental coincidences, the difference in counts between C_1, C_2, C_3 detectors and the C_4 detector represented the number of pions decaying in the flight along the path of length L. Knowing the average velocity and determining mean free path, the mean life time could the calculated. The value so obtained was 2.54×10^{-8} sec for π^+ mesons and 2.44×10^{-8} sec. for π^- mesons. Thus within the limits of experimental error, the mean lifetimes of the π^+ and π^- mesons were identical, with the average mean life = 2.5×10^{-8} sec.

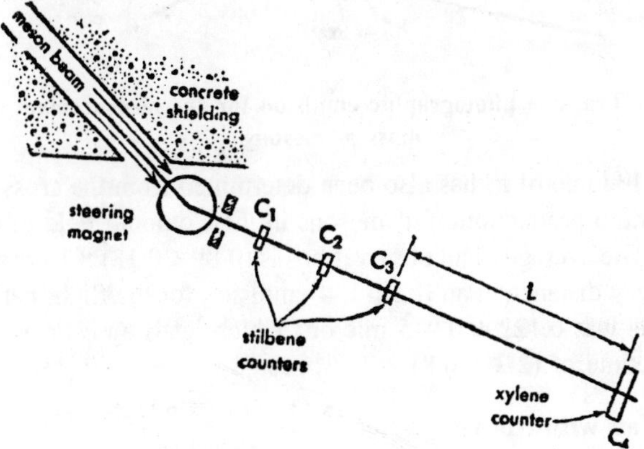

Fig. 8 : Measurement of life time of π-mesons.

Using a beam of 34 MeV positive pions from the Rochester 130" cyclotron, Kinsey, *et al.* (1964) found pion lifetime as (2.640 ± 0.008)

$\times 10^{-8}$ sec. By observation of decay in flight, Ayres et al; (1968) found that the lifetimes of the π^+ and π^- mesons are equal within standard deviations and suggested the value 2.6×10^{-8} sec. The decay of π^o goes through a virtual intermediate nucleon- anti-nucleon state.

$$\pi^o \xrightarrow[\text{Strong}]{} \underset{\text{Virtual}}{(n+\bar{n})} \xrightarrow[\text{EM}]{} \gamma + \gamma .$$

This two stage decay involves both strong and electromagnetic interactions, hence suggests a short mean life of the π^o. The accurate method of measuring the mean lifetime of π^o-mesons is to mesure the mean distance between the end of a track of a K^+ meson brought to rest in a photographic emulsion and the origin of a electron positron pair. The reactions are

$$K^o \rightarrow \pi^+ + \pi^o + 219 \text{ MeV}$$

$$\pi^o \rightarrow \gamma + e^+ + e^- \ (1.24\%)$$

R.G. Glasser (1960) found that the mean decay distance in an Ilford L-4 emulsion for the flight path of π^o was 0.122 ± 0.045 microns and hence the pion mean lifetime $(2.3 \pm 0.8) \times 10^{-16}$ sec. P. Stamer (1966) found lifetime as $(1.0 \pm 0.5) \times 10^{-16}$ sec.

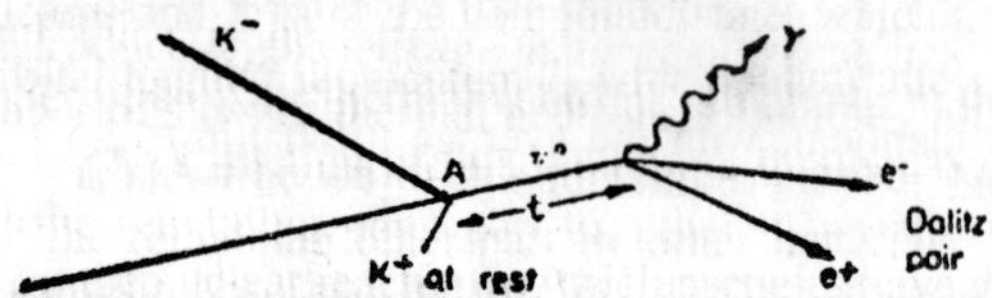

Fig. 9 : Track in photographic emulsion for the measurement of the mass π^o-mesons.

The lifetime of π^o has also been determined from the cross-section for the photo-production of π^o mesons in the Coulomb field of a heavy nucleus. The average of all observations is $(0.89 \pm 0.18) \times 10^{-16}$ sec. The mean decay distance in an Ilford L-4 emulsion for the flight path of the π^o-mesons was 0.122 ± 0.045 microns which leads an estimate for the mean lifetime of $(2.3 \pm 0.8) \times 10^{-16}$ sec.

Interaction with Matter

Because of their short life time ($\sim 10^{-16}$ sec.), it π^o-mesons move only a few atomic diameters before they decay and thus are not affected by the matter through which they pass. Positive pions are not able to surmount the nuclear Coulomb barrier and, therefore, undergo spontaneous decay, while negative pions are captured by nuclei. All of the energy of a captured pion is transmitted to the nucleus results to a nuclear disruption, *called a star*, because of its appearance in a particle track chamber or photographic emulsion. Menon, Muirhead and Rochat in 1950 investigated the stars produced by the capture of slow π^--measons in the nuclei of the elements composing nuclear emulsions. They studied more than 2500 stars and classified according to the number of prongs in each. About 46% of all visible stars were ascribed to the interaction of π^--mesons with light nuclei such as C^{12}, N^{14}, O^{16} and about 54% were with the heavy nuclei, such as silver and bromine. All the π^o-mesons absorbed in the emulsion do not produce visible stars.

By using available charged meson beams and hydrogen targets we can measure three cross-sections, corresponding to the reactions:

$$\pi^+ + p \to \pi^+ + p,\ \pi^- + p \to \pi^- + p,\ \pi^- + p \to \pi^o + n.$$

The first two reactions are purely elastic scattering and the last reaction is known as charge exchange scattering. The first accurate measurements upto 200 MeV π-meson indicated that the total cross-sections of reactions were in the ratio of 9 : 1 : 2.

The scattering cross-sections can be interpreted by assuming that in the strong interactions the total isospin is conserved as well as the third component. The total isospin of the π^+, p system is 3/2 and the component $T_3 = 1 + 1/2 + 3/2$. The π^-, p system has $T_3 = -1 + 1/2 = -1/2$ and is a mixture of T = 3/2 and T = 1/2 states. If the scattering amplitudes corresponding to the scattering in the T = 3/2 state and the T = 1/2 states are a(3/2) and a(1/2) respectively, then the cross-sections are:

$$\sigma_+ \propto \left|a\left(\frac{3}{2}\right)\right|^2 \qquad ...(1)$$

$$\sigma_- \propto \frac{1}{9}\left|2a\left(\frac{1}{2}\right)+a\left(\frac{3}{2}\right)\right|^2 \qquad ...(2)$$

$$\sigma_0 \propto \frac{2}{9}\left|a\left(\frac{3}{2}\right)-a\left(\frac{1}{2}\right)\right|^2 . \qquad ...(3)$$

If the π-p scattering predominantly occurs in the T = 3/2 state near 200 MeV, $a\left(\frac{3}{2}\right) >> a\left(\frac{1}{2}\right)$ and the cross-sections satisfy the ratios

$$\sigma_+ : \sigma_- : \sigma_0 = 9 : 1 : 2. \qquad ...(4)$$

From general collision theory, the maximum cross-section for elastic scattering is

$$\sigma_{max} = 4\pi\lambda\!\!\!^{-2} \frac{2J+1}{(2s_1+1)(2s_2+1)} \qquad ...(5)$$

In the case of π^+-p system s_1 the intrinsic spin of the pion is zero, s_2 the intrinsic spin of target proton is 1/2. For pions upto 300 MeV only s and p wave scattering is important, The d wave impact parameter is beyond the range of the *pion nucleon* force. The maximum value of total cross-section scattering of n π^+-mesons by protons was found 192 millibarns below approximately 200 MeV. At the peak of the resonance $\pi\lambda\!\!\!^{-2} = 24$ mb. These values with eqn (29) give us total angular momentum of the π-p system J = 3/2. This shows that the interaction is dominated by p-wave scattering

$$\left(L = J - s = \frac{3}{2} - \frac{1}{2} = 1\right).$$

Absorption of stopped π^--mesons : The fast charged pion slows down from an energy of about 100 MeV to about 1 keV in liquid hydrogen in 10^{-9} sec. In the further 10^{-12} sec, π-meson loses energy and becomes bound in a 1s orbit to a proton. From this orbit it may decay or react with the proton as

$$\pi^- + p \rightarrow \pi^o + n, \ \pi^o \rightarrow 2\gamma \qquad ...(6)$$

or

$$\pi^- + p \rightarrow n + \gamma. \qquad ...(7)$$

Panofsky, Aamodt and Hadley studied these reactions and found peak about 130 MeV describing reaction (8) and lower peak of mean energy of 70 MeV representing the energies of photons from the π^o-meson decaying as shown by reaction (8). The initial state has spin and parity J = $1/2^-$, since the π^- is captured from K-shell and the pion has odd parity. In reaction (9), the neutron state has J = $1/2^+$, the photon must, therefore, carry odd parity and unit angular momentum.

Intrinsic Spin

The reactions d + p = d + π^+ have been used to determine the spin of the π^+-meson (and therefore presumably of the π^--meson), using the measured cross-sections of the two reactions together with the *statistical mechanical principle of detailed balance.*

The differential cross-section for the reaction p + p → d + π^+ given in terms of transition matrix elements is

$$\frac{d\sigma_{pp\to d\pi}}{d\Omega} = \frac{1}{(2s_p+1)(2s_p+1)}\left(\frac{1}{q_{pp}}\right)^2 \Sigma|T_{ba}|^2 \quad ...(8)$$

where q_{pp} is the momentum in C.M. system of the initial state pp. Here T_{ba} is the matrix elements with a initial plane wave state and b the final state.

Similarly for the reaction d + π^+ → p + p, we have

$$\frac{d\sigma\pi_{d\to pp}}{d\Omega} = \frac{1}{(2s_d+1)(2s_\pi+1)}\left(\frac{1}{q\pi_d}\right)^2 \Sigma|T_{ab}|^2. \quad ...(9)$$

If both cross-sections refer to the same total energy and to the same angle of scattering in C.M. system, we contain the detailed balancing relation

$$\frac{d\sigma_{pp\to d\pi}}{d\Omega} = \frac{(2s_d+1)(2s_\pi+1)}{(2s_p+1)(2s_p+1)}\left(\frac{q_{\pi d}}{q_{pp}}\right)^2 \frac{d\sigma_{d\pi-pp}}{d\Omega}. \quad ...(10)$$

Using $S_p = 1/2$, $S_d = 1$ and the measured cross-sections and momenta in C.M. system. Cartwright, et al and Durkin, et al. suggested spin of the π^+-meson as zero.

It is assumed that the π-spin is the same as that of the π^+

The spin of the neutral pion cannot be determined directly. We know that π^o-meson decays into two photons and never into three photons. This implies that the intrinsic spin must be zero. Thus we see that all pions have spin zero and have no magnetic moment.

Parity

A negative pions are slowed down and captured into Bohr orbits in deuterium gas under pressure, we get following reactions :

$$\pi^- + d \to n + n \quad ...(11)$$

$$\pi^- + d \to n + n + \gamma \quad ...(12)$$

$$\pi^- + d \to n + n + \pi^o. \quad ...(13)$$

Reaction (13) is not observed. Panofsky measured the cross-section ratio

$$\frac{\sigma\left(\pi^- + d \rightarrow n + n\right)}{\sigma\left(\pi^- + d \rightarrow n + n + \gamma\right)} = 2.36 \pm 0.36 \qquad ...(14)$$

Slow π^- mesons are absorbed by the nuclei of hydrogen as $\pi^- + p \rightarrow \pi^o + n^o$ and $\pi^- + p \rightarrow n + \gamma$. The ratio of these two reactions has been determined by different workers. The latest value of this ratio is 1.53 ± 0.02.

To interpret these two ratios, there are following assumptions : *(1) The π^--mesons are absorbed by the nucleus from s-states of the mesic atoms formed in the capture of the π^- by the atoms of deuterium and hydrogen. (2) The proton and neutron have the same parity.*

In order for the final states to be consistent with Pauli's Principle, two neutrons can exist only in $^1S_0$; $^3P_{0,\,1,\,2}$; $^1D_2$ or $^3F_{2,\,3,\,4}$ etc. states. The total angular momentum of the initial state, i.e. the pion and deuteron system, is just the spin of the deuteron, J = 1. The parity, given by $P_\pi K_d (-1)^l$ is equal to P_π since P_d is even and $l = 0$. To conserve the total angular momentum and parity, the n + n system must be in a J = 1 state and must have same parity as that of the π^-meson. For L-even, the nucleon spins must be anti-parallel (S = 0) and for L-odd the nucleon spins must be parallel. Thus the only state with J = L + S = 1 is the $^3P_1$ state. This corresponds to an odd parity. Thus *the intrinsic parity of π-meson is negative.* This conclusion is consistent with the suppression since π-meson also has negative parity.

The similar reaction $\pi^+ d \rightarrow pp$ at rest cannot occur since Coulomb repulsion keeps if*- and d apart. A detailed analysis of the cross section suggests negative parity for π^+-mesons. The parity of the neutral pion is difficult to determine experimentally. It is again negative, can deduced using rare decay mode

$$\pi^+ \rightarrow e^+ + e^- + e^+ + e^-$$

and measuring distribution of angles between the planes of the two electron pairs.

K-MESONS

K-mesons are produced when high energy protons bombard a suitable target. The pions are also produced in this process. With increasing proton energy the fraction of K-mesons relative to pions increases. There are essentially four K-mesons, i.e. two charged mesons K^+ and K^-, which are particle and anti-particle of one another, and two uncharged mesons K^o and K^{-o}, which are also particle and anti-particle. These are obtained in laboratory by the interaction of high energy protons or pions with various atomic nuclei.

INTERACTIONS OF NUCLEAR RADIATION

Nuclear radiations are divided into three broad categories: (1) Charged heavy particles of mass comparable with nuclear mass and electrons, (2) electromagnetic radiation, γ-radiation, and (3) neutron radiation. A striking difference in the absorption of these types of radiation is that only charged particles have a range. In traversing a certain amount of matter a mono-energetic beam of heavy charged particles losses energy without changing the number of particles in the beam. After having crossed practically the same thickness of absorber they ultimately stopped. This minimum thickness of absorber is known as the *range of a particle.* As particles travel they lose energy. The energy lost per unit path length is known as the *specific energy* loss and its average *value stopping power of the absorpting substance.* The fluctuations in energy loss, because of the statistical nature of the collision processes along their path, also produce fluctuations in range. This small variation of the ranges is called *straggling.*

CHARGED NUCLEAR PARTICLES

An alpha particle, a proton, or another charged particle when moves through a matter ionizes or excites the atoms to which it comes sufficiently close. Coulomb force gives a sharp impulse to the atomic electrons as the particle moves swiftly by. The energy transferred to the electrons represents a loss of kinetic energy of the moving particle. The energy transfer to the positive heavy ions, released in ionisation, is negligible. Thus, we consider the phenomenon of the transfer of energy from a moving charged particle to the electrons of the surrounding medium.

Let us consider a particle of charge ze moving with a velocity v so that its closest distance of approach to a stationary electron is a. The velocity v is assumed to be so great that in most cases the electron has not appreciably changed its position before the particle has passed by. The transfer of momentum p to the electron or the net impulse given to the electron is obtained by integrating the force during the collision.

Thus,

$$p = \int_{-\infty}^{+\infty} \frac{ze^2}{4\pi \epsilon_0 r^2} \sin\theta \, dt \qquad ...(1)$$

where r is the distance of an electron. Motion of charged particle from the charged particle, can be with respect to
an electron, expressed as r = a/sin θ. By symmetry there is no component of momentum transferred parallel to the particle velocity.

x = cot – vt = a cot θ, and hence $dt = \left(\frac{a}{v}\right) \text{cosec}^2\theta \, d\theta$

Substituting values of r and dt in eqn. (1), we have

$$p = \frac{ze^2}{4\pi \epsilon_0 av} \int_0^{\pi} \sin\theta \, d\theta = \frac{ze^2}{2\pi \epsilon_0 av}.$$

Here v is taken outside the integral, as the change in it during the collision with single electron is very-very small. The kinetic energy transferred to the electron of mass m_0 is given by

$$T = \frac{p^2}{2m_0} = \frac{z^2e^4}{8\pi^2 \epsilon_0^2 m_0 a^2 v^2} \qquad ...(2)$$

Since there are 2πnZ a dadx electrons per length dx that have a distance between a and a + da from the charged particle, where n is the number of atoms per unit volume and Z is the atomic number of the medium. Hence the energy transferred from a charged particle to a cylindrical shell is given by

$$dE = T \, 2\pi nZ \, ada \, dx.$$

Thus the energy loss per unit path length of the particle is

$$-\frac{dE}{dx} = 2\pi nz \int_{a_{min}}^{a_{max}} Ta \, da = \frac{z^2e^4nZ}{4\pi \epsilon_0^2 m_0 v^2} \int_{a_{min}}^{a_{max}} \frac{da}{a}$$

$$= \frac{z^2e^4Zn}{4\pi \epsilon_0^2 m_0 v^2} \log_e \frac{a_{max}}{a_{min}} \qquad ...(3)$$

At first it seems natural to put $a_{min} = 0$ and $a_{max} = \infty$. This leads to the result that the rate of loss of energy is infinite. The lower limit a_{min} is determined by the uncertainty principle. In an elastic collision it is impossible to change the momentum of an electron by an amount greater than $2m_0v$ (assuming the heavy particle at rest and the electron impinging on it. Hence $a_{min} = h/2m_0v$. This is approximately the size of the electron as seen by the moving particle. The perturbing field from the charged particle will not change the state of an electron unless the time (a/v), is small compared with the period of oscillation (1/ν), hence a_{max} = v/ν, where ν is an average of the frequencies of oscillations of the electron in the atom.

Substituting the values of a_{min} and a_{max} in eqn. (3), we have

$$-\frac{dE}{dx} = \frac{z^2 e^4 Zn}{4\pi \epsilon_0^2 m_0 v^2} \log_e \frac{2m_0 v^2}{h\nu}. \quad ...(4)$$

The energy hν(= I) is an average *excitation energy* or binding energy for the atomic electrons of the medium and is an empirical constant which depends on and is related to the ionization potential of the stopping material. Typical values of I are given in the Table 1.1.

The mean ionization (excitation) potential is given by the empirical relation

$$I \simeq 9.1\, Z(1 + 1.9\, Z^{-2/3})\ \text{eV}. \quad ...(5)$$

In 1930, Bethe deduced an approximate quantum-mechanical relationship. A more exact quantum mechanical formula was derived in 1933 by F. Bloch. The Bethe formula, taking into account relativistic effects at high particle speeds, as

Table 1.1 : Average Excitation Potential

Substance	Z	I(eV)	K = I/Z
H	1	19	19
He	2	44	22
BE	4	64	16
C	6	77	12.8
N	7	88	12.6
Air	7.2	94	13.1
O	8	100	12.5
Al	13	166	12.7
Ar	18	230	12.8
Fe	26	300	11.5
Cu	29	371	12.8
Ag	47	586	12.5
Au	79	1017	12.8
Pb	82	1070	13.1

$$-\frac{dE}{dx}=\frac{z^2e^4Zn}{4\pi\in_0^2 m_0v^2}\left[\log_e\frac{2m_0v^2}{h\nu}-\log_e\left(1-\frac{v^2}{c^2}\right)-\frac{v^2}{c^2}\right] \quad ...(6)$$

The quantity –dE/dx is known as the *stopping power* of the slowing down medium. For all except the highest speeds the last two terms in the square bracket of above equation almost cancel each other. Since the term $\log_e (2m_0v^2/h\nu)$ is reasonably slowly varying with v, the main variation of stopping power with v is in the term outside the bracket.

Due to this inverse square term stopping power decreases rapidly in the particle energy range 1 to 10 MeV. There is a broad minimum in the stopping power at high energies, the stopping power increases slowly thereafter as v increases, because the second and third terms in the square brackets become important at high enough speeds.

It is also clear from equation (6) that the *stopping power does not depend on the mass of the particle but is only a function of its velocity and charge.* Thus the stopping power for an α-particle is four times the stopping power for a proton of the same velocity, except towards the end of the range.

When compounds or mixtures of elements are used as absorbers, the atomic stopping power of each atom is taken as independent of the presence of atoms of other type. The ratio of the stopping power of the material to the stopping power of air at STP for equal energy loss is known as the *relative stopping power* of the material. The relative atomic stopping power is the relative stopping power per atom and is defined as

$$S_a=\frac{\left(\frac{1}{N}\right)_m\left(\frac{dE}{dx}\right)_m}{\left(\frac{1}{N}\right)_{air}\left(\frac{dE}{dx}\right)_{air}} \quad ...(7)$$

The *relative electronic stopping power* is the relative stopping power per atomic electron and is defined as

$$S_e=\frac{\left(\frac{1}{NZ}\right)_m\left(\frac{dE}{dx}\right)_m}{\left(\frac{1}{NZ}\right)_{air}\left(\frac{dE}{dx}\right)_{air}} \quad ...(8)$$

$$S_m = \frac{\left(\frac{1}{\rho}\right)_m \left(\frac{dE}{dx}\right)_m}{\left(\frac{1}{\rho}\right)_{air} \left(\frac{dE}{dx}\right)_{air}} \quad \text{...(9)}$$

Limitations

The observed rate of energy loss of charged particles is in good agreement with the relation (6). Of course, perfect agreement is not to be expected for several important reasons:

1. If the speed of the incident charged particle is comparable with c the normally spherical field becomes distorted, shrining in the direction of motion and expanding laterally. This effect leads to an increase in the rate of energy loss for very high energy particles.

2. As charged particle moves through the medium, its electric field will temporarily polarize. Charges of unlike sign will be attracted toward the moving particle and like charges will be repelled. Thus polarization acts to reduce the rate of energy loss. This effect becomes significant only in dense media and at high energies and responsible for the phenomenon of *Cerenkov radiation*.

3. As the incident particle slows down, to a speed of the same order as that of the outer orbital atomic electrons, it will eventually pick up and carry along some of the less tightly bound electrons of the medium. This reduces the effective charge of the particle and therefore energy loss rate.

4. Eqn. (6) does not hold for velocities smaller than $(I/2m_0)^{1/2}$ as $-dE/dx$ becomes negative.

The Specific Ionization

The number of charge pairs created per millimeter of the charged particle's trajectory, depends upon the charge and energy of the incident particle. It rises to a maximum when the particle has lost nearly all its energy at the ends of its range. In the case of a beam of α-particles in air at 15°C and 1 atm, the peak corresponds to 6600 ion pairs per mm.

At this point the effective charge of the α-particle is not 2 but $Z \cong 1.5$ due to charge exchange process ($He^{++} + e^- \simeq He^+$). It begins to be appreciable at energies below $E_\alpha = 2$ MeV ($Z = 1.883$ at 1.7 MeV, 1.500 at 0.65 MeV dropping linearly with velocity to zero thereafter).

Corrections for Nonparticipating Electrons

In an absorbing atom, the K-electrons have velocity $vk = c(Z_{eff}/137)$ and $E_B = \frac{1}{2} m_0 c^2 (Z_{eff}/137)^2$. If the K-electrons participate in interactions with the incident particle of mass M and energy E, then

$$E = \frac{1}{2}Mv^2 = \frac{M}{4m_0} 2m_0 v^2 >> \frac{1}{8}Mc^2\left(\frac{Z-0.3}{137}\right)^2. \qquad ...(10)$$

It is because $Z_{eff} = Z - 0.3$ for K-electrons and the maximum energy transfer to the K-electrons, when $M >> m_0$ and $\beta^2 << 1$ is $2m_0v^2$. For the protons passing through Al, 1.0 MeV.

Bethe and coworkers derived a correction term C_K, due to the ineffectiveness of K-electrons and suggested the relation (6) as

$$-\frac{dE}{dx} = \frac{z^2 e^4 Zn}{4\pi \epsilon_0^2 m_0 v^2}\left[\log_e \frac{2m_0 v^2}{hv} - \log_e\left(1 - \frac{v^2}{c^2}\right) - \frac{v^2}{c^2} - \frac{C_K}{Z}\right] \qquad ..(11)$$

Similarly C_L, C_M, C_N, ...correction terms may nonparticipation of L, M, N, ...electrons respectively.

DELTA RAYS

We have so far considered only the ejection of electrons from the atoms of the stopping material through which the particle travels. The directly ejected electrons and the ionized atoms are known as primary ions. If one of these primary ions possesses sufficient energy to cause further ionization, it may eject further electrons from nearby atoms before being brought to rest. These electrons and ionized atoms are called *secondary ions.* The energetic electron is called *knock-on electron.* Evidence for these electrons can be seen in a cloud chamber, photographic emulsion or bubble chamber as short tracks extending out from the main track of the charged particle. These short range tracks are known as *delta rays.* Delta rays are visible prominently along the track of a particle moving at a speed of a few tenths c. Most of the δ-ray tracks are observed corresponding to the energy transfer of 100eV or more.

The primary ionization can be estimated by the number of collisions for which the energy transfer to an electron lies between $T_{max} = 2m_0v^2$ and $T_{min} = I$. Hence the number of collisions

$$= nZx\int_{T=2mv^2}^{T=I_0} 2\pi a da = \frac{nZd\times z^2e^4}{8\pi \in_0^2 m_0v^2}\left(\frac{1}{I_0}-\frac{1}{2m_0v^2}\right)$$

$$\simeq \frac{z^2e^4nZ}{8\pi \in_0^2 m_0v^2 I_0} \text{ ion pair/meter.} \qquad ...(1)$$

It is 6×10^4 for the 5 MeV protons passing through nitrogen at atmospheric pressure. As the observed value is 24×10^4, hence the difference can be accounted for by secondary ion production by the δ-rays ejected in the primary process.

RANGE AND STRAGGLING

The total range of a particle in a given material is the length of the flight path before the particle comes to rest and is given by the relation

$$R = \int_E^0 -dx = \int_E^0\left(-\frac{dE}{dx}\right)^{-1}(-dE) = \int_0^E\left(\frac{dE}{dx}\right)^{-1} dE \qquad ...(1)$$

In combining this with eqn. (1) we must remember that eqn. (1) fails at low energies. If, however, a range R_1 is measured for an energy E_1, which is higher than the energy for which eqn. (1) breaks down then the range for energy particles can be written as

$$R = R_1 + \int_{E_1}^{E}\left(-\frac{dE}{dx}\right)^{-1} dE = R_1 + R'. \qquad ...(2)$$

For the common non-relativistic case, we can write $E = \frac{1}{2}Mv^2$. Then eqn. (2) becomes

$$R' = \int_{E_1}^{E}\frac{dE}{-\frac{dE}{dx}} = \int_{v_1}^{v}\frac{d\left(\frac{1}{2}Mv^2\right)}{\frac{e^2z^2Zn}{4\pi \in_0^2 m_0v^2}\log_e\left(\frac{2m_0v^2}{hv}\right)}$$

$$\text{or } \int_{R_1}^{R} dx = \frac{\in_0^2 Mh^2v^2}{8z^2e^4m_0Zn}\int_{v_1}^{v}\frac{dy}{\log y} = \frac{M}{z^2}\frac{\in_0^2 h^2v^2}{8e^4m_0Zn}(Y-Y_1), \qquad ...(3)$$

where $y = \left(\frac{2m_0 v^2}{hv}\right)^2$, $Y = \int_0^v \frac{dy}{\log y}$. Hence Y and Y_1 are the function of v and v_1 respectively.

Thus, we see that the distance traversed for a given velocity loss in a specified medium is proportional to M/z^2 and to a comolicated function of v. Then for two particles which have different values of M/z^2 but have the same initial velocity every element of range dx will be proportional to M/z^2. Hence

$$\frac{dx_1}{dx_2} = \frac{\left(\frac{M}{z^2}\right)_1}{\left(\frac{M}{z^2}\right)_2}. \qquad ...(4)$$

If this relation is valid at any specified velocity, then it is valid over the entire range R. Hence for particles of equal initial velocity

$$\frac{R_1}{R_2} = \frac{\left(\frac{M}{z^2}\right)_1}{\left(\frac{M}{z^2}\right)_2}. \qquad ...(5)$$

For α-particles, Geiger discovered (in 1910) that in air at S.T.P.

$$R = 0.00318\ E^{3/2} \text{ meters}, \qquad ...(6)$$

where E is in MeV.

Brobeck and R. Wilson, for protons in the range from a few MeV to about 200 MeV gave an approximate range-energy relation

$$R = \left(\frac{E}{9.3}\right)^{1.8} \text{ meters}, \qquad ...(7)$$

where E is in MeV. At energies below about 2MeV, the exponent is close to 1.5.

It has been shown that the ranges R_v for protons and R_α for α-particles, travelling with the same velocity v, in air at S.T.P. are related as

$$R_p = 1.0072\ R_\alpha - 0.003 \text{ meter}. \qquad ...(8)$$

A fairly rough relation, accurate to about ± 15%, is provided by the Bragg-Kleeman rule

$$\frac{R_\rho}{\sqrt{A}} = \text{constant.} \qquad ...(9)$$

In the case of a mixture of various elements with atomic weights A_1, A_2, A_3,... and relative mass abundances n_1, n_2, n_3,..., the effective atomic weight is written as

$$A^{1/2} = n_1A_1^{1/2} + n_2A^{1/2} + n_3A_3^{1/2} + ...$$

For air at STP, $\sqrt{A_{air}} = 3.82$ and $\rho_{air} = 1.226$ kg/m³. Hence the Bragg-Kleeman rule takes the form

$$R = 0.32\left(\sqrt{\frac{A}{\rho}}\right)R_{air}. \qquad ...(10)$$

For the heavy charged fission fragments nuclear collisions occur more frequently than for the faster moving α-particles and energy losses through nuclear collisions become more important than ionization losses especially in the final portion of the range. The charge on the fission fragment decreases gradually as the fragment slows down by picking up electrons from its surroundings, thus state of ionisation varies along its path. In contrast with alpha particles which show an increasing rate of energy loss towards the end, the fission fragments lose most of their energy at the beginning of the track because of their high state of ionisation. The range of a fission fragment can be estimated in terms of the range of an α-particle of the same initial velocity by means of the approximate relation

$$\frac{R_f}{R_\alpha} = 3.73 \times 10^{-4}\frac{\left(\frac{A}{Z^{3/2}}\right)}{\left(\frac{v}{c}\right)^2}. \qquad ...(11)$$

In the above discussion we have only considered primary process of ionization in which the nuclear particle tears loose electrons from the atoms along the path. Due to their large energies (~ keV) many of them will knock out secondary electrons and the result is a *multiplication* of the number of ion pairs by a factor of 10 or more. Many events of atomic excitation take place in addition to these processes and a large number of photons are released in the following *de-excitations*. The total number of primary and secondary ion pairs produced in a given material is proportional to the energy loss of the particle approximately. If all ions produced by a particle coming to stop in a volume of air can be collected, the total charge is a measure of the particles energy. The ionization chamber operates on this principle.

Straggling

Identical charged particles, all having the same initial velocity, do not all have the same range. It is found that the ranges are distributed over a small interval about the mean range. This phenomenon is called *range straggling* and is due to several effects. Since the number of collisions is large and are independent to each other, hence the distribution may be expected roughly to be *Gaussian.*

(1) In computing the range, it is assumed that the particle follows a straight path. Actually, the particle follows a zigzag path and the measured range will be less than the actual range.

There are several processes which contribute to the dissipation of the energy of electrons passing through matter.

(a) Ionization : The energy loss by ionization may be treated in a manner similar to that used for heavy ions. It is the case of collisions between identical particles. For this Mott scattering formula may hold good at least at low energies. At higher energies, the electron scattering is termed as Moller scattering and the stopping power is expressed as

$$\frac{-dE}{dx} = \frac{e^4 nZ}{4\pi \epsilon_0^2 \, mev^2}\left[\log_e \frac{2m_e c^2}{I} - \left(3 + \frac{2}{\gamma} - \frac{1}{\gamma^2}\right)\log_e 2^{1/2}\right.$$

$$\left. + \frac{1}{6} - \frac{1}{8\gamma} + \frac{9}{16\gamma^2} + \log_e(\gamma - 1) + \frac{1}{2}\log_e(\gamma + 1)\right]$$

In the low energy limit $\gamma = (1 - \beta^2)^{-1/2} \to 1$, we get

$$\frac{-dE}{dx} = \frac{e^4 nZ}{4\pi \epsilon_0^2 \, m_e v^2}\left[\log_e \frac{2m_e v^2}{I} - 1.2329\right] \qquad ...(12)$$

It is same as eqn. (1) except the numerical term – 1.2329. At very high energies, $E_{kin} + m_0c^2 = m_0c^2\,\gamma$ or $E \simeq (\gamma + 1)$ m_ec^2 as $\gamma >> 1$ and we get

$$\frac{-dE}{dx} \simeq \frac{e^4 nZ}{4\pi \epsilon_0^2 \, m_e v^2}\left[\frac{1}{2}\log_e\left(\frac{E^2}{2m_e c^2 I^2}\right) + \frac{1}{16}\right] \qquad ...(13)$$

(b) Back scattering by Nuclei : Using Rutherford scattering theory, the cross section for electron scattering from the atomic nucleus of charge Ze through an angle q or greater is

$$\sigma = \int_{\pi}^{\theta} d\sigma = 2\pi \int_{\pi}^{\theta} \left(\frac{b}{2}\cot\frac{\theta}{2}\right) d\left(\frac{b}{2}\cot\frac{\theta}{2}\right)$$

$$= \frac{1}{2}\pi b^2 \cot^2\left(\frac{\theta}{2}\right), \qquad ...(14)$$

where $b = Ze^2/2\pi \in_0 m_e v^2$. The backward scattering corresponds to $\theta > \pi/2$. For incident electrons $v = \beta c$ and mass $m_e = m_0/(1 - \beta^2)^{1/2}$. Hence for the back scattering the cross section per nucleus

$$\sigma\left(\theta \geq \frac{\pi}{2}\right) = \pi Z^2 \left(\frac{e^2}{4\pi \in_0 m_0 c^2}\right)^2 \frac{1-\beta^2}{\beta^4}. \qquad ...(15)$$

Using Rutherford scattering the differential cross-section for elastic scattering of electrons into a solid angle $d\omega$, at a mean angle of θ in lab-system is given by

$$d\sigma = \frac{Z^2}{4}\left(\frac{e^2}{4\pi \in_0 m_0 c^2}\right)^2 \left(\frac{1-\beta^2}{\beta^4}\right) \text{cosec}^4 \frac{\theta}{2} d\omega \qquad ...(16)$$

Using relativistic Dirac theory and Born approximation Mc-Kinley and Feshback gave a relation for $d\sigma$, which is related with the Rutherford cross section as

$$d\sigma = d\sigma_{\text{Rutherford}}[1 - \beta^2 \sin^2 \theta/2 + \pi\beta\alpha Z (1 - \sin \theta/2) \sin \theta/2] \qquad ...(17)$$

where a is the fine structure constant

$$\left(= \frac{\mu_0 e^2 c}{2h} = 7.29735 \times 10^{-8}\right).$$

(c) Scattering by Atomic Electrons: For the energy Transfer greater than the binding energy of the struck electron, the problem can be assumed as a collision between two identical particles. The non-relativistic cross section per atom for scattering at a lab-angle θ or greater is given by

$$\sigma(\geq \theta) = 4\pi Z \left(\frac{e^2}{4\pi \in_0 m_e v^2}\right)^2 \cot^2 \theta. \qquad ...(18)$$

As each primary electron scattered at an angle θ is accompanied by a recoil electron at an angle $\pi/2 - \theta$. Hence

$$\sigma(\geq \theta) = 4\pi Z\left(\frac{e^2}{4\pi \in_0 m_e v^2}\right)^2 \left(\cot^2 \theta + \tan^2 \theta\right).$$

There is no back scattering in the lab-system and hence $\theta_{max} = \pi/2$. For a significant scattering, let $\theta = \frac{\pi}{4}$, hence

$$\sigma\left(\theta \geq \frac{\pi}{4}\right) \simeq 8pZ\left(\frac{e^2}{4\pi \in_0 m_e c^2}\right)^2 \beta^{-4}. \quad ...(19)$$

(d) **Bremsstrahlung:** This is the energy loss by radiation loss suffered by the electron when it is subjected to acceleration in the field of nucleus or of the atomic electrons. The rate of energy loss varies as the square of acceleration. In the case of radiating particles of mass m and charge z passing through a target of atomic number Z, the acceleration $\propto$ zZe^2/m.

The cross section for emission of bremsstrahlung

$$\sigma_{rad} \simeq \alpha Z^2\left(\frac{e^2}{4\pi \in_0 m_e c^2}\right)^2 \quad ...(20)$$

Using Born approximation and relativistic electrons, Heitler suggested different relations for different incident energy regions.

(i) For the nonrelativistic region ($E < m_e c^2$)

$$\sigma_{rad} = \frac{15}{3}\alpha Z^2\left(\frac{e^2}{4\pi \in_0 m_e c^2}\right)^2 \quad ...(21)$$

(ii) For the highly relativistic region, with no screening correction ($m_e c^2 \leq E \leq m_e c^2 Z^{-1/3}\alpha^{-1}$).

$$\sigma_{rad} = 4\left[\log_e\left(\frac{2E_{kin} + m_0 c^2}{m_0 c^2}\right) - \frac{1}{3}\right]\alpha Z^2\left(\frac{e^2}{4\pi \in_0 m_e c^2}\right)^2 \quad ...(22)$$

(iii) For the extremely relativistic region, with screening correction, ($E \geq m_e c^2 Z^{-1/3} \alpha^{-1}$)

$$\sigma_{rad} = 4\left[\log_e\left(138Z^{-1/3}\right) + \frac{1}{18}\right]\alpha Z^2\left(\frac{e^2}{4\pi \in_0 m_e c^2}\right)^2. \quad ...(23)$$

The total bremsstrahlung energy (in MeV per incident electron of K. E, E_c in MeV) is given by

$$I = KZE_e^2; \quad \text{for } E_e \geq 2.5 \text{ MeV}. \qquad ...(24)$$

where $K \simeq 7 \times 10^{-4}$. Thus the practical energy loss $\frac{I}{E_e}$ of 1 MeV electrons traversing a thick target of Pb is 6%.

The radiation loss per unit path length by the electrons of K.E. ($\geq m_0c^2$) is given by the relation

$$\frac{-dE}{dx} = n\,E\,\sigma,$$

$$-\frac{dE}{dx} = \frac{Ze^3 nE(Z+1.3)}{8\pi^2 \epsilon_0^2 m_0^2 c^5 h}\left[\log_e 183Z^{-1/3} + \frac{1}{13}\right] \qquad ...(25)$$

Here Z(Z + 1.3) is for Z^2 for higher Z-values.

The total path length S is measured along the actual path of the electron and is always considerably greater than the range R. For small energies, the mean path length S_{av}, in an absorber is given by

$$S_{av} \simeq \text{const.}\left(\frac{E_e^2}{NZ}\right).$$

The range R is roughly proportional to the total path length.

Since $\left(-\frac{dE}{dx}\right)_{rad}$ varies as Z^2, energy loss due to this factor is relatively more important for heavy elements than for light elements. The radiative energy loss is negligible for heavy charged particles like protons, mesons, deuterons and alpha particles as it is inversely proportional to the square of the mass of the incident charged particle. The energy loss through ionization is the most important factor for electron energies below 1 MeV, whereas energy loss in the form of radiation is predominant for energies above 1 MeV. In lead, radiation and collision losses are about equal at 6.9 MeV, in air equality is reached at about 83 MeV. An interesting fact is that the loss is proportional to the kinetic energy E of the incident particle (electron). It is clear from above relation that the radiation loss predominates an ionization loss at high speeds and loss can be written as

$$-\frac{dE}{dx} = \mu E,$$

where μ is a constant, independent of energy. On integration, we have

$$E = E_0 \; e^{-x/L}. \qquad ...(26)$$

where $L\left(=\frac{1}{\mu}\right)$ is a constant which can be found from eqn. (25) with $n = \frac{N\rho}{A}$. We thus find

$$L = \frac{1}{\mu} = \frac{8\pi^2 \in_0^3 h\, m_0^2\, c^5 A}{e^6 NZ(Z+1.3)\left[\log_e\left(183Z^{-1/3}\right)+\frac{1}{8}\right]} \qquad ...(27)$$

L, the *radiation length*, is the absorber thickness (measured in kg/m^2) needed to reduce the electron energy by radiation loss to l/e of its original value. It is 239 kg/m^2 for aluminium and 58 kg/m^2 for lead. The constant μ is a characteristic property of the absorber, known as the *linear absorption coefficient*. Its dimension is seen to be that of a reciprocal length. Mass absorption coefficient μ_m is defined as $\mu_m = \mu/\rho$, where ρ is the density of the absorption material. Experimentally it can be shown that μ_m is independent of A and increases slightly with Z.

Because of their excessive straggling and the large statistical fluctuations in the results of their collisions in the absorber, it is not possible to determine a range for β-particles in the precise way possible with the heavier charged particles. One obtains a practical range for β-particles, also called extrapolated range. This is the thickness of material required to reduce the observed β-particle intensity to the *background counting rate*. Empirical expressions for mean range (in gm cm^{-2}) as a function of kinetic energy E (in MeV) have been proposed by various workers.

N. Feather,	$R = 0.543\,E - 0.160$	$(E > 0.7)$
Bleuler and Zunti,	$R = 0.571\,E - 0.161$	$(1.2 < E < 2.3)$
L.E. Glendenin,	$R = 0.407\,E^{1.38}$	$(0.15 < E\; 0.8)$
	$= 0.542\,E - 0.133$	$(0.8 < E < 3)$
Katz and Penfold,	$R = 0.412\,E^{1.265 - 0.954 \log E}$	$(0.01 < E < 3)$
	$= 0.530\,E - 0.106$	$(2.5 < E < 20)$

The relative β-particle activity curve for continuous p-spectrum differs greatly from that for monoenergetic electrons. The thickness of the absorber required to stop the β-particles of highest energy, is called the maximum range. As it corresponds to a point of zero intensity, hence not very easy to determine experimentally. Feather and others suggested different methods for the determination of range and found the above relations useful also for the continuous spectrum. In this case, R will be the maximum range and E the maximum energy of the β-spectrum.

ABSORPTION OF GAMMA RAYS

The interaction of γ-rays with matter is markedly different from that of charged particles such as α- or γ-particles. The difference is apparent in the much greater penetrating power of γ-rays and in the absorption laws., Gamma-rays and X-rays, which are both electromagnetic radiations, have no definite range. Unliked charged particles, a well collimated beam of γ-rays shows the exponential absorption in matter. This is because photons are absorbed or scattered in a single event. If radiation of intensity I is incident upon an absorbing layer of thickness dx, the amount of radiation absorbed dI is proportional both to dx and I, so that

$$dI = -\mu I\, dx \text{ or } I = I_0\, e^{-\mu x}, \quad \text{...(1)}$$

where μ is a constant of proportionality which is a characteristic property of the medium, known as *linear absorption coefficient*. The mass absorption coefficient μ_m may be obtained by dividing the linear absorption coefficient by the density of the absorbing material.

The relation (44) gives the intensity (number of quanta per unit area per second) of the beam of initial intensity I_0, after traversing a thickness x of the homogeneous material.

There are a number of processes which can cause γ-rays to be scattered or absorbed. The wide-ranging choice requires that we restrict ourselves to those effects which are prominent at low energies, namely-photoelectric effect, the compton effect and the pair production. The minor effects which are of interest in special cases

(a) *Meson Production:* Mesons are produced if the γ-ray energy is above 150 MeV. The cross-section is very small (~ 10^{-3} barn/atom).

(b) *Photo Disintegration:* This process is allowed, if the γ-ray energy exceeds the separation energy of a neutron or proton (> 8 MeV). The cross-section is negligible compared with those for the Compton effect and for the pair production.

(c) *Delbruck Scattering:* It is due to virtual electron pair formation in the nuclear coulomb field and is also called *elastic nuclear potential scattering.* The cross-section is very small.

(d) *Nuclear Resonance Scattering:* In this process, the nuclear level is excited by an incident photons and de-excited by the re-emission of the excitation energy. The cross-section is very small.

(e) *Thomson Scattering:* Thomson assumed that the incident beam set each quasi-free electrons into forced resonant oscillation.

Using non-relativistic electro-dynamics he calculated the cross-sections for the re-emission of electro-magnetic radiation.

Consider a plane sine wave travelling in the x-direction, with the initial amplitude E_0, frequency v and displacement in the y-direction. The electric field at any instant t is given by

$$E = E_0 \sin \omega t = E_0 \sin 2\pi v t. \qquad ...(2)$$

The force on a free electron of mass m_e is eE and the equation of motion is

$$m_e \ddot{y} = eE = eE_0 \sin 2\pi v t.$$

The displacement $y = -\left[\frac{eE_0}{m_e}(2\pi v)^2\right] \sin 2\pi v t$. ...(3)

The instantaneous value of the equivalent dipole moment is ey. As the electrons are set into induced oscillation, the rate at which energy is taken from the incident plane wave per unit perpendicular area is known as incident intensity of radiation. It is the time average of the Poynting-vector $[S = (E \times B)/\mu_0$ or $S_x = E_y B_x/\mu_0 = E_v^2/c\mu_0]$.

$$\therefore \qquad \text{Intensity } I = |S_x| = \frac{E_0^2}{2c\mu_0}. \qquad ...(4)$$

The amplitude of an electric field E' due to an electric dipole (ey) at a distance R is given by the classical electrodynamics, as

$$E_0' = -\frac{\mu_0 e^2 E_0 \sin \Theta}{4\pi m_e R} = \frac{f(\Theta) E_0}{R}, \qquad ...(5)$$

where Θ is the angle between the direction of emission and the polarization vector of the incident wave E and $f(\Theta)$ is known as the *scattering amplitude.*

The time average of $|E'|^2$ induces a factor 1/2.

If the incident wave is unpolarized, we must find an average over all angles Θ in terms of scattering angle θ as

$$\left[\sin^2 \Theta\right]_{\text{average}} = \frac{1}{2}\left(1+\cos^2 \theta\right).$$

$$\left|S_{\text{out}}\right| = \left(\frac{\mu_0 e^2}{4\pi m_e}\right)^2 \frac{E_0^2}{R^2} \cdot \frac{\left(1+\cos^2 \theta\right)}{4c\mu_0}. \qquad ...(6)$$

The time averaged flux of radiation across a small area dA perpendicular to R is

$$\left|S_{\text{out}}\right| dA = \left(\frac{\mu_0 e^2}{4\pi m_e}\right)^2 \frac{E_0^2}{2c\mu_0} \frac{1+\cos^2 \theta}{2} \frac{dA}{R^2}$$

$$= \left(\frac{\mu_0 e^2}{4\pi m_e}\right)^2 \frac{E_0^2}{2c\mu_0} \frac{\left(1+\cos^2 \theta\right)}{2} d\Omega.$$

If the differentia] cross-section for re-emission of radiation (Thomson scattering) is $\sigma_T(\theta)$, then the scattered radiation

$$|S_{\text{out}}|\, dA = I,\ d\sigma_T = I\sigma_T(\theta)\, d\Omega,$$

hence $$d\sigma_T = \frac{1}{2}\left(\frac{\mu_0 e^2}{4\pi m_e}\right)^2 \left(1+\cos^2 \theta\right) d\Omega \qquad ...(7)$$

On integration we get the total scattering cross-section as

$$\sigma_T = \int d\sigma_T = \int_0^{\pi} \frac{1}{2}\left(\frac{\mu_0 e^2}{4\pi m_e}\right)^2 \left(1+\cos^2 \theta\right) 2\pi \sin \theta \, d\theta$$

$$= \frac{8\pi}{3}\left(\frac{\mu_0 e^2}{4\pi m_e}\right)^2 = \frac{8\pi}{3} r_0^2. \qquad ...(8)$$

This is the cross-section per electron. Hence the corresponding atomic cross section

$$a\sigma_T = Z\sigma_T = \frac{8}{3}\pi Z r_0^2. \qquad ...(9)$$

Because of the inverse square dependence upon the mass of the scattering body, the cross-section for Thomson scattering on heavier particles is vanishingly small.

(e) Rayleigh Scattering : It is the case of elastic scattering from the bound electrons, providing that the electrons do not receive sufficient energy to eject them from the atom, *i.e.,* the *bound electrons revert to their initial state after scattering.* The Rayleigh cross-section per atom by the bound electrons acting together is calculated by assuming that each electron scatters a wave with amplitude $f(\Theta)$[= scattering amplitude] given by Thomson and by coherent summation of these amplitudes. An integration over the electron distribution then leads to an *atomic scattering factor* f_θ. The differential cross-section for Rayleigh scattering is obtained in terms of $f(\theta)$ as

$$d\sigma_R = \left(\frac{\mu_0 e^2}{4\pi m_e}\right)^2 |f_\theta|^2 \frac{1}{2}\left(1+\cos^2\theta\right) d\Omega. \qquad ...(10)$$

For very small energies $f_\theta \to Z$ and

$$ds_R \sim r_0^2 Z^2 \left(1+\cos^2\theta\right). \qquad ...(11)$$

Its contribution is greater than that of Thomson scattering. For very small angles it exceeds even Compton scattering.

We shall now discuss briefly the various important processes.

1. **The Photoelectric Effect:** When the photon collides with an atom, it may impinge upon an orbital electron and transfer all of its energy to this electron by ejecting it from the atom. This process is known as *the photoelectric effect* and obeys the Einstein photo electric equation

   $h\nu = E_{kin} + E_B$,

   where $h\nu$ is the photon energy and E_B the binding energy of the ejected electron. Except for the absorbers of low Z, this effect is the predominant type of γ-ray interaction with matter for energies below about 0.1 MeV. The most likely interaction is one between the photon and the most tightly bound electrons, *i.e.,* the two K-shell electrons, which occur in about 80% of the photoelectric absorptions, if the incident photon is sufficiently energetic ($> E_{B, K}$).

The binding energy not only depends upon the value of Z but also upon the orbital electron shell. It decreases as one proceeds to outer shells according to the approximate relations:

$$E_{B,K} = R(Z-1)^2, \; E_{B,L} = \frac{1}{4}R(Z-5)^2, \; E_{B,M} = \frac{1}{9}R(Z-13)^2,$$

where R = Rydberg constant ≈ 13.61 eV.

The absorption curve as a function of incident energy shows sharp discontinuities corresponding to the initial energies equal to the ionization energies of the electrons in K, L M shell. As L, M... shells have subshells also, hence the curve has one edge corresponding to K-shell electron, and three nicks in the L-edge.

At low γ-ray energies, photo electrons are emitted in a direction perpendicular to the incident beam. At higher γ-ray energies, the angular distribution is more in the forward direction, *i.e., the position of the peak shifts to lower angles as energy increases.*

The absorption curve can be divided into three regions (a) vicinity of the absorption edge (< 0.1 MeV), (b) away from the edge > 0.1 MeV), and (c) relativistic region (hv > m_ec^2). The absolute probability of a photo-electric interaction is described by the atomic cross-section $\sigma_{pe,K}$, for K-electrons. The total photo-electric crosssection in the intermediate region (b), for both K-electrons together, using non-relativistic Born approximation, is given by

$$\sigma_{pe}, K = 4\sqrt{2\left(\frac{m_ec^2}{hv}\right)^{7/2}}\alpha^4 Z^5 \sigma_T, \qquad ...(12)$$

where $\alpha = \frac{1}{137}$ and $\sigma_T = \frac{3}{3}\pi\gamma_0^2$.

In the neighbourhood of the absorption edge (region a), it is necessary to use more exact form of electron wave functions. It gives us a correction function. In the relativistic region, the exponent (7/2) decreases with the increase of energy, until it reaches unity at very high energies.

$$\sigma_{pe,K} = 1.5\left(\frac{m_ec^2}{hv}\right)^{-1}\alpha^4 Z^5 \sigma_T. \qquad ...(13)$$

The Z^5 dependence means that for a given photon energy this process is much more important in heavy absorbers than light ones. Using empirical data, N.C. Rasmussen found the exponent of Z to vary from 4.0 to 4.6 with energy if varied from 0.1 MeV to 3 MeV.

Auger Effect: The ejection of an electron from a given shell through a photo-electric effect, creates a vacancy in that shell and leaves the atom in au excited state. The atom reverts to the lower state when an electron from the higher shell fills this vacancy. The amount of energy, equal to the difference in binding energy is emitted in the form of X-rays. If this energy is greater than the binding energy of the electron in the higher shell, the electron can be ejected from the higher shell. This electron is known as *Auger electron* and the effect as *Auger effect*, after the discoverer *Auger*. Thus if the K-shell is excited, the Auger electron may originate in the L-shell, if

$$E_{B, L} < E_{B, K} - E_{B, L}. \quad ...(14)$$

The energy of the L-Auger electron = $E_{B, K} - 2E_{B, L}$.

The relative probability of Auger emission and X-radiation is measured by the fluorescence yield

$$W_K = W \text{ (for the K-shell)} = \frac{\text{Number of K - quanta}}{\text{Number of K - shell vacancies}} \quad ...(15)$$

The correponding K-Auger yield is 1-W_K. The Z-dependence, by semi-empirical method, is given as

$$\frac{W_K}{(1-W_K)} = (-6.4 + 3.4\,Z - 0.000103\,Z^3)^4 \times 10^8. \quad ...(16)$$

This shows that for light atoms, the Auger effect is predominant.

2. **Compton Effect:** Rayleigh elastic scattering and the photo-electric absorption, both depend upon the reaction of an atom as a whole. As energy approaches 0.5 MeV, wavelength becomes shorter and there is a greater tendency for interaction to take place with individual electrons. Thomson theory of free electron scattering breaks down and the interaction can be regarded as the collision of a photon with an electron. This process is known as *Compton effect*. It is maximum around 1 MeV and is important between 0.5 McV and 10 McV. When a photon of energy hv strikes the perfectly free electron (at rest), the photon with diminished energy hv′ is scattered at an angle θ with the direction of incident photon and the electron recoils at an angle ϕ. From the laws of conservation of energy and momentum in directions both parallel and perpendicular to the incident photon, the change in wavelength for the scattered photon is given by the relation

$$\Delta\lambda = \lambda' - \lambda = \frac{h}{m_e c}(1 - \cos\theta), \quad ...(17)$$

where $\Delta\lambda$ is known as *Compton shift* and constant h/m_ec the *Compton wavelength* having value 2.42621×10^{-12} m. Compton shift in wavelength is independent of the incident wavelength and material of the scatterer. It is dependent on the angle 0. Since $\lambda = c/v$, hence

$$\frac{c}{\nu'} - \frac{c}{\nu} = \frac{h}{m_e c}(1 - \cos\theta) \text{ or } h\nu' = \frac{h\nu}{1 + (1 - \cos\theta)\xi},$$

where $\xi = \dfrac{h\nu}{m_e c^2}$. For very large incident photon energy ($h\nu \geq m_e c^2$), the energy of the backscattered photon approaches $\frac{1}{2} m_e c^2 = 0.25$ MeV at $\theta = 180°$ and 0.51 MeV at $\theta = 90°$. The kinetic energy of the recoil electron is given by

$$T = h\nu - h\nu' = h\nu \frac{(1 - \cos\theta)\,\xi}{1 + (1 - \cos\theta)\,\xi} \qquad ...(18)$$

The kinetic energy of the electron has its maximum value when $\theta = 180°$ and photon is scattered directly backward. The energy is zero for $\theta = 0$ and $\nu = \nu'$.

$$\therefore \qquad T_{max} = \frac{h\nu}{1 + \frac{1}{2}\xi}. \qquad ....(19)$$

The angles θ and ϕ can be related as

$$\cot\phi = (1 + \xi)\tan\frac{1}{2}\theta. \qquad ...(20)$$

The evaluation of the differential cross-section is complicated as it requires relativistic wave mechanics. Klein and Nishina, in 1929, derived a relation for the differential cross-section per electron in the case of a linearly polarized incident plane electromagnetic wave, given as

$$[s_0(\theta)]_{Pol} = \left[\frac{d\sigma_c}{d\Omega}\right]_{Pol} = \frac{1}{4} r_0^2 \left(\frac{\nu'}{\nu}\right)^2 \left(\frac{\nu}{\nu'} + \frac{\nu'}{\nu} + 4\cos^2\Theta - 2\right),$$

where Θ is the angle between the polarization directions of the incident and emergent ray. If the incident wave is unpolarized, the differential cross-section will be the mean value of the above equation averaged over all angle Θ, as

$$[\sigma_0(\theta)]_{unpol} = \frac{1}{2} r_0^2 \left(\frac{\nu'}{\nu}\right)^2 \left(\frac{\nu}{\nu'} + \frac{\nu'}{\nu} - \sin^2\theta\right)$$

$$= \frac{1}{2} r_0^2 \frac{-\xi \cos^2\theta + (\xi^2 + \xi + 1)(1 + \cos^2\theta) - \xi(2\xi + 1)\cos\theta}{[1 + \xi(1 - \cos\theta)]^2}.$$

For low energies $\xi \to 0$ and the above relation reduces to the Thomson cross-section, eqn. (50). Total cross-section is obtained by integration over all angles $(0 < \theta < \pi)$ as

$$\sigma_c = \int \sigma_c(\theta) d\Omega = \int_0^{\pi} \sigma_c(\theta) 2\pi \sin\theta \, d\theta$$

$$= 2\pi r_0^2 \left\{ \frac{1+\xi}{\xi^2} \left[\frac{2(1+\xi)}{\xi^2} - \frac{1}{\xi} \log_e(1+2\xi) \right] + \frac{1}{2\xi} \log_e(1+2\xi) - \frac{1+3\xi}{(1+2\xi)^2} \right\} \quad ...(21)$$

When ξ is small, above relation reduces to

$$\sigma_c \simeq \frac{8}{2} \pi r_0^2 (1 - 2\xi + 5.2\xi^2 - 13.2\xi^2 + ...) \quad ...(22)$$

when $\xi \to 0$, when $\xi \to 0$.

As the scattered energy is smaller than the incident energy by a factor $h\nu'/h\nu$. Hence the energy scattering cross-section

$$\sigma_s = \left(\frac{h\nu'}{h\nu} \right) \sigma_c. \quad ...(23)$$

The energy absorption cross-section $\sigma_a (= \sigma_c - \sigma_s)$ represents the probability for the recoil energy to be imparted to the electron in this process. The atomic Compton cross-section

$${}_a\sigma_c = Z\sigma_0. \quad ...(24)$$

The scattering is coherent when $h\nu = h\nu'$ and is incoherent when $h\nu' < h\nu$. In the former case the individual amplitudes for each of the atomic electrons are added and in the latter case the intensities are added. Thus the atomic differential Compton cross-section can be defined as

$${}_a\sigma(\theta) = {}_a s_{coh}(\theta) + {}_a\sigma_{incoh}(\theta). \quad ...(25)$$

From above relations it is clear that the total scattering coefficient per electron decreases with increasing photon energy and this decrease is quite slow at low values of the energy and for energies above 0.5 MeV, σ_c is roughly proportional to $(h\nu)^{-1}$. This decrease is much slow than does photoelectric absorption, even in heavy elements.

3. **Pair Production:** The third important mechanism by which electro-magnetic radiation can be absorbed is the production of positron-electron pairs. This is in effect the conversion of radiation energy into mass energy which was first observed by Anderson, in 1932, in cloud chamber photographs. This process can take place only when the photon energy exceeds $2m_0c^2$, which is the sum of the rest energies of the members of the pair. According to the principle of conservation of electric charge the created particles must have equal and opposite charges. Applying the principle of conservation of energy to this process, we must have that

$$E_\gamma = h\nu = 2m_0c^2 + E^+ + E^-. \qquad \text{...(26)}$$

Here we have neglected the small amount of energy carried away by the recoiling nucleus. This equation shows that the minimum gamma ray energy required to create an electron positron pair is equal to $2m_0c^2$ or 1.022 MeV. The photon energy in excess of this amount is shared almost equally between the two particles, with the positron receiving slightly more ($\simeq$ 0.0075 Z MeV) than the negatron because it is repelled by the nucleus while the negatron is attracted. Such differences tend to disappear as photon energy increases.

Dirac's relativistic wave equation gives us energy eigenvalues for the free electron as

$$E = \pm\,(p^2c^2 + m_0^2c^4)^{1/2}. \qquad \text{...(27)}$$

These values range from $-\infty$ to $-m_0c^2$ and then from $+m_0c^2$ to $+\infty$, with a gap $2m_0c^2$. Classically the negative energy states could be excluded as having no physical meaning. Quantum mechanics can however explain it.

The lowest state corresponding to an electron at rest, has total energy m_0c^2. Above this is a practical continuum of states corresponding to electrons with various kinetic energies. Dirac suggested that in the absence of an external field all negative energy states were filled with electrons. *The completely filled sea of electrons in the –ve region does not contribute to the total energy and momentum of the system.* The electrons in the +ve energy region cannot make the transition to the –ve region. The transition can take place in the other direction if a sufficient energy ($\geq 2m_0c^2$) is supplied to the system.

The removal of electron from the –ve region by the external field, which may be that of γ-radiation or that of charged particle, is the creation of electron ($\geq m_0c^2$) in the + ve region and a *hole* in the –ve region. The *hole* or *vacancy* is evinced as an observable particle *positron* e^+. It acquires an energy – (– E), a momentum – (– p) and a charge – (– e). The process inverse to pair production is called *annihilation.* In this process electron and positron disappear and energy $2m_0c^2$ is emitted as radiation.

The differential cross-section per nucleus for the creation of a positron of kinetic energy E_+ and a negatron of kinetic energy E_- is given by

$$d\sigma_{pp} = \alpha\left(\frac{e^2}{4\pi \in_0 m_0c^2}\right)\frac{Z^2 PdE_+}{h\nu - 2m_0c^2}, \qquad ...(28)$$

where P is a complicated function of hν and Z, varies between 0 (for hn ≤ $2m_ec^2$) and about 20 (for hν = ∞). The total pair production cross-section.

$$\sigma_{pp} = \alpha\left(\frac{e^2}{4\pi \in_0 m_0c^2}\right)^2 Z^2 \int_0^{h\nu-2m_0c^2} \frac{PdE_+}{h\nu - 2m_0c^2}$$

$$= \alpha\left(\frac{e^2}{4\pi \in_0 m_0c^2}\right)^2 Z^2 \overline{P}. \qquad ...(29)$$

For extremely relativistic cases the analytical integration of eqn. (28) gives

$$\sigma_{pp} = \alpha\left(\frac{e^2}{4\pi \in_0 m_0c^2}\right)^2 Z^2\left[\frac{28}{9}\log_e \frac{2h\nu}{m_0c^2} - \frac{218}{27}\right] \qquad ...(30)$$

For very high energy photons, screening correction is needed which reduces the effective nuclear charge. This screening is due to atomic electron cloud with in the radius $a = \frac{a_H}{Z^{1/3}}$. Using complete screening of the nuclear coulomb field by the field of the atomic electrons, we have

$$\sigma_{pp} = \alpha\left(\frac{e^2}{4\pi \in_0 m_0c^2}\right)^2 Z^2\left[\frac{28}{9}\log_e \frac{183}{Z} - \frac{2}{27}\right] \qquad ...(31)$$

It is independent of photon energy.

When pair creation occurs in the field of an electron, at an energy (> 4 m_0c^2), the total cross-section per electron (without screening).

$$ {}_e\sigma_{pp} = \alpha\left(\frac{e^2}{4\pi \epsilon_0 m_0c^2}\right)^2\left(\frac{28}{9}\log_e\frac{2h\nu}{m_0c^2}-11.3\right) \qquad ...(32)$$

and for the whole atom

$$\sigma_{pp} = Z\, {}_e\sigma_{pp}. \qquad ...(33)$$

Hence in an atom of any Z,

$$\frac{\text{Total cross-section for all Z-electrons}}{\text{Total cross-section for nuclear pair production}} = \frac{1}{CZ}.$$

The total linear attenuation coefficient μ, thus can be defined as

$$\mu = N(\sigma_R + \sigma_{pe} + \sigma_{pp}) + NZ\,\sigma_c, \qquad ...(34)$$

where N is the number of atoms per unit volume of the absorber.

Mass attenuation coefficient for gamma rays in lead as a function of energy. The coefficient for the photoelectric effect (μ_{pe}/ρ), the Compton effect (μ_c/ρ) and pair production effect (μ_{pp}/ρ) are shown separately by dotted curves.

The coefficients in cm^{-1} can be obtained by multiplying by the density of Pb = 11.35 gm./cm^3. The absorption by photoelectric effect predominates at small energies. The probability of photoelectric emission increases as $h\nu \to E_B$ or $\nu \to \nu_0$, the frequency of the *absorption edge*. Above the absorption edge, the probability of photoelectric emission varies roughly as $Z^5 E_\gamma^{-3.5}$.

If the photon energy is considerably below that of the K-line of an observer. There can be no *K-resonance* (represented by K-edge) absorption because of the lack of energy. In the energy region below 0.1 MeV, we get edges. At these edges cross-section shows discontinuous jump. The L_1, L_{11}, L_{111} edges are corresponding to electrons from L-orbit. Photoelectric effect decreases rapidly with increasing energy.

As it decreases, attenuation by the Compton effect becomes relatively more important, until these effects are equal. At energies (0.5 MeV to 1 MeV) most of the attenuation is caused by the Compton effect. Absorption by pair production starts at about 1 MeV and increases until at high energies, the absorption is almost completely by pair production only.

The removal cross-sections for γ-rays per atom of the absorber can be summarized as :

Attenuation process	Dependence on Z	Dependence on E
Compton	Z	1/E
Photoelectric	$Z^{4.5}$	1/E for $E > m_0c^2$ $1/E^2$ for $E < m_0c^2$
Pair Production	Z^2	$\log_e E$

An approximate classification of the energy regions (in MeV) for which each of the three processes predominates in the case of a light element (Al), a medium element (Cu) and a heavy element (Pb) can be summarized as:

Attenuation process	Al	Cu	Pb
Photoelectric	E < 0.05	E < 0.1	E < 0.5
Compton	0.05 < E < 15	0.1 < E < 10	0.5 < E < 5
Pair production	E > 15	E > 10	E > 5

Thus we see that the photoelectric effect dominates at low E and high Z, Compton effect dominates at medium energies E and low Z, and pair production effect dominates at high energies E and high Z.

Other important process is *diffraction* or *coherent scattering*. The example of it is the interference pattern produced when a photon beam is passed through a properly oriented crystal. Cross sections for coherent scattering are small and hence it is unimportant as a mechanism by which energy is removed from photon beam.

DECAY AND LIFETIME

Both neutral and charged K-mesons decay into ordinary particles. Assuming that $\Delta S = \pm 1$ in the decay, it is necessary to assign strangeness S = + 1 or S = – 1 to the K-mesons. For S = + 1, equation $Q = T_3 + B/2 + S/2$ implies that $T_3 = \pm 1/2$, corresponding to K^+ and K^o mesons. In addition, two corresponding anti-particles are expected for S = –1, these are K^- and $\overline{K^o}$ with $T_3 = -1/2$ and + 1/2 respectively.

There are six different ways that K^+-mesons commonly decay, in each case giving two or three less massive particles. The various decay modes of the K^+ are denoted by symbols, θ^+, τ, τ', $K_{\mu\gamma}$, $K_{\mu g}$ and $K_{\mu 3}$. This confusing terminology dates from the time when it seemed likely that the

θ^+, τ.... decays sprang from different parent particles. The notation $K_{\mu 3}$ represents a decay into three particles one of which is a muon. The decay reactions are :

$$K^+(\theta^+) \rightarrow \pi^+ + \pi^o + 219.3 \text{ MeV} \qquad (\sim 25\%)$$

$$K^+(\tau) \rightarrow \pi^+ + \pi^- + \pi^+ + 75.1 \text{ MeV} \qquad (\sim 6\%)$$

$$K^+(\tau') \rightarrow \pi^+ + \pi^o + \pi^o + 84.3 \text{ MeV} \qquad (2 \sim \%)$$

$$K^+(K_{\mu 2}) \rightarrow \mu^+ + \nu + 388.2 \text{ MeV} \qquad (\sim 58\%)$$

$$K^+(K_{\mu 3}) \rightarrow \mu^+ + \pi^o + \nu + 253.2 \text{ MeV} \qquad (\sim 5\%)$$

$$K^+(K_{e3}) \rightarrow e^+ + \pi^o + \nu + 358.3 \text{ MeV} \qquad (\sim 4\%)$$

The idea that the various empirical particles (θ, τ, etc.) are different manifestations of a single particle ran into difficulty. The difficulty was known as "tau-theta puzzle". Until 1956 it was assumed that parity was conserved in all nuclear and atomic processes. The two pion decay ($\theta^+ \rightarrow \pi^+ + \pi^o$) mode has even parity, but the parity of the three pion ($\tau^+ \rightarrow \pi^+ + \pi^+ + \pi^-$) mode is odd. There is no doubt that the theta and tau particles are both charged mesons with odd parity. Now difficulty is that the two identical particles end in states of different particles. This difficulty, known as "τ-θ *puzzle*", not yet been resolved but it led Lee and Yang in 1956 to consider the nature of the forces between particles and to suggest that weak interactions such as those involved in β-decay, parity would not be conserved.

An interesting situation arises in connection with the decay of the neutral K-mesons. Since K^o and $\overline{K^o}$ have different values of the hypercharge, + 1 and – 1 respectively, hence one cannot be converted into the other by the strong interaction. However, the K^o can be transformed into $\overline{K^o}$ by the *weak interaction*, For example :

$$K^o \rightarrow \pi^o + \pi^o \leftarrow \overline{K^o}$$

$$\rightarrow \pi^+ + \pi^- \leftarrow \overline{K^o}.$$

The transformation is possible only when CP symmetry invariance is applicable to the operation. Through the weak coupling between the K^o and its anti-particle the $\overline{K^o}$, it is possible to construct two neutral K-particles that are their own anti-particles. These states have wavefunctions

$$\psi\left(K_1^o\right) = \frac{\left[\psi\left(K^o\right) + \psi\left(\overline{K^o}\right)\right]}{\sqrt{2}} \qquad ...(1)$$

$$\psi\left(K_1^{\circ}\right)=\frac{\left[\psi(K^{\circ})-\psi(\overline{K}^{\circ})\right]}{\sqrt{2}}. \qquad ...(2)$$

Application of the CP operation to K_1°, i.e. converting K° to $\overline{K}^{\circ}$ and vice versa, changes it into K_1°, i.e. K_1° is unchanged, but K_2° is changed into–K_2°, i.e. the sign of the wavefunction is changed. When the CP operation is applied to the decay products, it can be shown that decay of K_1° into $\pi^+ + \pi^-$ or $\pi^{\circ} + \pi^{\circ}$ is permitted. However, the K_2° particle, cannot decay by these two pion modes, although it can decay into three pions. The decay modes are :

$$K^{\circ}(K^{\circ}) \rightarrow \pi^{\circ} + \pi^{\circ} + 227.8 \text{ MeV } (31\%)$$
$$\rightarrow \pi^{+} + \pi^{-} + 218.6 \text{ MeV } (69\%)$$

$$K^{\circ}\left(K_2^{\circ}\right) \rightarrow \pi^{\circ} + \pi^{\circ} + \pi^{\circ} + 92.8 \text{ MeV}(27.1\%)$$
$$\rightarrow \pi^{+} + \pi^{-} + \pi^{\circ} + 83.6 \text{ MeV}(12.7\%)$$
$$\rightarrow \pi^{\pm} + e^{\mp} + \nu + 229.3 \text{ MeV}(36.6\%)$$
$$\rightarrow \pi^{\pm} + e^{\mp} + \nu + 252.5 \text{ MeV}(26.6\%)$$

The decay of a neutral K-meson into two pions is accompanied by a greater decrease in mass, and hence is a large energy release, then in the three pion decays. The K_1°-particles decay rapidly and have a short life than the K_2°. These states are also represented by K_S and K_L, where S and L indicate the shorter and longer lives respectively.

Mass of Kaons

Bristol and Brown analyzed forty seven τ^+-decay events and from all ranges and angles found Q = 75.04 ± 0.16 MeV. Using m_{π} = 139.58 ± 0.02 MeV, it was suggested that

$$m_{K+} = 493.78 \pm 0.17 \text{ MeV}.$$

Decay of neutral K°-meson in the mode $\pi^+ + \pi^-$ provides one means for measuring its mass. The estimated mass is 498.1 ± 0.4 MeV for K°-meson. Crawford found that close to the associated production threshold the energy balances of the reactions

$$\pi^- + p \rightarrow \Lambda + K^{\circ}$$
$$\rightarrow \Sigma^- + K^+$$

are sensitive to the mass values of the K-mesons. They found

$$m_{K^o} - m_{K^+} = 4.8 \pm 1.1 \text{ MeV}. \quad ...(3)$$

Another approach is through the charge exchange reaction

$$K^- + p \rightarrow K^o + n.$$

Burnstein and Rubin were able to find mass difference

$$m_{K^o} - m_{K^-} = 3.90 \pm 0.25 \text{ MeV}. \quad ...(4)$$

An earlier similar measurement by Rosenfeld et al. gave an estimate of 3.7 ± 0.7 MeV for this quantity. The mass of K^o-meson thus calculated is

$$m_{K^o} = 497.7 \pm 0.3 \text{ MeV}.$$

The mass of the K^--menson has been determined by using the hyperon production reactions.

$$K^- + p \rightarrow \Sigma^+ + \pi^- + Q \quad ...(5)$$

$$K^- + p \rightarrow \Sigma^- + \pi^+ + Q'. \quad ...(6)$$

If the Q value of the eqn (5) is determined from the measured kinetic energies of the particles, the mass of the K^--meson can be found using the known mass of the Σ^+-hyperon.

$$m_{K^-} = 493.7 \pm 0.3 \text{ MeV (or } 965.8 \pm 1.5 \text{ } m_e).$$

K-meson (K^+, K^o) are strange particles and having strangeness quantum number +1.

Spin

The spin of the K-meson is zero, like of π-mesons. This can be obtained from any of the following arguments : (1) τ-decay does not involve a change of parity where as θ-decay does involve a parity change. This puzzle was resolved by Yang and Lee suggesting non-conservation of parity in weak interactions. The spectrum of the π^- in the 3π decay of K+-meson agrees with the theoretically predicted for a spiniess K-particle. (2)

The observation of the neutral decay of the K_1^o into π^o-mesons is consistent only with an even spin of the neutral K-meson. (3) The longitudinal polarization of the μ^+-meson arising from the K^+-decay has been measured. It is in the same direction as the μ^+ from the decay of the π^+. This implies that the K^+ is also a spiniess particle.

Parity

The decay of the K^+ meson into three pions ($K^+ \rightarrow \pi^+ + \pi^+ + \pi^-$) can give us information about the parity of K^+. Two like mesons $\pi^+ \pi^+$ must have even parity $P = P_\pi P_\pi (-1)^L$, hence in this ($\pi^+ \pi^+$) system L must be even. The intrinsic parity of the pion is negative. Hence if l is the orbital angular momentum of the π^--meson, then the parity of the three pion system is $(-1)^3 (-1)^L(-1)^l = (-1)^{l+1}$. If spin of the τ^+ particle is zero, and the parity is conserved in the τ^+ decay, then its parity will be odd.

In the two pion decay of θ^+ particle ($\theta^+ \rightarrow \pi^+ + \pi^+$), the total parity of the system is (–1) (–1) (–1)J. If parity is conserved in the θ-decay, the θ^+-particle parity is (–1)J. For a spiniess θ^+-particle, the parity is thus even. The even and odd parities of the two states of K^+-mesons were explained by Lee and Yang, who suggested that the τ^+ and θ^+ particles were identical particles and the parity conservation in K-meson decays was violated, as in β-decay.

The parity of the Kaon can only be determined in strong interactions. *All the Kaons ate found to have odd parity.*

Interaction with Matter

Charged K-mesons lose energy by ionization as they pass through matter and also have certain interaction with atomic nuclei. K^+-mesons undergo relatively little nuclear scattering or absorption and if brought to rest in matter undergo spontaneous decay as it would in the absence of matter. On the other hand, K^--mesons undergo quite strong nuclear scattering and absorption.

If they come to rest in matter, they form K-mesic atoms and are absorbed, producing large stars. Neutral kaons do not appear to have very strong interaction with nuclei and do not lose energy by ionization. On account of their short lives they decay in flight.

For K^--energies less than 300 MeV, the reactions are

Elastic $\qquad K^- + p \rightarrow K^- + p$

Charge exchange $\qquad K^- + p \rightarrow \overline{K}^o + n$

Inelastic
$$
\begin{aligned}
K^- + p &\rightarrow \Sigma^- + \pi^+ \\
&\rightarrow \Sigma^+ + \pi^- \\
&\rightarrow \Sigma^o + \pi^o \\
&\rightarrow \Lambda^o + \pi^o.
\end{aligned}
$$

Both elastic and inelastic scattering are characterized by rapidly increasing cross-sections as the K^--energy decreases to zero. A very sharp K^--p scattering resonance has been found in elastic and inelastic cross-sections at about 135 MeV.

The K^--mesons interact with the nuclear neutrons as :

$$K^- + n \rightarrow \Lambda^o + \pi^-$$

$$\rightarrow \Sigma^o + p^-$$

K^--mesons produce hypernuclei, when interact with O^{16} and He^4. The resonance states of K and Λ (K* and Λ*) have been observed due to the interaction of K^- with protons. η^o-mesons are also produced when K^- interacts with protons.

CP Violation in Neutral K-Decay

We have seen earlier that the 2-pion mode is allowed by CP for the K_1, but forbidden for the K_2. The most sensitive test of the hypothesis of CP conservation is to look for the CP-violating decay $K_2 \rightarrow \pi^+\pi^-$. Christenson, Cronin, Fitch and Turlay in 1964 have performed an experiment and have observed the CP-violating decay. It was found in that experiment that K_2 particles decayed into $\pi^+\pi^-$ if with a branching ratio

$$\frac{\Gamma(K_2 \rightarrow \pi^+\pi^-)}{\Gamma(K_2 \rightarrow \text{all charged modes})} \cong (2.02 \pm (0.01) \times (10^{-3}). \qquad ...(7)$$

This indicates that the violation is small. The charged particles are detected and their momenta measured, by the two spark chambers and magnet sandwiches. Most of the decays observed are the dominent K_2 modes (πμν and πeν). The π^--decays are recognized by computing $P_K = P_\pi^+ + P_\pi^-$ and requiring that with in experimental errors resultant momentum vector P_K lies along the beam direction and the invariant mass of two charged particles directed be equal to the K^o mass. A second experiment has observed the same effect. A constructive interference between the free K_2-decay into two pions ($\pi^+\pi^-$) and the two pions ($\pi^+\pi^-$) from K_1's regenerated in baryllium has been also observed.

This CP violation suggests that the long lived and short lived components of neutral K's, known as K_L and K_S, are not the K_1 and K_2 combinations defined by equations (39) and (40). The above experiment suggests that

$$\psi(K_L) = \psi(K_2) + \in\psi(K_1), \quad ...(8)$$

where $$|\in| = 2 \times 10^{-3}. \quad ...(9)$$

The CP violation also implies that it is possible to make an absulute distinction between matter and anti-matter. With CP violation, $\overline{K^o}$ and K^o which are particle and anti-particle pairs can be distinguished by their decays into two pions. These are mixtures of K_S and K_L. Since K_S and K_L are not very different from K_1 and K_2. Both K_S and K_L decay to two pions, so the time dependence of the 2-pion decay rate will involve are interference term, which has opposite sign for $\overline{K^o}$ and K^o.

η-Meson

Evidence of the existence of a neutral particle of mass ≈ 550 MeV disintegrating into 3π mesons was first reported by Pevsner and collaborators in 1961 and was named 11. It is a three-pion resonance or quasi-particles. It was observed in a deuteron bubble chamber exposed to a 1.23 BeV/e positive pion beam. The reaction is.

$$\pi^+ + d^+ \rightarrow p^+ + p^+ + \eta^o.$$

Its formation has been confirmed at Berkeley where this particle has been observed in the bombardment of protons by K^--mesons.

$$K^- + p^+ \rightarrow \Lambda^o + \eta^o.$$

Its chief modes of decay are

$\eta \rightarrow \gamma + \gamma$	(31.4%)	Electromagnetic interaction
$\eta \rightarrow \pi^o + \gamma + \gamma$	(20.5%)	Electromagnetic interaction
$\eta \rightarrow \pi^o + \pi^o + \pi^o$	(21%)	Weak interaction
$\rightarrow \pi^+ + \pi^o + \pi^-$	(22.4%)	Weak interaction

The η has a γ-width about 20 times larger than that of π^o. Considering that the γ-decay mode contributes roughly one half of the width, the mean life time η, obtained by comparison with that of π^o, should be 5×10^{-18} sec. This corresponds to the width of less than 1 keV.

Bastien et al. measured the invariant mass of the three pions in the reaction ($K^- + \pi \rightarrow \Lambda + \pi^+ + \pi^- + \pi^o$) and found the mass of η^o-meson as 549.9 ± 1.2 MeV. Delcourt et al. measured, with a quite different technique, the production of η-meson in the reaction $\gamma + p \rightarrow p + \eta$ just above threshold and estimated the mass 549.3 ± 2.9 MeV for η-mesons. The various published measurements are consistent. The best estimate derived is 548.7 ± 0.5 MeV.

As η, particle decays into two gamma rays or three pions, hence, it is necessary to assume the spin as 0, 2 or larger. Spin 2 or larger is excluded because it would result in marked anisotropies in the decay which are not observed. Thus it is better to assign $J = 0$ to the η-particle. Since the parity of two pions in a $T = 0$ state is necessarily even, hence the η-particle must have odd parity.

Hyperons

Hyperons are unstable particles, having masses greater than that of the nucleon. These particles are a second class of strange particles. The name hyperon was taken from Greek word hyper, "meaning above".

Λ-hyperon

The first hyperon was found in cosmic rays in 1947 by Rochester and Butler in England. They observed the tracks of two charged particles-emerging from a point in the gas, of a cloud chamber. The upside down V-shaped figure could best be explained by assuming that some unstable neutral particle shooting down from the sky had decayed to form positive and negative charged particles. The charged particles seen in the decay were identified as a proton and a negative pion.

This was the first evidence for the existence of the first and best known *hyperon. This neutral unstable particle was named as lambda and denoted by* Λ^o, *because of the characteristic appearance of its decay in a cloud chamber.* For its production in laboratories, Fowler, Shutt, Thorndike and Whittemore allowed a beam of 1.5 BeV π^--mesons to enter a diffusion cloud chamber containing hydrogen at a pressure of 18 atmospheres placed in a magnetic field pf 1100 gausses. The track of an incident pion can be seen at the end in the chamber. This end point is the vertex of a V, formed by the tracks of the negative pion and proton. To conserve energy and momentum in the production of Λ^o-hyperons, the necessity of the production of another neutral particle, probably K^o-meson, was also suggested. The reaction is

$$\pi^- + \pi \rightarrow L^o + K^o.$$

A-hyperon undergoes a following decay reactions :

$$\Lambda^o \rightarrow p + \pi^- \quad (67.7\%)$$

$$\rightarrow n^o + \pi^o \quad (31.6\%)$$

The mass of the Λ-hyperon has been obtained from the Q-value of the reaction $\Lambda^o \rightarrow p + \pi^-$. Both the pion and proton ranges, as well as the opening angles between the Λ-decay products were made and the Q-value so measured was 37.56 ± 0.13 MeV. Several other similar measurements have been made and the present best estimate of Q value is 37.60 ± 0.12 MeV. The corresponding mass

$$m_\Lambda = 1115.44 \pm 0.12 \text{ MeV}$$

The mean life of neutral lambda particle is 2.62×10^{-10} sec. The spin of Λ^o is 1/2. The magnetic moment of Λ^o, like the neutron is negative and equal to – (0.80 ± 0.07) nm.

The leptonic decay of the Λ^o-hyperon

$$\Lambda^o \rightarrow p + e^- + \overline{\nu_e} \quad (0.88 \times 10^{-5}\%)$$

has been observed to occur. Because of parity non-conservation in weak decay processes it is impossible to relate the parity of the Λ-hyperon to that of the πp system. By convention it is taken to have even parity. As Λ^o appears only in one charge state, neutral, hence it has isospin T = 0. The decay modes are permitted by the rule ΔS = 1.

Σ-hyperons

It is the family of hyperons with the greatest number of members. The first of these to be observed was the positive sigma hyperon denoted by the symbol Σ^+. It has two predominant decay schemes and three weak decays.

$$\Sigma^+ \rightarrow p + \pi^o + 116.1 \text{ MeV} \quad (\sim 51\%)$$

$$\rightarrow n + \pi^+ + 110.3 \text{ MeV} \quad (\sim 49\%)$$

$$\rightarrow n + \pi^+ + \gamma \quad (\sim 4 \times 10^{-7}\%)$$

$$\rightarrow \Lambda^o + e^+ + \nu \quad (\sim 2 \times 10^{-7}\%)$$

$$\rightarrow p + \gamma \quad (\sim 3 \times 10^{-5}\%)$$

The mean life of Σ^+-hyperon is 0.81×10^{-10} sec. The corresponding negative particle denoted by Σ^- was also observed within few months of the discovery of the Σ^+. Its decay scheme is

$$\Sigma^- \rightarrow n + \pi^- + 118.5 \text{ MeV (Mean life time } 1.65 \times 10^{-10} \text{ sec).}$$

So other wards evidence for the existence of a neutral Σ^- hyperon was also presented. Its decay scheme is

$$\Sigma^o \rightarrow \Lambda^o + \gamma + 80 \text{ MeV (Mean life time} < 10^{-11} \text{ sec).}$$

The Σ-hyperon masses come from measurements carried out on the products of K^- and Σ^- capture by protons and on the disintegration products of the hyperons themselves. When such measurements are combined, they yield $m_{\Sigma+} = 1189.39 \pm 0.14$ MeV.

The π^- and Σ^+ ranges in the reaction

$$K^- + p \rightarrow \pi^- + \Sigma^+ + Q$$

indicate a momentum balance and yield a Q-value of 103.0 ± 0.3 MeV. Using this value of Q and the mass of the Σ^+-particle, the mass of K^- -meson is calculated. Then from relation $K^- + p \rightarrow \pi^+ + \Sigma^- + Q'$ we get the mass of Σ^- as 1197.6 ± 0.5 MeV. This is considerably larger than the mass of Σ^+.

Burnstein et al observed the sequence $\Sigma^- + p \rightarrow \Sigma^o + n$ followed by the decay mode $\Sigma^o \rightarrow \Lambda^o + e^+ + e^-$. These events yielded a Σ^o momentum of 60.3 ± 0.9 MeV/c and $m(\Sigma^-) - m(\Sigma^o) = 4.7 \pm 0.1$ MeV or $m(\Sigma^o) = 1192 \pm 0.2$ MeV.

The less probable decay modes of Z^- are :

$$\Sigma^- \rightarrow n + \mu^- + \overline{\nu_\mu} \qquad (6.6 \times 10^{-5}\%)$$

$$\rightarrow n + e^- + \overline{\nu_\mu} \qquad (1.4 \times 10^{-5}\%)$$

$$\rightarrow \Lambda^o + e^- \overline{\nu_e} \qquad (7.5 \times 10^{-7}\%)$$

Σ^+, Σ^- and Σ^o all have intrinsic parity even and intrinsic spin 1/2 thus are fermions. They possess a magnetic moment. The experiments on the Σ^+-hyperon magnetic moment suggested the value 3.28 ± 0.58 nm. Because of the three charge states, Σ-hyperons have T = 1. The decay modes show that ΔS = 1. The Σ-parity can be related to that of the Λ by the analysis of the electromagnetic decay process $\Sigma \rightarrow \Lambda\, e^+e^-$

Σ^--hyperons undergo interactions with nucleons

$$\Sigma^- + p \rightarrow \Lambda^o + n.$$

Ω–hyperon

Before the discovery of Ω^--hyperon, all known particles had hypercharge values of + 1, 0, or – 1. Gell-Mann predicted the existence of such a particle, with a negative charge, as required to give it a hypercharge of –2. Furthermore, he suggested that it might be produced, as it actually was, by the interaction of a K^--meson at energies greater than 3.5 GeV with a proton. The expected strong interaction would be

$$K^- + p^+ \rightarrow \Omega^- + K^+ + K^o$$

Hypercharge – 1 + 1 – 2 + 1 + 1

Isospin 1/2 1/2 0 1/2 1/2

The experimental designed to produce and detect the Ω^--hyperon was carried out at Brookhaven National Laboratory in 1964. In this experiment short bursts of 5 GeV K^--mesons were allowed to enter a liquid hydrogen bubble chamber. The photograph obtained can be interpreted as : *A K^--meson enters from the bottom and interacts with a proton at the point a to yield a K^o-meson, a K^+-meson and the Ω^--hyperon.* The positive kaon is deflected to the left by the magnetic field whereas the Ω^--hyperon is deflected to the right. The Λ^o-hyper on decays at point b, after travelling a short distance during its lifetime, into Λ^o hyperon and π^--meson. Ξ^o-hyperon decays at point c, into Λ^o and π^o, but latter forms two gamma ray photons immediately. These photons then produce electron-position pairs at the points d and e. The Λ^o-hydrogen decays at point f into a proton and a π^--meson.

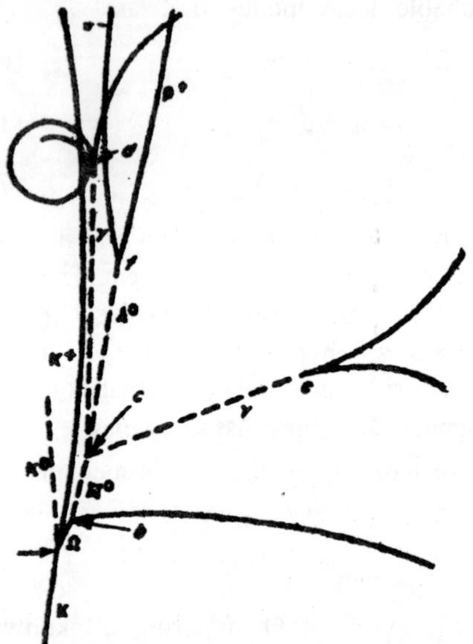

Fig. 10 : Photograph representing the discovery of Ω^--hyperon.

Gell-Mann noted that the Ω^--particle should decay by the weak interaction, which it does. Its decay modes are

$$\Omega^- \rightarrow \Xi^o + \pi^-$$

$$\rightarrow \Xi^- + \Lambda^o$$
$$\rightarrow \Lambda^o + K^-$$

From the momenta of decay products (Ξ^o and π^-), the mass of the Ω^- was measured to be 1686 ± 12 MeV. The mass was also measured by the decay mode $\Omega^- \rightarrow \Lambda^o + K^-$. The mass was 1674 ± 3 MeV. The mass was determined by decay mode $\Omega^- \rightarrow \Xi^- + \pi^o$. The averaged value used now is 1674 ± 3 MeV.

Ω^--hyperon is a baryon, having strangeness quantum number –3. hypercharge –2, spin 3/2 and parity even. It has a mean life time of 1.3×10^{-10} sec.

All these properties are summarised in Table 16.3.

HYPERNUCLEI

A hypernucleus or hyper fragment is a metatstable nucleus in which a bound hyperon replaces one of the nucleons. The symbol ${}_{\Lambda}B^{10}$ denotes a hypernucleus in which one of the neutrons of the ${}_5Be^{10}$ particle is replaced by a lambda hyperon. Actually the definition is misleading, the hypernucleus should rather be regarded as ${}_5Be^9$ which has captured a Λ^o-particle. The reason for this is that the Λ^o can occupy the same quantum state as a neutron already in the nucleus without violating the Pauli exclusion principle, since it is a different kind of particle. We shall restrict the discussion to Λ-hypernuclei. Because of the fast transformations strangeness conserving Σ or Ξ-hypernuclei have very short lifetimes and interact with nucleons as

$$\Sigma^+ + n \rightarrow \Lambda^o + p$$
$$\Sigma^- + p \rightarrow \Lambda^o + n$$
$$\Xi^- + p \rightarrow \Lambda^o + \Lambda^o$$
$$\Xi^o + n \rightarrow \Lambda^o + \Lambda^o.$$

The Polish physicists Danysz and Pniewski in 1952 deduced that a track observed in a nuclear emulsion block which had been exposed at a high altitude to cosmic radiation was due to hypernucleus of Beryllium. Hypernuclei of all the light elements from hydrogen (${}_{\Lambda}H^3$) to nitrogen (${}_{\Lambda}N^{14}$) have been reported. In addition, from a star produced in a photographic film bombarded by K^--mesons, evidence of a hypernucleus containing two Λ^o-hyperons was reported in 1963 at the CERN laboratory. The species formed was considered to be either ${}_{\Lambda\Lambda}Be^{10}$ or ${}_{\Lambda\Lambda}Be^{11}$. In 1966, a similar hypernucleus ${}_{\Lambda\Lambda}He^6$ was identified in U.S.A.

It is very difficult to obtain accurate determination of the average life-times of hypernuclei, but the average life-times are approximately slightly less than that of a free Λ^o-hyperon. There are usually several competing modes of decay of a given species of hypernuclei. The ${}_{\Lambda}He^4$ hypernucleus has the following decay modes:

$$
\begin{aligned}
{}_{\Lambda}He^4 &\rightarrow \pi^o + \text{nucleons} \\
&\rightarrow \pi^- + \pi + {}_{2}He^3 \\
&\rightarrow \pi + n + {}_{1}H^2 \\
&\rightarrow {}_{1}H^2 + {}_{1}H^2.
\end{aligned}
$$

The first two decay modes are called mesonic decays, in which hyperon may be regarded as undergoing its normal decay into $p + n^-$ or $n + \pi^o$. The nucleon thus produced may be expelled or it may remain in the product nucleus. Other two modes are called non-mesonic decays. The non-mesonic decay is equivalent to the interaction of a Λ^o-particle with proton or a neutron, i.e.

$$
\begin{aligned}
\Lambda^o + p &\rightarrow (\pi^- + p) + p \rightarrow n + p \\
\Lambda^o + n &\rightarrow (\pi^o + n) + n \rightarrow n + n.
\end{aligned}
$$

Since a Λ^o-hyperon does not normally interact with a nucleon, the interaction with a nucleon is stimulated by the remainder of the nucleus and the non-mesonic decay (dominant for hyper-nuclei with $Z > 3$ or 4) is referred to as stimulated decay.

The binding energy of the Λ^o-particle in the hypernuclei can be calculated from the observed kinetic energies of the decay particles, which can be derived from the lengths of the tracks in a photographic emulsion. In non-mesonic decay, one or more neutrons are generally emitted, they do not produce any track in the emulsion and the kinetic energies cannot be determined.

Hence, the binding energy of Λ^o-particle is obtained from mesonic decays. The values have been found to increase fairly regularly with the atomic number of the hypernucleus. The value is 0.2 MeV in ${}_{\Lambda}He^3$ and about 12 MeV in ${}_{\Lambda}C^{14}$ and ${}_{\Lambda}N^{14}$.

In this way it is clear that the existence of ${}_{\Lambda}H^2$ is not possible because in it the binding energy would be still smaller and in the heavier hypernuclei, Λ^o-hyperon is more strongly bound than a nucleon. Similar to with nucleon–nucleon force, the hyperon-nucleon force appears to be charge independent and dependent on the spin.

RESONANCE STATES

Most resonant or unstable states result from the strong interaction of a high energy meson (π or K) with a proton. Sometimes, resonant states also result from the interaction with a neutron or a deuteron. These states can be assigned mass, charge and spin consistent with the conservation laws. The earliest evidence for the formation of resonant states, also referred to as *resonance particles* or *resonances*, came in 1951 from a group working with Fermi at the Chicago University. In the course of studies of the scattering of positive and negative pions of various energies by protons, it was observed that the cross-sections increased rapidly at energies above about 60 MeV. In 1957, K.A. Brueckner at Indiana University suggested that a resonance interaction occurred between the proton and the pion, much as the resonance capture of a neutron by a nucleus was associated with a maximum in the cross-section. *It is for this reason that the term resonance or resonant state came into general use.* At the present time, the name is applied to a particle state of very short life (~ 10^{-23} sec) which decays into two or more particles. Because of their extremely short lifetimes, resonant states are described as unstable particles.

In spite of their extremely short lives, two general methods have been used for detecting resonances. *One is to measuse the scattering cross-section as a function of energy of the incident meson.* The plot of the results indicates the formation of a resonant state. The rest mass energy of this state can be calculated from the energy of the scattered particle corresponding to the maximum cross section. The average lifetime of the resonance, equal to the uncertainty in the time, $\tau = \Delta t = \hbar/\Delta E$, where ΔE is the uncertainty in energy which is equal to the full width of the peak at half the maximum height. Suppose the peak width at half maximum is found to be 100 MeV, the mean life of the resonant state is $6.6 \times 10^{-21}/100 = 6.6 \times 10^{-21}$ sec.

The scattering cross-sections of positive and negative pious by protons are shown in Fig. 11. The upper curve shows that the maximum of the resonant peak is at a pion energy of 195 MeV in the lab system. On converting it in the proton pion centre of mass system the rest mass energy of the resonant state is found to be 1238 MeV. The full width at half maximum is 120 MeV, corresponding to the life time of 5×10^{-21} sec. This short life indicates decay by the strong interaction. This particular resonance state was attributed to a state with an isospin of 3/2 by K.A. Brueckner, and was originally represented by the symbol $N_{3/}$

$_2$* indicating a *nucleon excited state with* T = 3/2. It is now designed as Δ(1238). Theory predicts that in the formation of an T = 3/2 resonance the maximum scattering cross-section for π^+-p should be three times greater than for π^--p. It is same as indicated by Fig. 11.

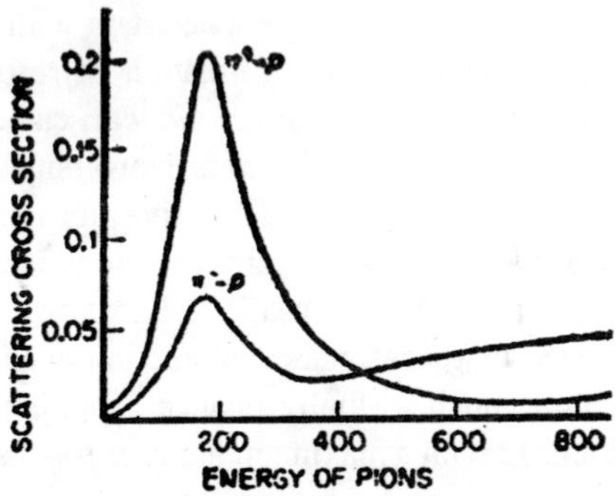

Fig. 11 : Scattering cross-ssctions of π^+, π^--mesons by protons.

When the cross-sections for scattering of positive and negative pions with energies in excess of 300 MeV by protons are measured, several additional resonant states are observed. Some of these are more prominent in the π^+ interactions, whereas other are in the π^+-interactions. These states can be assigned mass, charge and spin consistent with the conservation laws. The masses of some of well established resonant states are 1518, 1688, 1920 and 2360 MeV. These resonant states all have same mean lifetimes ~ 10^{-23} sec.

The second method for detecting resonances is based on a study of the energy distribution among the products, as determined from bubble chamber events. Suppose a high energy particle A interacts with a stationary particle B, in the bubble chamber to produce a particle C plus a resonant state R. The latter decays into two particles D and E in such a very short time that it leaves no track in the chamber.

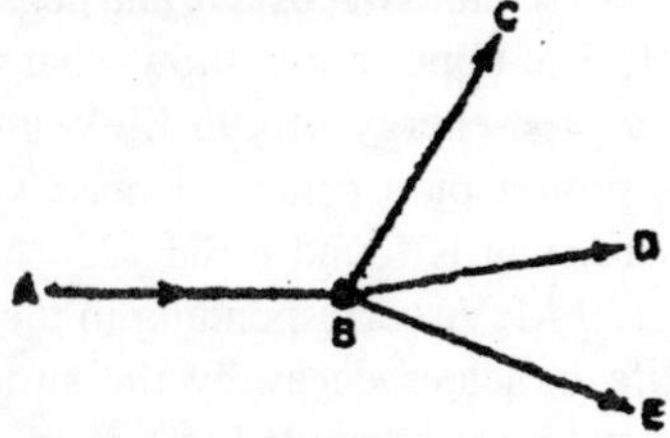

Fig. 12 : Bubble Chamber track.

If all these particles are charged, the tracks obtained in the bubble chamber will be as shown in Fig. 12. The problem is thus to distinguish between the postulated process.

$$A + B \rightarrow C + R \rightarrow C + D \rightarrow E \quad ...(1)$$

and the direct reaction

$$A + B \rightarrow C + D + E. \quad ...(2)$$

The problem can be solved by a kinematic analysis of the bubble chamber event made with the aid of a computer. The available energy will be distributed among the three particles in a random manner, as given by the law of conservation of momentum, in the case in which these particles are formed by a direct reaction. If large number of events of this type are examined, it will be found that the energy of any particle, say of C, has a spectrum of values ranging from zero to a maximum equal to the total energy available. If we consider another case in which a resonant state R is first formed together with a particle C. The available energy will be divided between them in a definite manner which depends only on the relative masses of these two. Thus in a large number of events, there will be many cases in which particle C has a specific energy value and a peak will be observed in the energy distribution curve. Energy of the resonance R can be calculated from the energy of the particle C corresponding to this peak. The mean life-time of the resonant state can be calculated by knowing the uncertainty in energy which is equal to the width of the peak at half maximum.

The problem that which of the three emergent tracks in the bubble chamber event represents particle C was solved by a group at the University of California, Berkeley in 1960. In the search for a hyperon-pion resonance analogous to the proton-pion resonance described earlier, the reaction was

$$K^- + p^+ \rightarrow \Lambda^o + \pi^+ + \pi^-.$$

Many hundreds of plates were analyzed by a computer. Some of the results suggested that the law of conservation of linear momentum was being violated and two resulting particles were indicated rather than three. Possibilities were

$$K^- + p^+ \rightarrow Y^{*+} + \pi^-$$

$$K^- + p^+ \rightarrow Y^{*-} + \pi^+,$$

where Y* was suggested a new resonance particle showing strong decay in 10^{-33} sec into

$$Y^{*+} \rightarrow \Lambda^o + \pi^+$$
$$Y^{*-} \rightarrow \Lambda^o + \pi^-.$$

Thus we see that $\Lambda^o\pi^+$ and $\Lambda^o\pi^-$ resonances are formed, the rest mass corresponding to the pion energy of 280 MeV is calculated to be 1385 MeV. This resonance state, originally designed Y* is now preferably indicated by the symbol Σ* (1385). The widths at half maximum of the energy peaks at 280 MeV are both 40 MeV. This shows that the Y* decays by the strong interaction. It is built from either $\Lambda^o\ \pi^+$ or $\Lambda^o\ \pi^-$, therefore, S = –1 and T = 1. Its baryon number is one and angular momentum is 3/2.

Evidence for an excited slate of a K-particle has been obtained from hydrogen bubble chamber events

$$K^+ + p \rightarrow K^{*+} + p;\ K^* \rightarrow \overline{K}^o + \pi$$
$$K^- + p \rightarrow K^{*-} + p$$
$$K^- + p \rightarrow \overline{K}^{*o} + m$$

The K* excited state has rest mass 892.4 MeV and isotopic spin half. It has a spin 1 and two values of strangeness S = ± 1 in the two charge states as for the K-meson. In addition to these resonances, large number of others have been observed.

Nambu in 1957 suggested the existence of a heavy neutral meson having zero isospin and unit angular momentum. Chew in 1960 had also shown that such a meson should exist on dynamical grounds as a 3π resonance, or bound state. The first experimental evidence for a three pion resonance, with T = 0 and I = 1, was found by a study of four pronged annihilation events when the 72 inch. Berkeley hydrogen bubble chamber was exposed to a high energy anti-proton beam. The over all process is

$$\overline{p} + p \rightarrow \pi^+ + \pi^- + \pi^o + \pi^+ + \pi^-.$$

In order to determine if a resonant state involving three pions are formed as an intermediate stage, three groups of pion triplets were studied, namely, singly charged $\pi^+\pi^+\pi^-$ and $\pi^-\pi^-\pi+$, doubly charged $\pi^+\ \pi^-\ \pi^o$ and $\pi^-\ \pi^-\ \pi^o$ and neutral $\pi^+\ \pi^-\ \pi^o$. About 800 bubble chamber events were analyzed and for each of the three groups a histogram was obtained of the number of triplets against the effective mass of the triplet. For the charged triplets, the histograms were in good agreement with the phase space distribution. The maximum is observed only in the case of neutral triplet. It follows, therefore, that a neutral resonant state is formed and that it decays into positive, negative and neutral pion. This resonant state is known as the omega (ω)-meson. Thus the above reaction can be written as two stage process.

$$\bar{p} + p \rightarrow \omega^o + \pi^+ + \pi^-$$

$$\omega^o \rightarrow \pi^+ + \pi^- + \pi^o.$$

In a bubble chamber study of the four pronged proton-antiproton annihilation events gave the evidence for a three pion resonance ω^o-meson. The state with charge zero shows a marked peack at a mass of 787 MeV but no such peak appears in the states with charge 1 or 2. The favoured state, therefore, has T = 0, I = 1 and parity odd. Gelfand *et al.* (1963) observed mass of 784.0 ± 0.9 MeV, its decay into 3 π with width 9.5 ± 2.1 MeV and a life time of $(0.69 \pm 0.15) \times 10^{-22}$ sec. Maglic *et al* (1961) observed mass of 187 MeV, resonance width <25 MeV and life time > 4×10^{-23} sec and the distance travelled before its decay into three pions > 13 fermis.

The ω-meson has also been observed in a liquid deuterium bubble chamber exposed to a high energy beam of π^+-mesons.

$$\pi^+ + d \rightarrow p + p + \pi^+ + \pi^- + \pi^o \text{ or } p + p + \omega^o.$$

This experiment is accompanied by less background events than in proton-antiproton annihilation.

ρ-meson

Another quasi-particle is the rho (ρ)-meson. It is a two pions resonance particle. It can exist in three charge states ρ^+, ρ and ρ^o. These are produced in a fairly large fraction of high energy antiproton-proton annihilations.

$$\bar{p} + p \rightarrow \rho + \pi + \pi.$$

The most frequent decay mode of the charged p-meson is

$$\bar{p} \rightarrow p^+ + \pi^o.$$

The most probable decay of ρ^o is

$$\rho^o \rightarrow \pi^+ + \pi^-.$$

Other decays have small branching ratios.

$$\rho^o \rightarrow \pi^+ + \pi^- + \pi^o$$

$$\rho^{o} \rightarrow \pi^{+} + \pi^{-} + \pi^{o} + \pi^{o}.$$

The observed rate of a dipion final state as a function of the total energy of the colliding beams (e^{+} and e^{-} beams) shows a maxima at about 750 MeV. By calculating momenta of $\pi^{+}\pi^{-}$ pair by track chamber methods, the mass of ρ-meson was calculated as 765 ± 4 MeV.

The width of ρ^{o}-meson is 100 MeV so that its life in sec it $\approx 10^{-33}$ sec.

As $\pi^{+}\pi^{-}$ states do not occurs hence $T \neq 2$. The neutral and singly charged are observed, hence isospin T = 1. The assignment of I = 1 is made by analyzing the angular distribution of the decay products of ρ-particles emitted parallel or anti-parallel to the direction of the incident pion. It follows negative parity, considering that two decay pions have zero intrinsic spin and even intrinsic parity.

The resonance state of η particle, named as η' or x^{o}, has mass 957.1 ± 0.6 MeV, spin 0, parity odd, isospin 0, hypercharge 0. The full width at half maximum is less than 4 MeV. Its decay modes arc

$$\eta' \rightarrow \eta + \pi + \pi \ (68\%)$$

$$\rightarrow \pi^{+} + \pi^{-} + \gamma \ (32\%).$$

SOLVED EXAMPLES

Example 1:

In an absorption experiment with 1.14 MeV γ-radiation from Zn^{65}, it is found that 25 cm of Al reduce the beam intensity to 2%. Calculate the half value thickness and the mass absorption coefficient of aluminium for this radiation.

Solution:

Half value thickness is that thickness of absorber that reduces the intensity at one half its original value. Thus

$$I = I_0 \exp.(-\mu_l x_{1/2})$$

or $$x_{1/2} = \log_e \frac{2}{\mu_l}$$

$$\frac{I}{I_0} = e^{-\mu_l x}$$

or $$\frac{2}{100} = e^{-\mu_l \times 0.25}$$

or $\mu_l = 15.64 \text{ metre}^{-1}$

$$\therefore \quad x_{1/2} = \frac{\log_e 2}{\mu_l} = \frac{0.693}{15.64} = 0.044 \text{ metre}$$

and mass absorption coefficient $\mu_m = \frac{\mu_l}{\rho} = \frac{15.64}{2700} = 0.00577 \text{ m}^2/\text{kg}.$

Example 2:

Compare the radiation loss with the ionization loss for I MeV β-particles in Pb. Calculate the β-energy for which these losses are equal to Pb.

Solution:

We know that

$$\frac{\left(\frac{dE}{dx}\right)_{rad}}{\left(\frac{dE}{dx}\right)_{ion}} = \frac{ZE}{800} = \frac{82\times 1}{800} = 10.25\times 10^{-2}$$

These losses are equal when

$$ZE = 800$$

or $$E = \frac{800}{82} = 9.75 \text{ MeV}.$$

Example 3:

Calculate the half value thickness for β-absorption in Al for the β-spectrum from Ra-E with E_{max} = 1.17 MeV for this spectrum.

Solution:

The half value thickness is that thickness of absorber that reduces the energy to its half value. Thus

$$E = E_0 e^{-\mu x}$$

or $$e^{-\mu x} = \frac{E}{E_0} = \frac{1}{2}.$$

$$\therefore \quad x_{1/2} = \log_e \frac{2}{\mu} = \log_e \frac{2}{\mu_{m\rho}}.$$

where $$\mu_m = \frac{22}{(E_{max})^{1.33}} = \frac{22}{(1.17)^{1.33}} = 17.8 \text{ cm}^2/\text{gm}.$$

Example 4:

A proton whose path has a radius of curvature of 250 cm. in a magnetic field of 1 tesla traverses a lead plate whose thickness is 40 gm/cm² along the path of the proton. What should the radius of curvature be after it emerges from the plate?

Solution:

We know the relation $Bq\nu = \dfrac{m\nu^2}{R}$

or $$\nu = \frac{BqR}{m}.$$

$\therefore\ \nu = 1 \times 1.6021 \times 10^{-19} \times 2.5/1.6725 \times 10^{-27}$

$= 2.395 \times 10^{8}$ m.

We know that the stopping power, *i.e.,* the amount of energy per unit length of path

$$-\frac{dE}{dx} = \frac{Zz^2ne^4}{4\pi \in_0^2 m_e\nu^2}\left[\log\frac{2m_e\nu^2}{I} - \log\left(1-\frac{\nu^2}{c^2}\right) - \frac{\nu^2}{c^2}\right].$$

For approximation we can neglect relativistic terms and we get

$$-dE = -M\nu\, d\nu = \frac{Zz^2ne^4}{4\pi \in_0^2 m_e}\cdot\frac{1}{\nu^2}\log\frac{2m_e\nu^2}{I}dx$$

or $$-\int\frac{\nu^3 d\nu}{\log\left(\dfrac{2m_e\nu^2}{I}\right)} = \frac{Zz^2ne^4}{4\pi \in_0^2 m_e M}\int dx.$$

For integration, let us assume $\log\left(\dfrac{2m_e\nu^2}{I}\right) = y$,

hence $$\left(\frac{2m_e\nu^2}{I}\right) = y \text{ and } \nu^2 = \left(\frac{I}{2m_e}\right)e^{\nu}.$$

$$\therefore\quad -\frac{I^2}{8m^2}\int\frac{e^{2y}}{y}dy = \frac{Zz^2ne^4}{4\pi \in_0^2 m_e M}\int dx.$$

In our problem $\int dx = x = 40\,\dfrac{gm}{cm^2} = \dfrac{40}{11.34}$ cm. The left hand side can be solved by standard integral table or by computer.

By solving the indefinite integral, we can get $v = 1.916 \times 10^8$ m. and hence R = 2 metre.

Example 5:

Calculate the range of a 9-MeV α-particle in aluminium if the relative stopping power of Al is 1700. Also calculate the thickness of Al that is equivalent in stopping power to 1 metre air.

Solution:

Range in air of an α-particle is given by the relation

$$R = 0.00318\, E^{3/2} = 0.00318\, (9)^{3/2}$$

$$= 0.08580 \text{ metre.}$$

$\therefore$ Range in aluminium

$$R_{AL} = R_{air}/S = 0.08586/1700$$

$$= 5.05 \times 10^{-5} \text{ metre.}$$

If ρ_{AL} is the density of A*l*, then the equivalent thickness in A*l* will be R_{AL}, ρ_{AL}, hence

$$R_{AL}\,\rho_{AL} = (R_{air}/S)\,\rho_{AL} = \left(\frac{1}{1700}\right)(2700) = 1.58 \text{ kg/m}^2.$$

Example 6:

Calculate the energy of the neutron produced when a slow negative pion is captured by a proton. Should neutron be treated relativistically?

Solution:

$$\pi^- + p^+ \rightarrow n^o + \gamma + Q.$$

$$139 + 938 \rightarrow 939 + h\nu + Q,$$

or $$Q = 138 \text{ MeV} = E_\gamma + E_n.$$

From conservation of momentum $m_n v_n = E_\gamma/c$

and $$\frac{E_n}{E_\gamma} = \frac{1}{2}\frac{m_n v_n^2}{c m_n v_n} = \frac{1}{2}\frac{v_n}{c} = \frac{1}{2}\frac{E_\gamma}{m_n c^2} = \frac{E_\gamma}{1878}$$

$$\frac{E_n}{E_\gamma + E_n} = \frac{E_\gamma}{1878 + E_\gamma} \quad \therefore \quad \frac{E_n}{138} = \frac{E_\gamma}{1878 + E_\gamma}$$

and $$\frac{E_n}{138} = \frac{138 - E_n}{1878 + 138 - E_n} = \frac{138 - E_n}{2016 - E_n}$$

or $E_n = 9$ MeV.

Using relativistic relations

$$m_n v_n = \frac{m_{0n} v_n}{\left(1-\beta^2\right)^{1/2}} = \frac{E_\gamma}{c}$$

and $E_n = m_{0n}c^2\ [(1 - b^2)^{-1/2} - 1]$

We get $E_n = 8.8$ MeV.

Example 7:

Prove that the photon energy shared by the particles in the pair production process is unequal.

Solution:

If a pair is created at a distance r from the centre of a nucleus of charge Ze, then the potential of each particle of the pair is $Ze^2/4\pi\epsilon_0 r$. As the electron is attracted and positron is repelled by the nucleus, hence the difference in their K.E. will be $2Ze^2/4\pi\epsilon_0 r$. In the process of pair production, the main contribution is from a region of the nuclear field for which r lies between $h/2\pi m_e c$ and $(h/2\pi m_e c)\ (h\nu/2m_e c^2)$. Hence the maximum energy of the average positron.

$$= \frac{2Ze^2}{4p\,\epsilon_0 \left(\dfrac{h}{2\pi m_e c}\right)} = \frac{Ze^2 m_e c}{\epsilon_0\, h}$$

$$= 0.0075 \text{ ZMeV.}$$

Example 8:

Compute the maximum energy of the Compton recoil electrons resulting from the absorption in Al of 2.19 MeV γ-rays.

Solution:

The maximum energy of the Compton recoil electrons is given by

$$E_{max} = \frac{E_\gamma}{1+\dfrac{m_0c^2}{2E_\gamma}} = \frac{2.19\times1.6021\times10^{-13}}{1+\dfrac{9.1091\times10^{-21}\times\left(3\times10^8\right)^2}{2\times2.19\times1.6021\times10^{-13}}}$$

$$= \frac{2.19\times1.6021\times10^{-13}}{1.1168} = 1.961 \text{ MeV.}$$

Example 9:

Discuss the following reactions using the conservation of isotopic spin : (i) $p + p \rightarrow p + p + \pi^o$, (ii) $\pi^o \rightarrow \gamma + \gamma$, (iii) $\Lambda^o \rightarrow \pi^- + p^+$, and (iv) $\pi^+ + p \rightarrow \pi^+ + p$.

Solution:

(i) $p + p \rightarrow p + p + \pi^o$

$$T = \quad \frac{1}{2} \quad \frac{1}{2} \quad \frac{1}{2} \quad \frac{1}{2} \quad 1$$

$$T_3 = +\frac{1}{2} + \frac{1}{2} + \frac{1}{2} + \frac{1}{2} \quad 0$$

We know that in all strong and electromagnetic reactions the third component of isotopic spin is conserved. As $Q = e\,(T_3 + B/2 + S/2)$ must be conserved, hence change of T_3 will be there when strangeness changes, which happens in the weak interactions. Secondly, in combining isotopic spins, we use the same rules as for ordinary spin. Hence vector sum of T on the right band side may be 2, 1, or 0. Again the total T_3 = l, ruling out T = 0. The conservation law can be satisfied with the final T = l.

(ii) $\pi^o \rightarrow \gamma + \gamma$

$T = 1$ no T

$T_3 = 0$

In this decay, T_3 is conserved hence the process does not involve the weak interaction. As the total isotopic spin changes, the decay occurs through the electro-magnetic forces whicn do not conserve isotopic spin.

(iii) $\Lambda^o \rightarrow \pi^- + p^+$

$$T = 0 \quad 1 \quad \frac{1}{2}$$

$$T_3 = 0 \quad -1 \quad +\frac{1}{2}$$

The decay of the Λ^o is a weak interaction which can bypass several conservative laws. The π^- is emitted in one direction from the plane defined by the Λ^o and the bombarding particle which produced it. Thus parity is not conserved. As T_3 does not conserve, hence the conservation of charge gives no conservation of strangeness.

(iv) $\pi^+ + p \rightarrow \pi^+ + p$

$$T = 1 \quad \frac{1}{2} \quad 1 \quad \frac{1}{2}$$

$T_3 = +1 \quad +\frac{1}{2} \quad +1 \quad +\frac{1}{2}$

In this strong reaction both T_3 and T must be conserved. In the case of the π^+ rection only one isotopic spin state is possible.

Example 10:

Allocate the isospin to the strange particles from following eqns :

(a) $p^- + p \rightarrow \Lambda^o + K^o$

(b) $p + p \rightarrow \Lambda^o + K^+ + p$

(c) $p^+ + n \rightarrow \Lambda^o + K^+$

(d) $p^- + p \rightarrow \Sigma^- + K^+$

(e) $p^+ + p \rightarrow \Sigma^+ + K^+$

(f) $p^+ + n \rightarrow \Xi^- + K^+ + K^+$

Solution:

(a) $p^- + \quad p \rightarrow \quad \Lambda^o + \quad K^o$

$T \quad 1 \quad \frac{1}{2} \quad 0$

$T_3 \quad -1 \quad +\frac{1}{2} \quad 0$

Hence T_3 must be –1/2 for K° produced with Λ° and T must have a value 3/2 or 1/2. The observed value fixes T = 1/2.

(b) $p + \quad p \rightarrow \Lambda^o + K^+ + p$

$T \quad \frac{1}{2} \quad \frac{1}{2} \quad 0 \quad \frac{1}{2}$

$T_3 \quad +\frac{1}{2} \quad +\frac{1}{2} \quad 0 \quad +\frac{1}{2}$

(c) $\pi^+ + \quad n \rightarrow \Lambda^o + K^+$

$T \quad 1 \quad \frac{1}{2} \quad 0$

$T_3 \quad +1 \quad -\frac{1}{2} \quad 0$

These reactions fix the value of T_3 for K^+ as +1/2 and T as 1/2.

(d) $p^- + \quad$ p ® $S^- + \quad K^+$

T	1	$\frac{1}{2}$	$\frac{1}{2}$
T_3	−1	$+\frac{1}{2}$	$+\frac{1}{2}$

This fixes the value of T_3 for Σ^- as −1 and of T as 1.

π^+ +	p → Σ^+ +	K^+
T 1	$\frac{1}{2}$	$\frac{1}{2}$
T_3 1	$+\frac{1}{2}$	$+\frac{1}{2}$

This fixes the value of T_3 for Σ^+ as + 1 and T as 1.

	π^+ +	n → X^- +	K^+ +	K^+
T	$1\frac{1}{2}$	$\frac{1}{2}$	$\frac{1}{2}$	
T_3	+1	$-\frac{1}{2}$	$+\frac{1}{2}$	$+\frac{1}{2}$

This fixes the value of T_3 for Ξ^- as −1/2 and of T as 1/2. The above results can also be obtained by Gell – Mann scheme.

EXERCISES

1. Estimate the straggling of 32 MeV α-particles in air.
2. Estimate the ranges in air of (a) 10 MeV H^3 ions, (b) double charge p 10 MeV He^3 ions at S.T.P.
3. Calculate the range in aluminium of a 6 MeV α-particle if the relative stopping power of A*l* is 1700. Also calculate its equivalent thickness in kg/m².
4. Complete the following reaction formula, taking account of conservation of charge, baryon number and strangeness (More than one answer is possible and more than one particle is necessary).

   $p + p \rightarrow \Xi^o +$
5. Which of the following reactions and decay processes strangeness and which interaction is responsible for each process?

   (a) $\pi^o \rightarrow \gamma + \gamma$

   (b) $\mu^- + p \rightarrow n + \nu_\mu$

   (c) $\Lambda^o \rightarrow p + \pi^-$

(d) $\Lambda^o + p \rightarrow S^+ + n$

(e) $p + n \rightarrow p + p + \pi^-$.

6. Calculate the cross-section of atomic photo-electric absorption coefficients for aluminium at 20, 60 and 100 MeV.
7. The radioisotope Na^{24} emits γ-rays energies 1.378 and 2.754 Me^2V in succession. After passing through 27.5 gm/cm^2 lead (density = 11 gm/cm), calculate their relative intensitites. The linear absorption coefficients are 48 and 62 m^{-1} respectively for harder and softer components.
8. Calculate the range of 1 MeV β-particles in A*l*. Compare the radiation loss with ionization loss for these particles in A*l*.
9. What thickness of aluminium foil will reduce the energy of 40 MeV He^4 ions to 32 MeV? What energy loss will a 20 MeV deuteron suffer in passing through the foil?

3

COLLECTIVE NUCLEUS MODELS

INTRODUCTION

The shell model has been most successful in explaining a number of nuclear features. The deviations of magnetic moments from the Schmidt curve make this model less acceptable. The measured quadrupole moments are several times larger than can be attributed to the odd nucleon even in nuclei with just one nucleon more or less than a closed shell, where the single particle model should be at its best. Related to it is the observation that E_2 transitions are often much faster than would be expected for transition between single particle states.

J. Rainwater, the American physicist, in 1950, suggested that these discrepancies might be overcome in odd-A nuclei by considering the polarization of the even-even core by the motion of the odd nucleon. The nuclear core, consisting even nucleons, thus have a spheroidal rather than a spherical shape. This distortion would make an additional contribution to the quadrupole moment and to quadrupolar transition rate. The idea of the deformed nuclear core has been developed by A. Bohr and B. Mottelson. The individual nucleons are imagined to move in orbits as before in a potential distribution determined by the remaining nucleons. It is now suggested that the entire shell configuration can undergo periodic oscillations in shape. This collective motion of the nucleons influences the individual particle orbits because it changes the potential of the region in which these particles move. Because of the stability of the core, the collective motion is small and the independent particle characteristics are prominent, for the nuclei consisting of almost closed shells.

The scheme of nuclear energy levels which results from the collective motion of the nucleons in the core and interplay between the motion of losely bound surface nucleons depends upon the strength of the coupling between them. When the coupling is strong the energy states resemble with the linear molecules. Corresponding to the *rotation, vibration, and electronic energy* states of a molecule, we have *rotational, vibrational and nucleonic energy* states in the nuclei. The rotations and vibrations arise from the motion of the nuclear core and the nucleonic states arise due to the motion of the loosely bound nucleons. The total energy is expressed as :

$$W = E_{rot} + E_{vib} + E_n. \quad ...(1)$$

Mathematically this means that the Hamiltonian is composed of three additive parts containing, (1) rotational co-ordinates, (2) vibrational co-ordinates and (3) nucleonic co-ordinates. The wave function is then the product of three wave functions each containing the respective co-ordinates.

The vibrational states of nuclei are formed by flexings of the nuclear surface and are of complex nature. Nuclear rotational motion is also some what more complex in that it is not rigid body rotation but a rotation of the shape of the deformed surface enclosing A free particles. The collective motion now becomes a vibration about the equilibrium shape, and a rotation of the nuclear orientation which maintains the deformed shape. Let us now discuss these states one by one.

VIBRATIONAL STATES

The nucleus is considered as an incompressible liquid drop and is described in terms of the radius vector specifying the nuclear surface. The general shape of the nuclear surface can be written as

$$R(\theta, \phi) = R_0\left[1 + \sum_{\lambda,\mu} \alpha\lambda\mu \; Y\lambda\mu^*(\theta, \phi)\right], \quad ...(1)$$

where R_0 is the nuclear radius if it was spherical, $Y\lambda_i$ are the spherical harmonics representing successive modes of surface standing waves produced by surface disturbances, $\alpha\lambda\mu$ are deformation parameters determining the nuclear shape, and θ and ϕ are polar angles. The subscript μ takes the values $-\lambda$ to $+\lambda$, hence there are $2\lambda + 1$ modes of deformation of order λ. The mode with $\mu = 0$ (for all λ values) represents an axially symmetric nuclear shape.

Any collective motions are expressed by letting $\alpha\lambda\mu$ vary in time. The kinetic energy of the nuclear mass is of the form

$$T = \frac{1}{2} \sum_{\lambda,\mu} B\lambda \, |\alpha\lambda\mu|^2 \quad ...(2)$$

In the case of the irrotational flow of a constant density fluid, Rayleigh's method gives

$$B\lambda = \frac{rR_0^5}{\lambda} \quad ...(3)$$

where ρ is the density of nuclear matter. The potential energy for collective motion is

$$V = \frac{1}{2}\sum_{\lambda,\mu} C\lambda|\alpha\lambda\mu|^2, \quad ...(4)$$

where Cλ are the deformability coefficient, given as

$$C\lambda = (\lambda - 1)(\lambda + 2)\, S\, R_0^2 - \frac{3}{2\pi}\frac{Z^3e^2}{R_0}\frac{\lambda - 1}{2\lambda + 1}. \quad ...(5)$$

where S is the surface tension.

Eqns. (3) and (4) show that the total Hamiltonian H is given by

$$H = E_0 + \sum_{\lambda,\mu}\left[\frac{1}{2}B\lambda|\alpha\lambda\mu|^2 + \frac{1}{2}C\lambda|\alpha\lambda\mu|^2\right]. \quad ...(6)$$

Hence the classical frequency of oscillation

$$\omega\lambda = \left(\frac{C\lambda}{B\lambda}\right)^{1/2}. \quad ...(7)$$

This shows that the frequency ωλ = 0 for λ = 0 and λ = 1. The λ = 0 motion implies change of density and is not an oscillator. The λ = 1 motion implies translational motion of the whole system only. The energy eigenvalues corresponding to Hamiltonian H are the harmonic oscillator energies, given by

$$E = E_0 + \sum_{\lambda\mu}\left(n\lambda\mu + \frac{1}{2}\right)\hbar\omega\lambda, \quad ...(8)$$

where nλμ is the number of oscillators or phonons in the λμ-mode of oscillation. The phonon of type λμ carries angular momentum quantum number λ, with Z-component μ and parity $(-1)^\lambda$.

According to the collective model, the electromagnetic radiation field produced during a transition from the first excited vibrational state to the ground state results from a rearrangement of the nuclear charge from a spheroidal to a spherical distribution. Under this process the angular momentum changes ($\Delta\lambda = 2$) and the electromagnetic radiation emitted has the characteristics of E 2 transition. Since pairing forces depress the energy of the ground state of even-even nuclei relative to the levels predicted by shell model, no low energy individual particle states exist and states formed by collective motions can more easily be observed. The first excited level of most even-even nuclei is formed by the λ = 2 quadrupole surface vibrational state and is generally known as a one phonon state. The next excited state always appears to have +ve parity and even angular momentum. Level scheme of Cd^{114} shows that the third, fourth and fifth levels (0^+, 2^+ and 4^+) are formed by the coupling of two quadrupole surface vibrations each with $\lambda = 2$. This triplet set of levels is known as two phonon state. Even three phonon states have also been identified experimentally. For the octupole oscillations begin with 0^+, 3^-, since the phonons would have $\lambda = 3$. The 3^- state is expected to occur very close to the second excited state of the quadrupole oscillation.

For odd nuclei a certain amount of insight can be gained by considering the odd nucleon (in ground state) coupled to the one or more vibrational collective excitations of the even-even core. The shell model predicts a $d_{3/2}$ ground state for the 79th proton. Experimental observations have found a level at 77 keV ($1/2^+$ level) and a level at 409 keV $\left(1\frac{1^-}{2}\text{ level}\right)$.

But there are three other observed levels at 268, 279 and 548 keV, that are explained easily by the shell model. Braunstein and deShalit have proposed that these are core excited vibrational states formed by the coupling of the angular momentum of the 2^+ vibrational level of the core (Pt^{196}) to the angular momentum of the ground state of Au^{197}.

Let us confine attention to quadrupole shape $\lambda = 2$. The eqn. of the nuclear surface is

$$R(\theta', \phi') = R_0\left[1 + \sum_{\mu=-2}^{2} \alpha^*_{2\mu}\, Y_{2\mu}(\theta'\, \phi')\right], \qquad ...(9)$$

where the deformation parameters $\alpha_{2\mu}$ and the polar co-ordinates θ', ϕ' are in lab system. Since $\delta R = R(\theta', \phi') - R_0$ is real, hence

$$\alpha_{2-\mu} = (-1)^{\mu} (\alpha_{2\mu})^*. \qquad ...(10)$$

In the rotation from one set of co-ordinates to another by Euler angles Θ Φ Ψ, the spherical harmonics transform according to

$$Y_{\lambda\mu}(\theta, \phi)\ \sum_{\nu} Y_{\lambda\nu}(\theta', \phi')\, D^{\lambda}_{\mu\nu}(\Theta, \Phi, \Psi) \qquad ...(11)$$

The function D^{λ} constitute a unitary matrix, and are found in the study of angular momentum.

$$\therefore \alpha_{2\mu} = \sum_{\nu} a_2 \nu D^{2*}_{\mu\nu}(\Theta, \Phi, \Psi). \qquad ...(12)$$

Since the body axes are principal axes, the products of inertia are zero, which implies that

$$a_{20} = \beta \cos\gamma,\ a_{21} = a_{2,-1} = 0 \text{ and}$$

$$a_{22} = a_{2,-2} = \sqrt{\frac{1}{2}}\ \beta \sin\gamma,$$

where β and γ are new parameters. The deformations δR along the principal axes j = 1, 2, 3 are obtained as

$$\delta R_1\left(\frac{\pi}{2}, 0\right) = \left(\frac{5}{4\pi}\right)^{1/2} \beta R_0 \cos\left(\gamma - \frac{2\pi}{3}\right)$$

$$\delta R_2\left(\frac{\pi}{2}, \frac{\pi}{2}\right) = \left(\frac{5}{4\pi}\right)^{1/2} \beta R_0 \cos\left(\gamma - \frac{4\pi}{3}\right)$$

$$\delta R_3(0, \phi) = \left(\frac{5}{4\pi}\right)^{1/2} \beta R_0 \cos\gamma$$

or in general

$$\delta R_j \left(\frac{5}{4\pi}\right)^{1/2} = \beta R_0 \cos\left(\gamma - \frac{2\pi j}{3}\right). \qquad ...(13)$$

For γ = 0, the nucleus would be prolate (cigar shaped) spheroid with the 3 axis as its symmetry axis. For γ = 2π/3 and 4π/3, the nuclei would be prolate spheroids with 1-and 2-axes as symmetry axes. For γ = π, π/3, 5π/3, the nuclei are oblate spheroids. If γ ≠ nπ/3, the nuclear shape is that of an ellipsoid with three unequal axes. The kinetic energy for collective oscillations, eqn. (53),

$$T = \frac{1}{2} B_2 \left(\beta^2 + \beta^2 \gamma^2\right) + \frac{1}{2} \sum_{j=1}^{3} g_j \omega j^2 \qquad ...(14)$$

where B_2 is the mass parameter for collective quadrupole oscillations, ω_j is the angular velocity of the principal axes with respect to the space fixed axes and g_j are effective moments of inertia, given by

$$g_j = 4B_2\beta^2 \sin^2\left(\gamma - \frac{2\pi j}{3}\right) = \frac{15}{4\pi} g_{rigid} \beta^2 \sin^2\left(\gamma - \frac{2\pi j}{3}\right) \qquad ...(15)$$

where g_{rigid} is the moment of enertia of a rigid sphere of radius R_0 and is having value $(16\pi/15)\,B_2$. Above eqn. shows that the moment of inertia for collective rotation about the symmetry axis is zero, *i.e.,* for $\gamma = 0$ or π, $\gamma_3 = 0$. For the other axes, its values are equal, *i.e.,*

$$g_1 = g_2 = g = 3B_2b^2 = \left(\frac{45}{16\pi}\right) g_{rigid}\,\beta^2. \qquad ...(16)$$

Since β is small, very little of the nuclear matter is actually taking part in the effective rotation.

ROTATIONAL STATES

The observable rotational motion is possible if the nucleus is pictured to be a fluid drop or to have any form with a definite surface. This rotational effect can be either rigid in which case particles actually move in circles around the axis of rotation, or wavelike in which case particles perform oscillatory motions and only the geometrical shape of the drop changes. If we retain the shell model as a reasonably good picture of the independent motions of individual nucleons, within a nucleus, it is difficult to picture the nucleus as a rigid body rotating around an axis. Thus a wave like rotation seems to be a more logical explanation, but such wave like rotations can be observed only in deformed nuclei because the apparent motion must then be solely a surface phenomenon. If the rotating nuclear system were a rigid structure, the energy associated with rotation would be purely kinetic and would have value $\frac{1}{2}g_0\omega^2$. Here g_0 is the moment of inertia which for a rigid system is given by the relation

$$g_0 = \sum_p m_p r_p^2, \qquad ...(1)$$

According to the collective model, moments of inertia of nuclei can be determined from the energies of their rotational levels. Rotational energy levels of an axially symmetric nucleus can be described by the three constants of motion : I, the total angular momentum; K, the projection of I on the nuclear symmetry axis (z-axis); and M, the projection of I on a space fixed axis (z'-axis). The collective rotational angular momentum R is perpendicular to the symmetry axis. For a wave like rotation there can be no rotation about the symmetry axis. The quantum number K is, therefore, a constant for each set of rotational levels and represents an intrinsic angular momentum for that band.

If g_3 and g are the moments of inertia for rotations about symmetry axis 3 (*i.e.,* z-axis) and about an axis ⊥ to it, and I_1, I_2 and I_3 are the components of the total angular momentum operator along the body fixed axes, the Hamiltonian is given by

$$H = \sum_{i=1} \frac{\hbar^2}{2g_i} I_i^2 = \frac{\hbar^2}{2g}\left(I_1^2 + I_2^2\right) + \frac{\hbar^2}{g_3} I_3^2$$

$$= \frac{\hbar^2}{2g}\left(I^2 - I_3^2\right) + \frac{\hbar^2}{2g_3} I_3^2. \qquad ...(2)$$

For such an Hamiltonian eigen functions are the D-functions, which are the transformation functions for the spherical harmonics under finite rotations. We thus have

$$I^2 D^I_{MK} = I(I+1) D^I_{MK}$$

$$I_3 D^I_{MK} = K D^I_{MK}$$

$$I'_z D^I_{MK} = M D^I_{MK} \qquad ...(3)$$

$$\text{and } H D^I_{MK} = \left\{\frac{\hbar^2}{2g}\left[I(I+1) - K^2\right] + \frac{\hbar^2}{2g_3} K^2\right\} D^I_{MK}$$

and the energy eigen values are

$$E = \frac{\hbar^2}{2g}\left[I(I+1) - K^2\right] + \frac{\hbar^2}{2g_3} K^2. \qquad ...(4)$$

In the case of symmetry both the model and experimental facts indicate that g_3 quite small. This leads to low lying rotational states for which K = 0, and the energy expression thus becomes

$$E = \frac{\hbar^2}{2g} I(I+I). \qquad ...(5)$$

States with K ≠ 0, other rotational states may occur if one or more of the nucleons are excited to a higher shell model state so that the intrinsic angular momenta no longer cancel by pairs. For non-zero K, I takes all the values K, K + l, K + 2, etc.

Thus we see that these low lying rotational levels be characterized by the sequence of states with I = 0^+, 2^+, 4^+, etc. with even parity throughout and with energies proportional to I (I + 1). It may be seen that the ratio of excitation of successive states is

$$\frac{E_4}{E_2} = \frac{10}{3}, \frac{E_6}{E_2} = 7, \frac{E_8}{E_2} = 12, \text{ etc.,}$$

and these characteristic values are verified for many even-even nuclei. As an example, Fig. 12 shows the energy level diagram of $_{72}Hf^{180}$. The excitation energies are measured with great precision with a bent crystal spectrometer. Substituting experimental value of energy for the 2^+ level in equation (5), we obtain $\frac{\hbar^2}{2g} = 15.55$. Using this value of $\frac{\hbar^2}{2g}$ the energies of other levels are calculated and are listed to the extreme right of each level in the diagram. The agreement is quite good, but there is a symmetric difference which increases with increasing excitation energy. This difference may be explained as resulting from an increase in the moment of inertia with increasing I because of the actions of the centrifugal force. When a correction for this effect is included, the equation (5) then becomes

$$E = \frac{\hbar^2}{2g} I(I + 1) - BI^2 (I + 1)^2. \qquad ...(6)$$

The agreement with experimental value is now extremely good.

Thus we see that collective model is quite successful in interpreting the pattern of exited states of even-even nuclei. For odd-even nuclei, the situation is complicated by the fact that the motion of the core and the motion of the odd nucleon must be coupled and thus the motion of the odd nucleon is no longer adequately described by the shell model states which correspond to a static spherical core.

Furthermore, the excitation energy of the first rotational state of the deformed nuclei shows a smooth variation with A, whereas magic and near magic nuclei show no rotational spectra. From these data and with the help of eqn. (6) we can obtain effective moment of inertia of the various nuclei. The values of moments of inertia help us to understand the kind of rotational motion that occurs. If R be the mean radius and ΔR the difference between the major and minor semiaxes of the deformed well, then we can define deformation parameter by using the relation $R = R_0 [1 + \beta (5/4\pi)^{1/2} \left(\frac{3}{2}\cos^2\theta - \frac{1}{2}\right)$ as

$$\beta = \frac{4}{3}\left(\frac{\pi}{5}\right)^{1/2} \Delta\frac{R}{R} = 1.06\frac{\Delta R}{R_0}. \qquad ...(7)$$

The nuclear deformation is connected with the electric quadrupole moment as

$$Q_0 \frac{3}{\sqrt{(5\pi)}} Z_e R_0^2 \beta [1 + 0.36\,\beta + ...]. \qquad ...(8)$$

To find the value of Q, multiply intrinsic quadrupole moment Q_0 by a projection factor, which is a function of I and K By using the coupling scheme, a complicated calculation gives

$$Q = Q_0 \frac{3K^2 - I(I+1)}{(I+1)(2I+3)}$$

which in the ground state (K = I) reduces to

$$Q = Q_0 \frac{I(2l-1)}{(I+1)(2I+3)}. \qquad ...(9)$$

The collective model has its roost convincing successes, ascribing the Q values to a surface deformation of the nuclear core. This model goes a long way to explain the individual deviations from the Schmidt curves. Prof. Rainwater, Prof. A, Bohr and Prof. Mottelson shared the Noble prize in Physics for 1975 for their work on collective model of nuclei.

UNIFIED MODEL

According to this model, nuclei are deformed away from a spherical shape because the nucleus is not rigid structure and nucleons outside closed shells can set up tensions in the closed shell core and thereby establishing polarization of the nucleus. If the forces between the external nucleons and the core are repuilsve, the prolate spheroid is formed. On the other hand, oblate spheroid is formed when the forces are attractive. Although small effects of polarization of the core can be observed with only one or two nucleons outside closed shell.

Strength of polarization is appreciable when we have several nucleons. It is because the first nucleon generates a potential that tends to align other particles with their orbital planes as nearly coincident as possible with their orbital planes of the first particle. After several nucleons have been added, the polarization becomes strong enough for deformation of the nucleus to take place, even in its ground state.

Nilsson has calculated the expected changes in the relative positions of the shell model energy levels and found that, upon deformation of the nuclear surface, each level splits into (2j + l)/2 levels, a different level for each other value of m_j. In the theory of harmonic oscillator, we have assumed all three component frequencies to be identical. Nilsson assumed that a deformation δ could be defined in terms of the unequal frequencies in the component directions of the harmonic oscillator.

Using a term proportional to 1.1 and a spin orbit term proportional to 1. s in the Hamiltonian of an anisotrople oscillator, Nilsson suggested the interaction of one nucleon with the nuclear field as

$$H = H_0 + C\ 1.s + D\ 1.1, \qquad ...(10)$$

where
$$H_0 = -\frac{\hbar^2}{2M}\nabla^2 + \frac{M}{2}\left(\omega_x^2 + \omega_y^2 y^2 + \omega_x^2 z^2\right) \qquad ...(11)$$

Nilsson used the case of cylindrical symmetry and introduced a single parameter of deformation, as

$$\omega_x^2 = \omega_0^2\left(1 + \frac{2}{3}\delta\right) = \omega_y^2,\ \omega_z^2 = \omega_0^2\left(1 - \frac{4}{3}\delta\right).$$

Thus the dependence af ω_0 on δ is given by

$$\omega_0(\delta) = \omega_0(0)\left[1 + \frac{4}{3}\delta^2 - \frac{1}{2}\frac{6}{7}\delta^3\right]^{-1/6}. \qquad ...(12)$$

The parameter δ is related to the parameter β by the relation

$$\delta \simeq \frac{3}{2}\sqrt{\frac{5}{2\pi\beta}}\,. \qquad ...(13)$$

Let us introduced a term Hδ representing the coupling of the particle to the axis of deformation, having value

$$H\delta = -\delta\hbar\omega_0 \frac{4}{8}\left(\frac{\pi}{5}\right)^{1/2} r^2\, Y_{20} \qquad ...(14)$$

∴ Total Hamiltonian $H = H_0 + H\delta + C\ 1.s + D\ 1.1$

$$= H_0 + k\hbar\omega_0(0)R, \qquad ...(15)$$

where
$$k = \frac{1}{2}\frac{C}{\hbar\omega_0}(0),\ R = \eta U - 21.s = -\mu 1.1 \qquad ...(16)$$

with $\mu = 2\frac{D}{C}$, $\eta = \frac{\delta}{k}\frac{\omega_0(\delta)}{\omega_0(0)}$ and $U = -\frac{4}{3}\sqrt{\frac{\pi}{5}}r^2 Y_{20}$ ...(17)

Here η is a deformation parameter and μ its independent of deformation μ determines the depression of higher levels, k determines the spin-orbit spliting. If N represents the total number of oscillator quanta, and Ω is the quantum number corresponding to the operator j; then the eigenvalues of the total Hamiltonian are given by

$$E^{N\Omega} = \left(N + \frac{3}{2}\right)\hbar\omega_0(\delta) + k\hbar\omega_0(0)r^{N\Omega}, \qquad ...(18)$$

where, $E^{N\Omega}$ represents the eigenvalues of operator R. The changes in the energy levels of selected single particle nuclear states, as a function. The hydrodynamic picture of the nucleus also explains other observed rotational bands besides the ground state rotational band in deformed even-even nuclei. These bands are formed from the coupling of nuclear rotations to a variety of a vibrational oscillations. The unified model yields quadrupole moments that agree with experiment much better.

SUPER CONDUCTIVITY MODEL

According to this model the individual nucleons retain their shell model states. The quantum numbers *l*, j and ± m_j, are used to describe the pairing of nucleons. With in two closed shells the nuclear shell model predicts a determinable arrangement for individual nucleon states. For example the $2p_{3/2}$, $1f_{5/2}$, $2p_{1/2}$ and $1g_{9/2}$ states according to this order are predicted between the closed shell at N = 28 and that at N = 50. If individual nucleons are noninteracting, as suggested by shell model, the first nucleon or two beyond a closed shell would be expected to be descriable by the lowest energy state configuration beyond the closed shell. However, quantum mechanics allows some admixture of levels. For example, in the nuclide $_{28}Ni^{58}$, the two neutrons would be expected to be $p_{3/2}$ fermions. They are in this state in most of the time (about 66%) but according to calculations they are expected to spend about 28% of their time as $f_{5/2}$ neutrons, 3% of their time as $p_{1/2}$ neutrons and 3% of their time as $g_{9/2}$ neutrons.

A missing nucleon (hole) can create the same effect as an extra nucleon. Thus a nucleus with two nucleons less than a closed shell can be treated in a manner analogous to a nucleus with two extra nucleons. For example, the nuclide $_{82}Pb^{200}$ has 4 neutron pairs (or 8 holes) less than the doubly closed shell nuclide Pb^{208}. The neutron levels immediately between the closed shell at N = 126 and that at N = 82 are in descending order, $3p_{1/2}$, $2f_{5/2}$, $3p_{3/2}$ $1h_{9/2}$ and $2f_{7/2}$. We would expect the ground state of Pb^{200} to appear as two $p_{1/2}$ holes and six $f_{5/2}$ holes. The state is thus described as a $p_{1/2}$ $f_{5/2}^{-6}$ configuration. Negative number indicates the number of holes. Enough energy is required to break a pair from a stable nuclear system consisting solely of pairs. After the pair is broken, each separate particle (hole) goes into a state that can be described by shell model configuration. Therefore, the particles of these unpaired state take on many characteristic of the non-interacting particles of the shell model. However, they are still coupled to the remaining paired particles by longer range nuclear forces.

Vibrational and rotational states that are observed in energy gap can be explained by this model by assuming that the long range part of the nuclear force is a quadrupole force that can interact between particles in different orbits as well as in the same orbit.

The formalism of the super-conductivity model of the nucleus gives approximately the same levels as the other models. This model is more satisfactory than others. Unfortunately, no single model has been found that can correlate all or even a major fraction of the existing experimental data. No doubt, models such as the unified and super-conductivity models have shown considerable success in predicting the basic features of nuclear levels, but we have to do more in this field.

FERMI GAS MODEL

In this model the nucleus is taken to be composed of a degenerate Fermi gas of neutrons and protons. The gas is considered degenerate because all the particles are crowded into the lowest possible states in a manner consistent with the requirements of the Pauli principle. For each type of particle, the gas may be characterized by the K.E. of the highest filled state, the *Fermi energy.* The nucleons move freely within a spherical potential well of the proper diameter with depth adjusted so that the Fermi energy raises the highest lying nucleons up to the observed binding energies.

Let us neglect for the moment the electrostatic charge of the protons and suppose that our nucleus has N = Z = A/2. The well is filled separately with nucleons of each type, allowing just two particles of a given type with opposite spin to each cell in phase space of volume h^3. According to the assumption of Fermi statistics, the number of neutron states per unit momentum interval is

$$\frac{dN}{dP} = \frac{2.4\pi^2 V}{(2\pi\hbar)^3} = \frac{Vp^2}{\pi^3\hbar^3}. \quad ...(1)$$

Here V is the volume of the nucleus. If $p_f[= 2ME_f)^{1/2}]$ is the limiting momentum, below which all the states are filled, then the number of neutrons occupying momentum states upto this maximum momentum is given by

$$N = \int_0^{p_f} \frac{dN}{dp} dp = \frac{V}{3\pi^2\hbar^3} p_f^3 \quad ...(2)$$

or
$$p_f = \left(3\pi^2\right)^{1/3} \hbar \left(\frac{N}{V}\right)^{1/3}. \quad ...(3)$$

$$\therefore \text{ Fermi energy } E_f = \frac{p_f^2}{2M} = \left(3p^2\right)^{2/3} \frac{\hbar^2}{2M}\left(\frac{N}{V}\right)^{2/3}, \quad ...(4)$$

For a spherical well of radius $1.2 \times 10^{-15} A^{1/3}$ m filled with N neutrons, in a crude approximation Z = N = A/2, we have $E_f = 30$ MeV. Since it is known experimentally from photo-disintegration experiments that the binding energy of the least bound neutron (ionization potential) of average heavy nuclei is 8 MeV. Hence the neutron gas is contained in a potential energy well of depth ≈ 38 MeV.

For protons the situation is similar except for the potential barrier. However, a nucleus contains fever protons than neutrons and hence the maximum K.E. of the protons is less than that of the neutrons. If the electrostatic repulsion between the protons is considered, the symmetry is destroyed and the model must involve two separate gases contained in two different wells and having different energy levels. The levels are occupied to the same height if the nucleus is β-stable.

Otherwise a sequence of β-decays will occur until this situation is reached. The well well depth for the proton gas is considerably less than that for the neutrons, which can be explained by the fact that the neutrons feel only the attractive nuclear forces, but the protons feel in addition the repulsive Coulomb potential of their mutual electrostatic interactions.

Thus for stable nuclei there will be an excess of neutrons because of the greater depth of neutron well.

The Fermi gas model is not useful for the prediction of the detailed properties of low lying states of nuclei observed in the radioactive decay processes.

LIQUID DROP MODEL

In this model the finer features of nuclear forces are ignored but the strong internucleon attraction is stressed. The essential assumptions are:

1. The nucleus consists of incompressible matter so that $R = R_0A^{1/3}$;
2. The nuclear force is identical for every nucleon;
3. The nuclear force saturates. Thus one might inquire whether a nucleus can be represented as a crystalline aggregate of nucleons. But it gives that the zero point vibrations of the nucleons about their mean rest positions would be too violent for stability. Hence the individual nucleons must be able to move about within the nucleus much as does an atom of a liquid and one might, therefore, think of a nucleus as being like a *small drop of liquid.* Such a model is thus known as *liquid drop model.*

The idea that the molecules in the drop of liquid correspond to the nucleons in the nucleus is confirmed due to following similarities:

1. The nuclear forces are analogous to the surface tension force of a liquid;
2. The ucleons behave in a manner similar to that of molecules in a liquid;
3. The fact that the density of nuclear matter is almost independent of A shows resemblance to liquid drop where the density of a liquid is independent of the size of the drop;
4. The constant binding energy per nucleon is analogous to the latent heat of vaporisation;
5. The disintegration of nuclei by the emission of particles is analogous to the evaporisation of molecules from the surface of liquid;
6. The energy of nuclei corresponds to internal thermal vibrations of drop molecules;
7. The formation of compound nucleus and absorption of bombarding particles are correspond to the condensation of drops.

Semi-empirical formula gives no information about any other properties of nuclei than their energies and the Z/A ratio. In estimating the properties of the excited states of nuclei, one has to consider deformation of a spherical drop giving rise to periodic oscillations of the surface. A spherical drop (a nucleus) of radius R_0 is deformed. If R (θ, ϕ) be the distance of the deformed surface from the centre at an angle θ, ϕ, the difference can be expressed as

$$q(\theta, \phi) = R(\theta, \phi) - R_0 \sum_{l=0}^{\infty} \sum_{m=-l}^{l} q_{lm} Y_{lm}(\theta, \phi). \quad ...(5)$$

To make the problem simpler, let us consider cylindrically symmetric deformations for which m = 0 and q_{l0} oscillate harmonically in time as

$$q_{l,0} = q_l \cos(\omega_l t). \quad ...(6)$$

The characteristic frequency ω_l is determined by the dynamics of the vibration. The surface tension opposes the surface deformation. Since the liquid is incompressible, the waves at the surface implies motion with the liquid. The wavelength of surface vibrations of liquid drop is given by the relation $\lambda\!\!\!^{-} = \frac{R}{l}$. The mass μ participating in the vibration is equal to the mass of the outer shell of thickness $\lambda\!\!\!^{-}$, given by

$$\mu \simeq M\left(\frac{3\lambda\!\!\!^{-}}{R}\right). \quad ...(7)$$

The kinetic energy is the vibration $T = \frac{1}{2}\mu q_l^2$. ...(8)

Due to the wave of amplitude q_l and wavelength $\lambda\!\!\!^{-}$, the plane surface S increases by an amount

$$\Delta S \simeq \frac{1}{2}\left(\frac{q_l}{\lambda\!\!\!^{-}}\right)^2 S. \quad ...(9)$$

The change in surface energy of the drop ΔSα is equal to the potential energy. Hence we get

$$\Delta E_3 = \frac{1}{2}\left(\frac{q_l}{\lambda\!\!\!^{-}}\right)^2 S\alpha = \frac{1}{2}K q_l^2. \quad ...(10)$$

From relations (8) and (10), we get the frequency

$$\omega_l = \left(\frac{K}{\mu}\right)^{1/2} \left(\frac{4\pi\alpha l^3}{3M}\right)^{1/2}. \quad ...(11)$$

The specific properties of the spherical surface change the l^3 into $l(l-1)(l+2)$. The surface tension coefficient α can be calculated in terms of surface energy $E_3 = a_3 A^{2/3}$ as

$$4\pi R^2\alpha = E_3 = a_s A^{2/3} \text{ or } \alpha = \frac{a_s}{4\pi R_0^2}. \qquad ...(12)$$

Hence relation (11) becomes

$$\omega_l \simeq \left[l(l-1)(l+2)\left(\frac{a_s}{3R_0^2} MA \right) \right]^{1/2}. \qquad ...(13)$$

This gives us too high value of the excitation energy. The frequency ω_l is reduced by the Coulomb effect, as

$$\omega_l = \left[l(l-1)\left\{ (l+2)0\frac{10\gamma}{2l+1} \right\} \frac{a_s}{3r_0^2\, MA} \right]^{1/2}, \qquad ...(14)$$

where γ is the ratio between the Coulomb energy $E_c = \frac{3}{5}\frac{Z^2e^2}{4\pi\epsilon_0}R$ and the surface tension energy $E_s = 4\pi\alpha\, R^2 = a_s\, A^{2/3}$. Thus equation leads to somewhat smaller frequencies for heavier nuclei, but is insufficient to represent the actual level distances.

Relation (14) shows that the frequency becomes imaginary where γ is larger than a certain limiting value γ_c. Its smallest value is 2 for $l = 2$. Hence condition for stability against surface deformation is

$$g\gamma_c\left(= \frac{E_c}{E_s} = 0.0474\frac{Z^2}{A} \right) < 2 \text{ or } \frac{Z^2}{A} > 42.2 \cdot \qquad ...(15)$$

We, therefore, expect in this model that the nuclei near the limit of Z^2/A split into two parts by the additional supply of small amounts of energy.

We summarise the results: The liquid drop model not only gives atomic masses and binding energies accurately, but also predicts α-and β emission properties. The binding energy formula does not include closed shell effects, but it can be used to provide a base line from which shell effects can be calculated.

This model is able to explain certain features of nuclear fission, but is not very successful in describing the actual excited states as it gives to large level distances. *This model is the forerunner of the collective model of nuclear structure.*

It forms the basis of Bohr's theory of the compound nucleus formation in nuclear reactions.

SHELL MODEL

This is a great departure from the atomic model where the emphasis is on the motion of the electrons in the field provided by the nucleus. Now questions are: Can be nucleons exist in well ordered quantum controlled nuclear shells? Is there any evidence for the grouping of nucleons into shells? Can quantum numbers similar to n, *l*, s, j be applied to the nucleus? For certain numbers of neutrons or protons, *called magic numbers,* nuclei exhibit special characteristics of stability reminiscent of the properties shown by noble gases among the atoms.

Nuclei in which either N or Z is equal to one of these magic numbers (2, 8, 20, 28, 50, 82, 126) show certain particulars that are not understandable in terms of the liquid drop model. We will see that the magic numbers of the nucleons can be explained with a shell model of the nucleus. Evidently protons and neutrons in the nucleus are not all equivalent as had been assumed in introducing the liquid drop model. We will see that this model is capable of explaining not only the magic numbers but also many other nuclear properties such as spin, magnetic moment, and energy levels.

1. Evidence for the Existence of Magic Numbers

1. It is found that some isotopes are spontaneous neutron emitters when excited above the nucleon binding energy by a preceding β-decay. These are : O(N = 9), Kr (N = 51) and Xe(N = 83).
2. The particularly weak binding of the first nucleon outside a closed shell is shown by the unusually low probabilities for the capture of neutrons by nuclides having N = 50, 80 and 126.
3. Very similar relations exist among the energies of beta-ray emissions. These energies are abnormally large when the neutron or proton number of the final nucleus assumes a magic value.
4. Alpha decay energies are rather smooth functions of A for a given Z but show striking discontinuities at N = 126. This represents the magic character of the number 126 for neutrons.
5. Mayer in 1948 suggested that nuclei with a magic number of nucleons are especially abundant in nature.
6. ${}_2He^4$, and ${}_8O^{16}$ are particularly stable, can be seen from the binding energy curve. Thus we see that numbers 2, 8, indicate stability.

7. No more than five isotopes occur in nature for any N except N = 50, where there are six and N = 82, where there are seven. Neutron numbers of 82, 50, therefore, indicate particular stability.
8. Above Z = 28, the only nuclides of even Z which have isotopic abundances exceeding 60% are Sr^{88} (N = 50), Ba^{138} (N = 82) and Ce^{140} (N = 82).
9. Sn (Z = 50) has ten stable isotopes, more than any other element, while Ca (Z = 20) has six isotopes. This indicates that elements with Z = 50 and Z = 20 are more than usually stable.
10. The Schmidt theory of magnetic moments for odd A nuclides shows that the ground states of these nuclides change from even parity to odd parity or vice versa at the numbers A = 4, 16, 40, when the nucleon numbers are 2, 8 and 20 respectively.
11. The asymmetry of the fission of uranium could involve the structure of nuclei, which is expressed in the existence of the magic numbers.
12. The binding energy of the next neutron or proton after a magic number is very small.
13. The doubly magic nuclei (Z and N both magic numbers) ${}_2He^4$, ${}_8O^{16}$, ${}_{20}Ca^{40}$ and ${}_{82}Bb^{208}$ are particularly tightly bound.
14. Nuclei with the magic proton numbers 50 (Sn) and 82 (Pb) have much smaller capture cross-sections than their neighbours.
15. The electric quadrupole moments of nuclei show sharp minima at the closed shell numbers, indicating that such nuclei are nearly spherical.

2. Extreme Single Particle Model

In this model it is assumed that the nucleons in the nucleus move independently in a common (mean) potential, determined by the average motion of all the other nucleons. Most of the nucleons are paired so that a pair of nucleons contributes zero spin and zero magnetic moment. The paired nucleons thus form an inert core. The properties of odd A nuclei are characterized by the unpaired nucleon and of odd-odd nuclei by the unpaired proton and neutron. In order to understand some of the properties of nuclei including the magic numbers two cases, *infinite squares well and harmonic oscillator potentials* are considered. In these case we can obtain an exact solution. They provide two contrasting view points. The square well has an infinitely sharp edge where us the harmonic oscillator potential diminishes steadily at the edge. The addition of the spin orbit potential eliminates some of the difficulties experienced with the above two potentials.

The starting point of all shell models is the solution of the Schrodinger equation for a particle moving in a spherically symmetrical central field of force. The eigenstates available to a nucleon of mass M moving in a (mean) spherically symmetrical potential V(r) are determined by the solutions of the equation

$$\left(\nabla^2 + \frac{2M}{\hbar^2}\{E - V(r)\}\right)\phi(r) = 0, \quad ...(16)$$

where E is the energy eigenvalue. In such a relation reduced mass m is replaced by M, as in a heavy nucleus m is practically equal to the nucleon mass M. The general solution of this equation can be written as

$$\phi_{nlm}(r, \theta, \phi) = u_{nl}(r)\, Y_{lm}(\theta, \phi), \quad ...(17)$$

where $u_{nl}(r)$ is the radial function and $Y_{lm}(\theta, \phi)$ are the spherical harmonics. The set of quantum numbers n, *l*, m determines an eigenstate corresponding to an eigen-value E_{nl}. The radial wave-function $u_{nl}(r)$ is a solution of the equation

$$\frac{1d}{r^2 dr}\left(r^2 \frac{du_{nl}}{dr}\right) + \frac{2M}{\hbar^2}\left[E_{nl} - V(r) - \frac{l(l+1)\hbar^2}{2Mr^2}\right]u_{nl} = 0 \quad ...(18)$$

(a) Square-well of Infinite Depth

We will here first treat the manageable problem of calculating the position of the various energy levels in an infinitely deep square well of radius R. For simplicity let us assume that the potential is zero inside the well and infinite outside. Outside and at the boundary of the well the radial wave function u_{nl} (r) vanishes. The radial solutions are regular at the origin and inside the well are the spherical Bessel functions ji(k_{nlr}).

$$\therefore \qquad u_{nl}(r) = j_l(k_{nlr}) = \frac{A}{\sqrt{(k_{nlr})}} J_{l+1/2}(k_{nlr}), \quad ...(19)$$

where A is a constant, $J_l + {}_{1/2}(k_{nlr})$ is a Bessel function and k_{nl} is the wave number can be denned by the equation

$$k_{nl^2} = \frac{2M}{\hbar^2}[E_{nl} - V(r)], \quad ...(20)$$

where E_{nl} is the total – ve energy and V(= – U) is the well depth. It is better to measure energies from the bottom of the well and then we have

$$k_{nl^2} = \left(\frac{2M}{\hbar^2}\right)E'_{nl}, \qquad ...(21)$$

where E'_{nl} is positive, measured from the bottom of the well.

The permitted values of k_{nl} are selected by the boundary condition. In the simple case of a well of infinite depth, the wave function has to vanish at the nuclear boundary. Thus at r =.R

$$u_{nl}(R) = j_l(k_{nl}\ R) = 0.$$

Each l-value has a set of zeros and each of them corresponds to a value of km, given by $k_{nl}R$. Thus the eigen-value $k_{nl}R$ is the nth zero of the lth spherical Bessel function. The number n, giving the number of zeros of the radial part of the wavefunction (not counting the origin), is known as the *radial quantum number*. It differs from the principal quantum number of atomic spectroscopy, since the latter counts all the total wavefunction, angular as well as radial and is of major importance for specifying the energy of the corresponding state.

A graph of spherical Bessel function for l = 0, 1 and 2. There is succession of zeros at $k_{nl}R$ = x, numbered serially n = 1, 2...and these values differ for different l. The order of levels in a spherical square well of infinite depth is given by the order of the zeros in the Bessel functions. We indicate such levels in order of increasing energy. The level energies are given by the relation

$$k'_{nl} = \frac{k_{nl}^2\hbar^2}{2M} = \left(\frac{\hbar^2}{2MR^2}\right)x^2. \qquad ...(22)$$

For R = 8 × 10^{-15} m, the quantity $\frac{\hbar^2}{2MR^2} = 0.34$ MeV. The spectroscopic symbols s, p, d,..., represent the states l = 0, 1, 2 ... Each shell model level actually consists of $2l + 1$ sub-states, each with a different angular wavefunction Y_l, m (θ, ϕ). Because of the two different orientations of the nucleon spin, a level can have $2(2l + 1)$ protons. Since neutron and protons are two different particles, hence this level can also have $2(2l + 1)$ neutrons in addition to $2(2l + 1)$ protons.

The first energy level corresponds to x = 3.14, where $j_0(x) = 0$. This $l = 0$ state the written as 1s. It consists 2(2.0 + 1) = 2 particles. The next level corresponds to x = 4.49, where $j_1(x) = 0$. This $l = 1$ state is written as 1p It consists 2(2.1 ± 1) = 6 particles, thus giving altogether a total of eight particles in the shell. The other levels in sequence are 1d[l = 2, x = 5.76, no. of nucleons 10], 2s[l = 0, x = 6.28, number 2], 2s[l = 3, x = 6.99, number 14],...

These are listed. The left hand side represents the number of particles upto any particular level. Letter before terms s, p, d, f.... .represents the order of zero in the Bessel function, *e.g.,* 1s for first zero in the Bessel function $j_0(x)$, 2s for second zero in this function. Here we see that shell closes at total particle numbers 2, 8, 18, 20, 34, 40, 58,.... These are not the nuclear magic numbers. The situation is not altered by calculating the levels for a finite rather thus infinite well. The level order is not essentially different although the excitation alters. This explains the pronounced stability of He^4, O^{16} and Ca^{40}. However, there is no indication of the occurrence of gaps at the magic numbers 28, 50, 82, 126.

(b) Harmonic Oscillator Potential

We now consider a particle moving with simple harmonic motion isotropically bound in three dimensions. A particle of mass M, held to a fixed point by a restoring force kt (where k is a force constant), will vibrate about this fixed point with frequency $\nu = \left(\frac{1}{2\pi}\right)\sqrt{\left(\frac{k}{M}\right)}$ or angular velocity $\omega = \sqrt{\left[\left(\frac{k}{M}\right)\right]}$. Hence the potential energy function V(r) of the oscillating particle

$$V(r) = \int_0^r krdr = \frac{1}{2}kr^2 = \frac{1}{2}M\omega^2 r^2 . \qquad \text{...(23)}$$

The energy eigen value corresponding to the eigen function

$\psi_{nlm}(r, \theta, \phi) = u_{nl}(r)\, Y_{lm}(\theta, \phi)$ is given by

$$E_{nl} = \hbar w\left(2r + l - \frac{1}{2} = \left(\Lambda + \frac{3}{2}\right)\right)\hbar\omega \qquad \text{...(24)}$$

with n = 1, 2, 3,..., l = 0, 1, 2,..., and $\Lambda = 2n + l - 2$. ...(25)

When the angular dependence of the wavefunctions is examined, it is found that for each Λ value, there is a degenerate group of levels with different l values such that $l \leq A$ and even (odd) corresponds to even (odd) values of Λ. Thus the sequence of single particle levels for the 'harmonic oscillator well consists of bands of degenerate levels, each band separated by energy $\hbar\omega$ from the next. The degeneracy corresponding to each l value is 2 (2l + 1) as before. However, the eigenstates corresponding to the same value of Λ (= 2n + l) are also degenerate. Since 2n = $\Lambda - l + 2$ is even for Λ even or odd, the degenerate eigenstates are

$$(n, l) = \left[\frac{1}{2}(\Lambda+2), 0\right], \left[\frac{\Lambda}{2}, 2\right] ... [2, \Lambda-2], [1, \Lambda] \quad \text{for } \Lambda\text{=even}$$

$$= \left[\frac{1}{2}(\Lambda+1), 1\right], \left[\frac{1}{2}(\Lambda-1), 3\right] ... [2, \Lambda-2], [1, \Lambda] \quad \text{for } \Lambda\text{=odd.}$$

∴ Number of neutrons or protons with the eigen value E_Λ.

$$N_\Lambda = \sum_{k=0}^{\Lambda/2} 2[2(2k)+1] \qquad \text{for even } \Lambda$$

$$= \sum_{k=0}^{(\Lambda=1)/2} 2[2(2k+1)+1] \qquad \text{for odd } \Lambda$$

$$= (\Lambda + 1)(\Lambda + 2) \quad \text{in either case.} \qquad ...(26)$$

and the total number of particles for all levels upto Λ is

$$\sum_\Lambda N_\Lambda = \frac{1}{3}(\Lambda+1)(\Lambda+2)(\Lambda+3). \qquad ...(27)$$

The single particle level scheme predicted by the infinite harmonic oscillator well is given in the Table 1.

Table 1: Level scheme predicted by the infinite harmonic oscillator.

Λ	Degenerate States E_Λ	l values	nl for each Λ	$(\Lambda + 1)(\Lambda + 2)$	$\frac{1}{3}(\Lambda+1)(\Lambda + 2)(\Lambda + 3)$
0	$\frac{3}{2}\hbar\omega$	0	1s	2	2
1	$\frac{5}{2}\hbar\omega$	1	1p	6	8
2	$\frac{7}{2}\hbar\omega$	0, 2	2s, 1d	12	20
3	$\frac{9}{2}\hbar\omega$	1, 3	2p, 1f	20	40
4	$1\frac{1}{2}\hbar\omega$	0, 2, 4	3s, 2d, 1g	30	70

Thus we see that in the harmonic oscillator potential, the levels appear in groups such as 1s; 1p; 1d. 2s; 1f, 2p; etc. These grouped levels are degenerate, occupying the same energy state, whereas if the infinite square well potential, the degeneracy is split. The level are of even parity when the oscillator number Λ is even and of odd parity when the oscillator number Λ is odd. It is clear from the above results that the closure of a shell for the harmonic oscillator potential occurs corresponding to neutron or proton numbers 2, 8, 20, 40, 70, 112 and 168, whereas the square well potential suggests magic numbers at 2, 8, 18, 20, 34, 40, 58, 68, 70, 92, 106, 112, 138 and 156. Experimentally observed values are 2, 8, 20, 50, 82 and 126. Thus the truth may lie in between these two potentials.

(c) Spin-Orbit Potential

Several attempts have been made to modify the potentials to yield the observed magic numbers. Mayer and Haxel, Jensen and Suess in 1949, suggested that a non-central component should be included in the force acting on a nucleon in a nucleus. It is corresponding to the interaction between the orbital angular momentum and the intrinsic angular momentum (spin) of a particle. The magnetic moment is associated with the spin angular momentum and the magnetic field is induced due to the orbital angular momentum. This magnetic field has an effect on the magnetic moment. The interaction energy

$$W = -\vec{\mu}_s \vec{B} = -f(r)s.1, \qquad \text{...(28)}$$

where s and 1 denote the spin and orbital angular momentum vectors respectively and f(r) is a potential function. Thus the potential which determines the single particle wave function will be V(r)–f(r) S.I. where V(r) and f(r) depend only on the radial distance and the size of the nucleus. Because of the strong coupling, the two vectors combine to a total angular momentum J for this particle. Since $s = \frac{1}{2}$, there are only two possible ways of combining s and *l*, resulting in the *stretch case* j = *l* + s and the ***jacknife** case* j = *l* – s. We can find the product s.1 by the cosine rule **applied** to the triangle formed by the vectors s, I and j, as follows:

$$1 = \frac{1}{2}\left[j^2 - 1^2 - s^2\right] = \frac{1}{2}\left[j(j+1) - l(l+1) - s(s+1)\right]$$

$$= \frac{1}{2}l \text{ for } j = l + \frac{1}{2}$$

$$= -\frac{1}{2}(l+1) \text{ for } j = l - \frac{1}{2}. \qquad ...(29)$$

The energy shifts from the central value are

$$\Delta E_{nl} = -\frac{1}{2} l \int \frac{dr}{\psi_{nl}} (r)^{/2} f(r)$$

$$= \frac{1}{2}(l+1) \int \frac{dr}{\psi_{nl}} (r)^{/2} f(r) \text{ for } j = l - \frac{1}{2} \qquad ...(30)$$

and the total spin-orbit energy splitting is

$$\Delta(\Delta E_{nl}) = \Delta E_{nl}\left(j = l - \frac{1}{2}\right) - \Delta E_{nl}\left(j = l + \frac{1}{2}\right),$$

$$= \left(l + \frac{1}{2}\right) \int \frac{dr}{\psi_{nl}} (r)^{/2} f(r). \qquad ...(31)$$

Thus we see that the spin orbit interaction splits each of the higher single particle levels. The term with $j = l + \frac{1}{2}$ lies lower in energy and is thus more tightly bound, than the term with $j = l - \frac{1}{2}$. The splitting increases with l and can become so large that for a given n the term with the largest l value and $j = l + \frac{1}{2}$ can slide down to energies as low as those of the multiple with quantum number (n – 1). The effect of such an spin-orbit coupling is analogous to the effect of the magnetic coupling that causes the fine structure splitting in atomic physics, but such effects are much too weak to give necessary splitting. There is an evidence for the existence of a strong spin-orbit force between nucleons from high energy polarisation experiments. Sequence of energy levels is like. Spectroscopic notation is used here to denote the terms. The letters (s, p, d,... etc.) determine l and the subscripts the j, the total angular momentum quantum number. The number in front of the letter is 1 if the symbol following appears for the first time, 2 if it appears for the second time and so on. Spin-orbit level scheme also shows that the odd-parity $1h_{11/2}$-level is grouped with the even-parity $3s_{1/2}$ level, although the energy difference between these is small and they are markedly different in their spins. Similarly $1g_{9/2}$ and $2p_{1/2}$; $1i_{13/2}$ and $3p_{1/2}$ are bounded together.

The lowest state $1s_{1/2}$ holds two nucleons. There is thus a closed shell at 2. In the next state the interaction is not yet strong enough to separate the $1p_{s/2}$, and $1p_{1/2}$ levels by an amount comparable with well spacing so the next closed shell occurs, when all the 6p states are filled, at 2 + 6 = 8. The state $ld_{5/2}$ is definitely lower than $1d_{3/2}$. but it still forms part of the same shell. The states $1d_{5/2}$, $1d_{3/2}$ and $2s_{1/2}$ are sufficiently close to constitute a single shell. The next shell therefore, closes at 8 + 12 = 20.

For the state $1f_{7/2}$, the l value $l = 3$, is high enough to lower the state below all the other N = 3 states, but not enough to make it join the group from N = 2 , Thus this shell closes at 20 + 8 = 28. The next shell comprises $1f_{5/2}$, $2p_{3/2}$, $2p_{1/2}$ and $1g_{9/2}$. The last coming down because of spin-orbit coupling from the next higher group of levels. As it contains 22 sub-levels, this shell closes at 28 + 22 = 50.

Similarly, the next shell is made up of the 32 sub-levels of $1g_{7/2}$, $2d_{5/2}$, $2d_{3/2}$, $3s_{1/2}$ and $1h_{11/2}$, closes at 50 + 32 = 82 and the one after that with 44 sub-shells of $1h_{9/2}$, $2f_{7/2}$, $2f_{5/2}$, $3p_{3/2}$, $3p_{1/2}$ and $1i_{13/2}$ at 82 + 44 = 126. In this way we see that shell closures occur at particle numbers 2, 8, 20, 28, 50, 82, 128 exactly as required by experiments

3. Single Particle Model

In this model, the nucleus with mass number A is considered to consist of filled shells containing the maximum number of neutrons and protons allowed by the Pauli principle and unfilled shells containing "loose" particles. In the extreme single particle model, all the k-particles with the same (n, l, j) have the same energy. In the single particle model the particle state is the superposition of the wave functions of states whose energies are close to one another; such a procedure is called configuration mixing.

In the discussion of the light nuclei (A £ 50), the i-spin plays an important role. The exclusion principle requires that the wavefunction be antisymmetric under inter change of all co-ordinates of any pair of nucleons. Thus for the unfilled shell containing only protons, the angular momentum wavefunction must be anti-symmetric in all pairs For the shell containing neutrons and protons both, the angular momentum function must be symmetric or anti-symmetric for exchange of a given pair, depending on the symmetry of isospin function for the same exchange.

The total spin J of the nucleus is determined by assuming that j of the individual nucleon is a good quantum number. The total spin of k_p loose protons is J_n and of k_n (=k – k_n) loose neutrons is J_n.

$$J = J_p + J_n. \qquad ...(32)$$

Let us consider the case of two particles with the same value of j in the state of configuration (n, l, j) The total angular momentum (J, M) is

$$Y_j M(r_1, r_2) = u_n(r_1)\ u_n\ (r_2) \sum_m < jjm\ M - m/JM > \chi_j^m(1)\chi_j^{M-m}(2) \quad ...(33)$$

First inchanging the particles (1) and (2), then changing the summation index by introducing in m' = M-m und dropping the prime, we get

$$\Psi_J^M(r_1, r_2) = u_n(r_1)u_n(r_2)\sum_m < jjm\ M - m/JM > \chi_j^m(2)\chi_j^{M-m}(1)$$

$$= u_n(r_1)\ u_n(r_2)\sum_{m'} < jj\ M - m'\ m'/JM > \chi_j^{M-m'}(2)\chi_j^{m'}(1)$$

$$= u_n(r_1)\ u_n\ (r_2)\sum_m < jj\ M - m\ m/JM > \chi_j^{M-m}(2)\chi_j^{m'}(1)$$

$$= u_n(r_1)\ u_n\ (r_2)\sum_m (-1)^{2j-J} < jjm\ M - m/JM > \chi_j^m(1)\chi_j^{M-m}(2)$$

$$= (-1)^{2-j} Y_j M\ (r_1, r_2), \quad ...(34)$$

where we have used the following property :

$$< j_1 j_2\ m_1 m_2/JM > = (-1)^{j+j2-J} < j_1 j_2\ m_2 m_1/JM >.$$

Since j is half an odd integer, hence 2j = odd and eqn. (34) shows that the antisymmetric functions exist only for even J. It must be clear that this conclusion applies only to the same type of particles, *i.e.*, either protons or neutrons.

Table 2

Nucleus	Unfilled level	J	J	T	Possibility
$_1D^2$	$1d_{1/2}$	1/2	0	1	Yes
			1	0	Yes
$_8O^{18}$	$1d_{5/2}$	5/2	0, 2, 4	1	Yes
			1, 3	0	No
$_9F^{18}$	$1d_{5/2}$	5/2	0, 2, 4	1	Yes
			1, 3	0	No
$_{16}S^{31}$	$1d_{3/2}$	3/2	0, 2	1	Yes
			1	0	No
$_{17}Cl^{31}$	$1d_{3/2}$	3/2	0, 2	1	Yes
			1	0	Yes
$_{20}Ca^{42}$	$1f_{7/2}$	7/2	0, 2, 4, 6	1	Yes
			1, 3 5	0	No
$_{21}Sc^{42}$	$1f_{7/2}$	7/2	0, 2, 4, 6	1	Yes
			1, 3, 5	0	Yes

In light nuclei, when a shell is filled by both neutrons and protons together, the concept of isotopic spin plays an important role. As each nucleon possesses an isotopic spin t and the total isotopic spin is T, hence the wave function with the isospin formalism is given by

$$\Psi_J^M(r_1, r_2) = (-1)^{2j-J}(-1)^{2t-T}\Psi_J^M(r_1, r_2) \quad ...(35)$$

This shows that for the existence of the wavefunction, J must be even for T = 1 and be odd for T = 0. Table 2 shows how the last two nucleons determine the nuclear spin.

Let us consider an example of three nucleons (Two neutrons and one protons say) in $j = \frac{5}{2}$ level. The two neutrons together can have an antisymmetric state with J_n = 0, 2, or 4. To each of these J_n values the $j = \frac{5}{2}$ of the proton can be added vectorially, to obtain all possible values of J. For the case of three nucleons in the $j = \frac{5}{2}$ level the possible values for the 3 m_j to give the total $M = \Sigma m_j = \frac{5}{2}$, not all three m_j being the same, are : $\left(\frac{5}{2}, \frac{5}{2}, -\frac{5}{2}\right)$, $\left(\frac{5}{2}, \frac{1}{2}, -\frac{1}{2}\right)$, $\left(\frac{3}{2}, \frac{3}{2}, -\frac{1}{2}\right)$ and $\left(\frac{3}{2}, \frac{1}{2}, \frac{1}{2}\right)$. The first three contain $+m_j$, $-m_j$ pairs, but the last do not. The first three sets thus occur, while the last two do not. To explain this, we consider two other quantum numbers, the *seniority* s and the *reduced isotopic spin* t. The seniority quantum number s is defined as the number of particles left in the shell after removal of all nucleon pairs that have J = 0 and T = 1 (like above $+m_j$, $-m_j$ pairs, first three in the above example). The reduced isotopic spin t is defined as the isotopic spin of these s-nucleons. The above definition shows that s is necessarily less than 2j, its smallest possible is 0 for even number of nucleons k and is 1 for odd k.

In the even Z-odd N nucleus, the protons couple to J = 0 and neutrons to $J = J_n$, thus the resultant angular momentum of the whole nucleus is $J = J_n$. Similarly, in an odd Z-even N nucleus $J = J_v$ and in an even-even nucleus J = 0. For odd-odd nuclides, the total angular momentum J is given by *Nordheim's* rules can be stated as :

(a) Strong rule

For $j_p = l + \frac{1}{2}$ and $j_n = l - \frac{1}{2}$ or the reverse, $I(= J) = |j_p - j_n|$.

(b) Weak rule

For both $j + l + \frac{1}{2}$, or both $j = l - \frac{1}{2}$, $|j_v - j_n| \leq I(= J) \leq j_n + j_p$.

The above rules can also be written as: For $N(j_p + j_n - l_p - l_n) = 0$, $I = |j_p - j_n|$ add for $N = \pm 1$, I is either $|j_p - j_n|$ or $j_n + j_p$.

4. Individual (Independent) Particle Model

In this model, all the A particles in the nucleus are taken into account, hence is also named as *many particle shell model.* Each nucleon moves independently of all other nucleons in a common potential field, hence, is considered as independent particle. The wave function $\Psi^v(r)$ for one configuration of the nucleus is the properly antisymmetrized Slater determinant of the single particle wave functions Ψ^v for all the particles.

$$\Psi^v(r) = \begin{vmatrix} \phi^{v1}(r_1) & \phi^{v1}(r_2)\ldots\phi^{v1} & (r_A) \\ \phi^{v2}(r_1) & \phi^{v2}(r_2)\ldots\phi^{v2} & (r_A) \\ \vdots & \vdots & \vdots \\ \phi^{v}A(r_1) & \phi^{v}A(r_2)\ldots\phi^{v}A & (r_A) \end{vmatrix} \quad \ldots(36)$$

The Ψ^v are solutions of the Schrodinger equation

$$H_1\Psi^v = E_1\Psi^v, \quad \ldots(37)$$

where
$$H_1 = \sum_{r=1}^{A}\left[-\frac{\hbar^2}{2m}\Delta_i^2 + V_1(r_i)\right] \quad \ldots(38)$$

Hence $V_1(r_i)$ is the single particle potential in which each nucleon moves. The Hamiltonian of A nucleon system

$$H = \frac{\hbar}{2m}\sum_{i=1}^{A}\nabla_i^2 + \sum_{i<j=2}^{AV_{ij}(r_{ij})} \quad \ldots(39)$$

where V_{ij} (r_{ij}) two nucleon interaction between nucleons i and j. The actual wavefunction Ψ of the nucleus is obtained by the equation

$$H\Psi = E\Psi. \quad \ldots(40)$$

In practice $\Psi = \Sigma a_v \Psi^v$ ...(41)

The nuclear wavefunction Ψ and the energy eigenvalue E are determined by diagopalizing the matrix $(\Psi^\mu \mid H \mid \Psi^v)$, *i.e.* by variational procedure of minimizing E.w.r.t. the coefficients of the linear combination, eqn. (41); There are two commonly used sets of basic functions, described as the L–S and j–j coupling schemes.

L–S Coupling. If the interaction is central, then for k-particles, the total orbital angular momentum $L = l_1 + l_2 + l_3 + ... l_k$ and the total intrinsic spin $S = s_1 + s_2 + ... + s_k$ are both constants of motion, and are coupled together provided the nuclear spin-orbit interactions can be considered weak, compared to the central force experienced by each nucleon. The single particle wave function is of the form

$$\phi \text{ (nlstmi } m_s \; m_t) = u_{nl} (r) \; Y_t^{ml} (\theta, \phi) \; \gamma_{-s}^{ms} \; \chi_l^{mt} \qquad ...(42)$$

where n is the total quantum number, l, s, t, are the orbital angular momentum, spin and isospin and m_i, m_s, m_t their Z-components.

For two particle states ϕ, the multiplicities of the states are 2T + l and 2S + 1, where $T = t_1 + t_2$ and $S = s_1 + s_2$ The totally anti-symmetric states thus are

^{21}L, ^{13}L with L = 0, 2, ... 2l (even)

^{11}L, ^{33}L with L = 1, 3, ... 2l – 1 (odd),

Here superscripts are the values of 2r + l and 25 + 1.

Actually the quantum numbers quoted are insufficient to identify these states completely. We shall denote extra quantum number by [λ], defined as

$$\lambda_1 > \lambda_2 ... \geq \lambda_p \text{ and } \Sigma\lambda_i = k \qquad ...(43)$$

PREDICTIONS OF THE SHELL MODEL

(1) Stability of the closed shell nuclei

This scheme clearly reproduces all the magic numbers.

(2) Spins and Parities of nuclear ground states

The shell model has been very successful in predicting the ground state spin of a large number of nuclei. According to this model the neutron and proton levels fill independently. There are following rules for the angular momenta and parities of nuclear ground states.

(i) Even-even nuclei have total ground state angular momentum l = 0^+. There is no known exception to this rule.

(ii) With an odd number of nucleons, *i.e.,* a nucleus with odd Z or odd N, the nucleons pair off as far as possible so that the resulting orbital angular momentum and spin direction are just that of the single odd particle. There are some exceptions to this rule and they will be discussed below.

(iii) An odd-odd nucleus will have a total angular momentum which is the vector sum of the odd neutron and odd protons-values. The parity will be the product of the proton and neutron parity, *i.e.*, parity = $(-1)^{ln+lp}$. The extreme single particle-model is unable to make predictions concerning this since from the model it is not known how the last neutron and the last proton couple their j's.

The rules proposed by L. W. Nordheim are very helpful. If for the two odd nucleons $j_1 + j_2 + l_1 + l_2$ = an even number, the resultant angular momentum $I = |j_1 - j_2|$. If this sum is an odd number the I is large, approaching $j_1 + j_2$.

We expect from first rule that the angular momentum is zero not only for $_2He^4$ and $_8O^{16}$ but also for $_{38}Sr^{88}$, $_{76}Os^{192}$, $_{92}U^{238}$ and all the other even nuclei. Some actual examples of odd even nuclei are now presented. Consider the nucleus $_6C^{13}$. The six protons and six of the seven neutrons are paired up in the contiguraton $(1s_{1/2})^2\ (1p_{3/2})^4$. The odd neutron is in the $1p_{1/2}$ level and the entire nucleus in its ground state is characterized by the $1p_{1/2}$ designation. The ground state angular momentum indicated as the subscript in $p_{1/2}$ is 1/2, a value which is observed experimentally. For nucleus $_7N^{13}$, the unpaired particle is a proton with spin 1/2. As a second example consider $_8O^{17}$ and $_9F^{17}$. The shells are filled according to $(ls_{1/2})^2\ (1p_{3/2})^4\ (1p_{1/2})^2\ (1d_{5/2})^1$. If it is O^{17}, the last unpaired nucleon is a neutron and has a spin 5/2 : if it is F^{17}, the last particle is a proton with spin 5/2. Thus the model predicts 5/2 which is also the observed value for the ground state spin for each of these nuclei.

In this way, numerous examples can be cited where the extreme single particle model has been successful in predicting the spins. The occasional discrepancies, found chiefly among the heavy nuclei, can be understood by modifying the-rules and stating that *if the high spin shell comes after low spin shell the high spin shell fills faster, pairing its particles before the low spin shell can be fille l completely*. Following this rule, we may write for $_{33}As^{76}$ and $_{28}N^{61}$:

$$(1s_{1/2})^2 \mid (1p_{3/2})^4\ (1p_{1/2})^2 \mid (1d_{5/2})^6\ (2s_{1/2})^2\ (1d_{3/2})^4\ |(1f_{7/2})^8|\ (2p_{3/2})^3\ (1f_{5/2})^2.$$ instead of writing

$$(1s_{1/2})^2 \mid (1p_{3/2})^4\ (1p_{1/2})^2 \mid (1d_{5/2})^6\ (2s_{1/2})^2\ (1d_{3/2})^4\ |(1f_{7/2})^8|\ (2p_{3/2})^4\ (1f_{5/2})^1$$

Experimentally both nuclei have spin 3/2 rather than 5/2 as is indicated by the model. Similar kinds of exceptions have been observed for neutron numbers 57, 59, and 61. These exceptions have been eliminated by using the above rule.

The assumption that like nucleons outside the closed shells pair off so as to cancel their angular momenta is really not at all self- evident. It has been found that the above assumption is correct except when the $1d_{5/2}$, $1f_{7/2}$ and $1g_{9/2}$ levels are filled to the extent of three nucleons or three holes. In these cases the angular momenta would be $\frac{3}{2}, \frac{5}{2}, \frac{7}{2}$ instead of the predicted values $\frac{5}{2}, \frac{7}{2}, \frac{9}{2}$. For example, $_{11}Na^{23}$ has angular momentum $\frac{3}{2}$ instead of $\frac{5}{2}$ $_{25}Mn^{55}$ has $\frac{5}{2}$-instead of $\frac{7}{2}$ and $_{34}Se^{79}$ has $\frac{7}{2}$ instead of $\frac{9}{2}$. In these three nuclides the total angular momentum is due to the three nucleons outside the closed shell, can be explained by single particle model.

(3) Magnetic moments of nuclei

In an odd nucleus, the total angular momentum of the nucleus is equal to the angular momentum j of the last unpaired nucleon. Thus we see that magnetic moment of the nucleus is produced by the odd nucleon only. The orbital angular momentum with numerical value $\sqrt{[l(l+1)]}$ and the spin s with numerical value $\sqrt{[s(s+1)]}$ couple to a total angular momentum j with numerical value $\sqrt{[j(j+1)]}$, in units of $\hbar$. The magnetic moment associated with spin angular momentum s is given by vector $\mu_s = g_s$ s.

Similarly the magnetic momentum associated with orbital angular momentum I vector $\mu_l = g_l$ I.

Hence vector μ = sum of the components of the vectors g_l I and g_s s along the j vector

$$= g_l \sqrt{[l(l+1)]} \cos(\text{I}.\text{j}) + g_s \sqrt{[s(s+1)]} \cos(\text{s}.\text{j}) \qquad ...(45)$$

By applying he cosine rule to the triangle formed by the vectors I, s and j, the above relation can be written as

$$\mu = g_l \sqrt{[l(l+1)]} \frac{j(j+1)+l(l+1)-s(s+1)}{2\sqrt{[l(l+1)\,j(j+1)]}}$$

$$+g_s \sqrt{[s(s+1)]} \frac{j(j+1)+s(s+1)-l(l+1)}{2\sqrt{[s(s+1)\,j(j+1)]}}$$

$$= \frac{j(j+1)+l(l+1)-s(s+1)}{2\sqrt{[j(j+1)]}} g_l + \frac{j(j+1)+s(s+1)-l(l+1)}{2\sqrt{[j(j+1)]}} g_s$$

Since for a single particle, the spins s = 1/2 and there are two possible cases.

(i) I parallel to s (Stretch case) $I = j + s = l + 1/2$.

(ii) I antiparallel to s (*Jackknife case)* $I = j = l - s = l - 1/2$.

These relations define two curves, for μ versus I, with the values $I = l \pm 1/2$, for each class of odd even nucleus. The values of μ are known as the *Schmidt values* and the curves are known as *Schmidt lines.* When we substitute into equations (46) and (47) the g factors which correspond to single nucleons are

$g_l = 1$ and $g_s = 5.58$ for protons

and $g_l = 0$ and $g_s = -3.82$ neutrons,

we obtain the predicted relationship between μ and I on the single particle model. Thus eqns (46) and (47) reduce to :

Odd proton : $\mu = I = 2.29$ for $I = l + 1/2$

$$\mu = I = -2.29 \frac{I}{(I+1)} \quad \text{for } I = l - \frac{1}{2} \qquad ...(48)$$

Odd neutron : $\mu = -1.91$ for $I = l + \frac{1}{2}$

$$\mu = 1.91 \frac{I}{(I+1)} \quad \text{for } I = l - \frac{1}{2} \qquad ...(49)$$

To each value of I, there are two values of μ, depending on whether $I = l + \frac{1}{2}$ or $I = l - \frac{1}{2}$. These two values of μ constitute the to *Schmidt limits.* Equs. (48) and (49) are plotted for nuclei of odd nucleons. The resulting lines represent the magnetic moments predicted by the extreme single particle model. All measured values lie within the regions bounded by the curves. The values lie mainly in two broad bands roughly parallel to the two Schmidt curves. In a large number of cases the value lies nearer one of the curves than another, and in nearly all these cases, the *l* value so indicated is the shell model value. These deviations are evidence of the approximate nature of this model. The most important reason for this discrepancy is that our shell model is based on a spherically symmetric potential, whereas in reality nuclei are ellipsoidal. Another possible reason is that it is not at all certain that the magnetic moments of the nucleons in the nuclear matter are the same as in the free states.

(4) Electric quadrupole moment

The shell model also makes predictions about the electric quadrupole moment of a odd A, odd Z nuclide, on the assumption that this is due to the unpaired proton. The quadrupole moment for a given I is given by

$$Q = -\frac{2I-1}{2I+2}\left(r_{r.m.s}\right)^2, \qquad \text{...(50)}$$

where $l = l \pm s$ and $(r_{r.m.s.})^2$ is the mean square radius of the orbit. The observed values are usually much larger than obtained from eqn. (50) and are not negative always. A further puzzling feature is that the quadrupole moments of odd A – odd N nuclides are not significantly smaller than those of odd A – odd Z nuclides, as required if the quadrupole moment is largely due to odd proton. For a uniform charge distribution

$r^2_{r.m.s} = \frac{3}{5} R_C^{\ 2}$, where Coulomb radius $R_C = 1.2\ A^{1/3}$ fermi.

(5) Nuclear isomerism

The single particle model has succeeded in explaining the phenomenon of nuclear isomerism. It was mentioned previously in the spin orbit level scheme that sometimes a high spin state is depressed and is grouped with the low spin state, opposite in parity but differing in energy by only a small amount. If, now a gamma ray is emitted from an excited state that has a high spin value to a lower state with low spin value, then there is a large spin change involved in this low energy transition, which leads to a low probability and thus long life.

The nucleus ${}_{19}K^{41}$ decays by gamma rays of 1.3 MeV with a half life of 6.6×10^{-9} sec. The spin orbit level scheme is $({}^1s_{1/2})^2\,|(1p_{3/2})^4\,(1p_{1/2})^2|\,(1d_{5/2})^6\,(2s_{1/2})^2\,(1d_{3/2})^5$. The extreme single particle model interprets this gamma ray to be M_2 arising in transition from the $f_{7/2}$ excited state to the ground state $d_{3/2}$ with a change of parity.

The shell model also predicts that almost all of the isomeric states with long life are found for nuclei with N or Z near the end of a shell. These nuclei are found in four distinct groups, known as *islands* of *isomerism*, given by Z or N between 19 and 27, 39 and 49, 69 and 81, or between 111 and 125. The level scheme indicates that in these regions the following levels lie close together.

(i) $19 \le N$ or $Z \le 27$ $1f_{7/2}$ and $1d_{3/2}$ levels

(ii) $39 \le N$ or $Z \le 49$ $1g_{9/2}$ and $2p_{1/2}$ levels

(iii) $69 \le N$ or $Z \le 81$ $1h_{11/2}$ and $3s_{1/2}$ levels

(iv) $111 \le N$ or $Z \le 125$ $1i_{13/2}$ and $3p_{1/2}$ levels.

(6) Beta decay

One other application of the shell model is that of beta decay. Theory of β-decay shows that the life time can be understood in terms of relative parity and angular momenta of the states involved. The nuclear shell model is able to predict the spins and parities of these states, of the unstable nuclei which decay by β-emission. Hence β-decay study is the most powerful method of getting the detailed order of orbital momenta with in the shells.

(7) Stripping reactions

Stripping reactions can be explained by single particle model.

SOLVED EXAMPLES

Example 1:

Show that the wavefunction inside the spherical box ($V = 0$, $r < R$; $V = \infty$, $r > R$) of a particle of spin 1/2, energy E and angular momentum l is

$$Y_{nlm} = \frac{1}{\sqrt{2}} J_{l+1/2}\left(k_{nl}\right) P_l^m (\cos\theta) e^{im\phi}$$

where $k_{nl}^2 = \frac{2ME}{\hbar}$. *By considering the zeros of* $J_{l+1/2}(k_{nl}R)$ *for* $k_{nlR} <$ *9.5, show that there are 92 particles in the box with energy less then* $E_{max} = \frac{45\hbar^2}{MR^2}$. *Show also that according to the eqn. (2) there are 120 particles with energy less than* E_{max} *and that the maximum energy for 92 particles is* $37.5 \frac{\hbar^2}{MR^2}$.

Solution:

General solution of the Schrodinger eqn. is given by

$$\psi_{nlm}(r) = u_{nl}(r)\ \psi_{lm}\ (\theta, \phi) = u_{nl}(r)\ P_l^m\ (\cos\theta)\ e^{lm\phi}.$$

Using eqn. (10), we get

$$\psi_{nlm} = \frac{A}{\sqrt{(k_{nlr})}} J_{l+1/2}(k_{nlr}) P_l^m (\cos\theta)\, e^{lm\phi}.$$

That for k_{nl} R < 9.5, the order of the zeros in the Bessel functions, is ^{1}s, ^{1}p ^{1}d, ^{2}s, ^{1}f, ^{2}p, ^{1}g ^{2}d, ^{1}h, ^{3}s. The number of particles upto these levels is 2 + 6 + 10 +2 + 14 + 6 + 18 10 + 22 + 2 = 92. The energy is given by

$$E_{nl}\ \frac{k_{nl}^2\hbar^2}{2M} = \left(\frac{\hbar^2}{2MR^2}\right)(k_{nl}R)^2$$

$$= (9.5)^2\left(\frac{\hbar^2}{2MR^2}\right) = 45.125\left(\frac{\hbar^2}{MR^2}\right).$$

From eqn. (2), we get N

$$= V\frac{p_f^3}{3\pi^2\hbar^3} = \frac{4}{3}\pi R^3\frac{(2ME_f)^{3/2}}{3\pi^2\hbar^3}.$$

∴ No. of particles with energy less than E_{max}

$$= \frac{4R^3\left(\frac{2M\times 45\hbar^2}{MR^2}\right)^{3/2}}{9\pi\hbar^3} = 120$$

and the Fermi energy for 92 particles

$$= (3\pi^2)^{2/3}\frac{\hbar^2}{2M}\left(\frac{92}{\frac{4}{3}\pi R^3}\right)^{2/3} = \frac{37.5\hbar^2}{MR^2}.$$

Example 2:

The first excited state of W^{182} is 2^+ and is 0.1 MeV above the ground state. Estimate the energies of the lowest lying 4^+ and 6^+ states of W^{182}?

Solution:

Energy expression to low lying rotational states is

$$E = I(I + 1)\,\frac{\hbar^2}{2g}.$$

For 2^+ state $0.1 = \left(\frac{\hbar^2}{2g}\right) 2 \times 3$ or $\left(\frac{\hbar^2}{2g}\right) = \left(\frac{0.1}{6}\right) = 0.01667.$

Hence energies corresponding to the 4^+ and 6^+ states are:

$E(4^+) = 0.01667 \times 4 \times 5 = 0.333$ MeV

$E(6^+) = 0.01667 \times 6 \times 7 = 0.700$ MeV.

Example 3(a):

Predict the characteristics of the ground state of ${}_8O^{17}$, ${}^{18}S^{33}$, ${}_{29}Cu^{63}$ *and* ${}_{83}Bi^{209}$ *values.*

Solution:

Shell model terms of the nuclides are given as :

For ${}_8O^{17}$ – $({}^1s_{1/2})^2$; $({}^1p_{3/2})^4$ $({}^1p_{1/2})^2$; $({}^1d_{5/2})^1$, hence I = 5/2, l = 2 and thus even parity.

∴ Magnetic moment of this nucleus due to odd neutron $({}^1d_{5/2})$ = – 1.91 and quadrupole moment

$$Q = -\frac{3}{5}\frac{2I-1}{2I+2}R_e^2 = -\frac{3}{5}\times\frac{4}{7}\times\left(1.2\times 17^{1/3}\times 10^{-15}\right)^2$$

= 0.0326 barn.

For ${}_{16}S^{33}$ – $({}^1s_{1/2})^2$; $({}^1p_{3/2})^4$ $({}^1p_{1/2})^2$; $({}^1d_{5/2})^6$ $({}^2s_{1/2})^2$ $({}^1d_{3/2})^1$, hence , l = 2 and thus even parity. The magnetic moment of this nucleus due to odd neutron ${}^1d_{3/2}$ is $1.91\left(\frac{3}{2}/\frac{5}{2}\right) = 1.146$.

The quadrupole moment

$$Q = -\frac{3}{5}\times\frac{2}{5}\times\left(1.2\times 33^{1/3}\times 10^{-15}\right)^2 = -0.0355 \text{ bern}.$$

For ${}_{29}Cu^{63}$ – [Shells upto 28 protons]; $({}^2p_{3/2})^1$, hence I = 3/2, l = 1 and thus odd parity. The magnetic moment of this nucleus due to odd proton ${}^2p_{3/2}$ is $\frac{3}{2} + 2.29 = 3.79$. The quadrupole moment

$$Q = -\frac{3}{5}\times\frac{2}{5}\times\left(1.2\times63^{1/3}\times10^{-15}\right)^2 = -0.0547 \text{ barn.}$$

For $_{83}Bi^{209}$ – [Shells upto 82 protons]; $(lh_{9/2})^1$, hence I = 9/2, l = 5 and thus odd parity. The magnetic moment of this nucleus due to odd proton $(lh_{9/2})$ = 4.5 – 2.29 × 9/11 = 2.63. The quadrupole moment $Q = -\frac{3}{5}\times\frac{8}{11}\times(1.2\times209^{1/3}\times10^{-15})^2 = 0.176$ barn.

Example 3(b):

Write the normalized nucleon wave functions for $s_{1/2}$, $p_{3/2}$ and $p_{1/2}$ states. Hence find the nuclear density for He^4, C^{12} and O^{16} nuclei.

Solution:

For a single particle of spin 1/2 and orbital angular momentum l, j = l + 1/2 and l – 1/2. The wave function can be written as

$$\Psi_{n,\,1/2,\,l,\,j\,mi} = \sum_{ms=-1/2}^{1/2} C\left(\frac{1}{2}, l, j; m_i, m_j\right)\chi_{1/2,\,ms}\,\Psi_{n,\,l,\,mj-ms},$$

where $\Psi_{n,\,l,\,mj-ms} = \frac{u_{nl}(r)}{r}Y_{l,\,mj-ms}(\theta,\phi)$

and $\chi_{1/2,\,1/2} = \chi_{1/2,\,-1/2} = \chi$.

$$Y_{n,\frac{1}{2},\,l,\,j,\,mj} = \left[\left(\frac{l+mj+\frac{1}{2}}{2l+1}\right)^2 \chi_+ Y_{l,\,mj-1/2} \pm \left(\frac{l\mp mj+\frac{1}{2}}{2l+1}\right)^{1/2} \chi_- Y_{l,\,mj+1/2}\right]\frac{u_{n,\,l}(r)}{r},$$

Here $u_{nl}(r) \sim r^{l+1}\exp.\left(-\frac{r^2}{2a^2}\right)$ with $a = \left(\frac{\hbar}{M\omega}\right)^{1/2}$.

Hence we can write the normalized nucleon wave functions for $s_{1/2}$, $p_{3/2}$ and $p_{1/2}$ states as

$$\Psi_{s_{1/2}} = \left(\frac{1}{\pi a^6}\right)^{1/4}\exp.\left(\frac{-r^2}{2a^2}\right)Y_{0,\,0}\chi_\pm$$

$$= \left(\pi a^2\right)^{-3/4} \exp.\left(-\frac{r^2}{2a^2}\right)\chi_{\pm}.$$

$$\Psi p_{3/2} = 2\left(\frac{4}{9\pi a^{10}}\right)^{1/4} r\ \exp.\left(\frac{-r^2}{2a^2}\right) \begin{cases} \chi_+ Y_{1,1} \text{ for } m_j = +\frac{3}{2} \\ \chi_- Y_{1,-1} \text{ for } m_j = -\frac{3}{2} \\ \sqrt{\frac{2}{3}}\chi_+ Y_{1,-1} + \sqrt{\frac{2}{3}}\chi_{-1} Y_{1,0} \\ m_j = \frac{1}{2} \\ \sqrt{\frac{1}{3}}\chi_+ Y_{1,-1} + \sqrt{\frac{2}{3}}\chi_{-1} Y_{1,0} \\ \text{for } m_j = -\frac{1}{2}. \end{cases}$$

The nucleon density of a nucleus $\rho(r) = \sum_{k=1}^{A} |\Psi_k(r)|^2$.

Hence for He^4, consisting 2 protons and 2 neutrons all in $s_{1/2}$ states, the nucleon density

$$\rho(r) = 4\left(\frac{1}{\pi a^2}\right)^{3/2} \exp.\left(\frac{-r^2}{a^2}\right)$$

C^{12} – nucleons consists of 2 protons and 2 neutrons in $s_{1/2}$ states and 4 protons and 4 neutrons in $p_{3/2}$ state. The nucleon density is thus given by

$$\rho(r) = r\left(\frac{1}{\pi a^2}\right)^{3/2} \exp.\left(\frac{-r^2}{a^2}\right) + 8\times\frac{2}{3}\left(\frac{1}{\pi a^{10}}\right)^{1/2}\frac{4}{3}$$

$$\left(Y^2{}_{1,1} + Y^1{}_{1,0} + Y^2{}_{1-1}\right)^2 \exp.\left(\frac{-r^2}{a^2}\right)$$

$$= 4\left(\frac{1}{\pi a^2}\right)^{3/2}\left(1 + \frac{4r^2}{3a^2}\right)\exp\left(\frac{-r^2}{a^2}\right)$$

Similarly for O^{16}, which consists of 12 nucleons as in C^{12} and 2 protons and 2 neutrons in $p_{1/2}$ state, the density is

$$\rho(r) = 4\left(\frac{1}{\pi a^2}\right)^{3/2}\left(1+2\frac{r^2}{a^2}\right)\exp.\left(\frac{-r^2}{a^2}\right).$$

Example 4:

On the basis of the extreme single particles hell model, what would be the expected ground state spectroscopic configurations of the following nuclei :

C^{11}, Sc^{45}, Ni^{61}, Ge^{73}, In^{109}, Ta^{181}, Tl^{203}, Am^{241}?

The measured spins are respectively, $\frac{3}{2}, \frac{7}{2}, \frac{3}{2}, \frac{9}{2}, \frac{7}{2}, \frac{1}{2}, \frac{5}{2}$. If you find any discrepancy between your expectations and experiments, try to give some explanation.

Solution:

Order of energy states, as suggested by Mayer and Jenson, is $^1s_{1/2}$, $^1p_{3/2}$, $^1p_{1/2}$; $^1d_{5/2}$, $^2s_{1/2}$, $^1d_{3/2}$; $^1f_{7/2}$; $^2p_{3/2}$, $^1f_{5/2}$, $^2p_{1/2}$, $^1g_{9/2}$; $^1g_{7/2}$ $^2d_{5/2}$, $^2d_{3/2}$, $^3s_{1/2}$, $h_{11/2}$; $^1h_{9/2}$, $^2f_{7/2}$.

For $_6C^{11}$, $(^1s_{1/2})^2$ $(^1p_{3/2})^3$.

Hence I = 3/2, l = 1 and thus parity – 1 or odd.

For $_{21}Se^{45}$, $(^1s_{1/2})^2$; $(^1p_{3/2})^4$ $(^1p_{1/2})^2$; $(^1d_{5/2})^8$ $(^3s_{1/2})^3$ $(^1d_{3/2})^4$; $(^1f_{3/2})^1$, hence I + 7/2, l = 3 and thus odd parity.

For $_{28}Ni^{61}$, [Shells upto 28 neutrons]; $(^2p_{3/2})^4$, $(^1f_{5/2})^1$.

Hence I = 5/2, l = 3 and thus odd parity. The experimental value 3/2 of I can be explained by stating that if the high spin shell hours after low spin shell, the high spin shell fills faster, pairing its particles before the low spin shell can be filled completely, *i.e.*, I = 3/2 due to the unfilled level $^2p_{3/2}$ and hence I = 1 and parity odd.

For $_{32}Ge^{73}$, [Shells upto 28 neutrons];

Hence $\quad I = \frac{9}{2}$, l = 4 and this even parity.

For $_{49}In^{109}$, [Shells upto 28 protons];

$$(^1g_{7/2})^8\ (^2d_{5/2})^6\ (^2d_{3/2})^4\ (^3s_{1/2})^2\ (^1h_{11/2})^3.$$

Hence I = 11/2, l = 5 and thus odd parity. The experimental value 7/2 can be explained by standing that the total angular momentum is due to the three nucleons outside the closed shell, *i.e.*, level $g_{7/2}$ just outside the closed shell.

For $_{81}TI^{203}$, [Shells upto 50 protons];

$$({}^1g_{7/2})^8\ ({}^2d_{5/2})^6\ ({}^2d_{3/2})_2{}^4\ ({}^3s_{1/2})^2\ ({}^4h_{11/2})^{11}.$$

Hence I = 11/2, l = 5 and thus odd parity. The experimental value 1/2 can be explained by stating that the high spin shell fills faster when its comes after low spin level, *i.e.*, I = 1/2 due to unfilled level ${}^3s_{1/2}$.

For $_{95}Am^{241}$, [Shells upto 82 protons] : $({}^1h_{9/2})^{10}\ ({}^2f_{7/2})^3$.

Hence I = 7/2, l = 3, and thus odd parity. The experimental value 5/2 can be explained by stating that when ${}^1d_{5/2}$, ${}^1f_{7/2}$ and ${}^1g_{9/5}$ levels are filled to the extent of three nucleons or three holes the angular momentum are $\frac{3}{2}, \frac{5}{2}$ and $\frac{7}{2}$ respectively.

The disagreement with shell model levels can easily be explained by Nilsson potential.

Example 5(a):

Determine the harmonic oscillator frequencies ω appropriate to the nuclei O^{17} and Ni^{60}.

Solution:

The total single particle energies in a nucleus of mass number A are

$$E = M\omega^2 A\,(r_{rms})^2 = \frac{3}{5} M\omega^2 A R_c^2, \qquad ...(1)$$

with the Coulomb radius $R_c = 1.2\ A^{1/3}$ fermi. Assuming an equal number of neutrons and protons and that all energy states upton E_{Λ_0} are occupied, we have

$$A = \sum_{\Lambda=0}^{\Lambda_0} 2N\Lambda = \frac{2}{3}(\Lambda_0+1)(\Lambda_0+2)(\Lambda_0+3)$$

$$= \frac{2}{3}(\Lambda_0+2)^2 + \text{... terms of higher } \Lambda_0 \qquad ...(2)$$

$$\text{and } \frac{E}{\hbar\omega} = \sum_{\Lambda=0}^{\Lambda_0} 2N_\Lambda\left(\Lambda+\frac{3}{2}\right) \simeq \frac{1}{2}(\Lambda_0+2)^4 - \frac{1}{3}(\Lambda_0+2)^3 + \text{...} \qquad ...(3)$$

$$\simeq \frac{1}{2}\left(\frac{3}{2}A\right)^{4/2}. \qquad ...(4)$$

Solving eqns. (1) and (4), we get

$$\omega = \frac{5\times 3^{1/3}\hbar\, A^{1/3}}{2^{7/3} M R_0^2} = \frac{5\times 3^{1/2}\times 1.0545\times 10^{-34}}{2^{7/3}\times\left(1.67\times 10^{-27}\right)\left(1.2\times 10^{-15}\right)} A^{-1/3}$$

$$= 6.3 \times 10^{22} A^{-1/3}.$$

For O^{17} nucleus $\omega = 6.3 \times 10^{22} \times (17)^{-1/3} = 2.45 \times 10^{22}$

For Ni^{60} nucleus $\omega = 6.3 \times 10^{22} \times (60)^{-1/3} = 1.61 \times 10^{22}$.

Example 5(b):

Estimate the Fermi energies of neutrons and protons in the centre of U^{238}. Assume the density of nuclear matter in the centre of the ${}_{92}U^{238}$ nucleus to be 2×10^{38} nucleus cm^{-3}.

Solution:

Fermi energy of neutron $E_f = (3\pi^2)^{2/3} . \dfrac{\hbar^2}{2M}\left(\dfrac{N}{V}\right)^{2/3}$

$\therefore$

$$E_f = \left(3\times 3.14^2\right)^{2/3} \frac{\left(1.0545\times 10^{-34}\right)^2}{2\times 1/.67482\times 10^{-27}}\left(\frac{146}{238/2\times 10^{44}}\right)^{2/3}$$

$$= 48.99 \text{ MeV}.$$

Similarly the Fermi energy of proton E_f = 35.99 MeV.

EXERCISES

1. Suppose that we try to interpret the ${}_2He^3$ nucleus as a system consisting of a deuteron bound to a proton. From the known deuterium magnetic moment (0.857 μ_n), and proton moment (2.79 μ_n), calculate the magnetic moment of ${}_2He^3$, expected on this model. Also calculate the Schmidt model prediction for ${}_2He^3$. Which agrees better with the experimental value of –2.13 μ_n.

2. Calculate the magnetic moments and quadrupole moments of Mg^{25}, K^{39} and Pb^{207}.

3. Give the expected ground state spins and parities for the nuclei Li^6, Al^{28}, K^{40}, C^{60} and Bi^{306}.

4. What do you expect the ground state spins and parities to be for Al^{27}, Ar^{30}, Zr^{39}, Co^{55} and Pt^{195}.

5. Compare the difference in Fermi energies of neutrons and protons in U^{238} with the Coulomb repulsion experienced by a proton approaching Pa^{237}.

6. Find the position in energy of the $s_{1/3}$ states in a square well potential of depth V_0 = 45 MeV, as a function of $R = R_0 A^{1/3}$.

4

THEORY OF NUCLEAR FISSION WITH BOHR AND WHEELER

INTRODUCTION

The first thorough theoretical treatment of this process was earned out by Bohr and Wheeler in 1939. They applied a simple form of analysis (Legendre polynomial expansion) to express the radius r making angle θ with the axis of maximum deformation

$$r = R\left[1 + \sum_{l=0}^{\infty} \alpha_l P_t(\cos\theta)\right]$$

$$= R\,[1 + \alpha_2 P_2 (\cos\theta) + \alpha_3 P_3 (\cos\theta) + ...] \qquad ...(1)$$

where R is the radius of the spherical nucleus α_2, α_3, are the deformation parameters. Here $\alpha_0 = \alpha_1 = 0$, as the centre of mass of the drop is assumed to remain unchanged.

The surface energy of a spherical drop $E_s^{phere} = 4\pi R^2 T$, where T is the surface tension. Hence surface energy of the deformed drop

$$E_s = 4\pi R_0^2\, A^{2/3}\, T\left[1 + \alpha_2\left(\frac{3}{2}\cos^2\theta - \frac{1}{2}\right) + ...\right]^2$$

$$= 4\pi R_0^2\, A^{2/3}\, T\left[1 + \frac{2}{5}\alpha_2^2 + ...\right]$$

$$\therefore \quad \Delta E_s = E_s^{sphere}\left[\frac{2}{5}\alpha_2^2 + ...\right] \qquad ...(2)$$

The Coulomb energy of a spherical drop $E_C^{sphere} = \frac{3}{5} Z^2 e^2 / 4\pi \epsilon_0 R$ hence that of the deformed drop

$$E_e = \frac{3}{5}\frac{Z^2e^2}{4\pi \epsilon_0 A^{1/3}R_0}\left[1+\alpha_2\left(\frac{3}{2}\cos^2\theta - \frac{1}{2}\right)+...\right]^{-1}$$

$$\therefore\ \Delta E_e = E_c^{sphere}\left[-\frac{1}{5}\alpha_2^2 - ...\right] = -E_c^{sphere}\left[\frac{1}{5}\alpha_2^2 + ...\right] \quad ...(3)$$

Thus the total energy variation

$$\Delta E = \Delta E_s + \Delta E_c = \frac{1}{5}\alpha_2^2\left[2E_s^{sphere} - E_c^{sphere}\right] \quad ...(4)$$

If it is +ve, i.e., $2E_s^{sphere} > E_c^{sphere}$, the drop is stable to small distortions. Fissions may occur spontaneously if ΔE = –ve or $E_s^{sphere} < \frac{1}{2}E_c^{sphere}$

$\therefore\ 4\pi R_0^2 A^{2/3} T < 3Z^2e^2/40\pi\varepsilon_0 A^{1/3} R_0$ or $Z^2/A > 45$

The ratio $\frac{E_c^{sphere}}{2E_s^{sphere}}$ known as *critical parameter*, represented by χ. Thus when $\chi < 1$, the nuclear is stable against spontaneous fission. It is possible to estimate the degree of distortion of a nucleus in the critical state by equating the *critical or threshold* energy E_{th} to the total energy variation ΔE. From semi-empirical data $4\pi R_0^2T$ = 13 MeV, hence for U^{238}, E_s^{sphere} = 520 MeV and E_c^{sphere} = 830 MeV and thus $\alpha_2^2 = 1/7$. The energy that has to be imparted to the nucleus in order to reach this critical shape, the threshold energy is given as

$E_{th} = 4\pi R^2T\ f(\chi) = 4\pi R_0^2\ A^{2/3}\ T\ f(\chi) = 17.8$ A2/3 $f(\chi)$ Mev ...(5)

This energy can be calculated by neglecting the second order change in energy due to the neck joining the two fragments

$$E_{th} = 2\left(4\pi R_0^2\right)T\left(\frac{1}{2}A\right)^{2/3} - 4\pi R_0^2 A^{2/3}T + 2\times\frac{3}{5}\times\left(\frac{1}{2}Ze\right)^2 /$$

$$4\pi \epsilon_0 R_0\left(\frac{1}{2}A\right)^{1/3} + \left(\frac{1}{2}Ze\right)^2/8\pi \epsilon_0 R_0\left(\frac{1}{2}A\right)^{1/3} - \frac{3}{5}(Ze)^2/4\pi \epsilon_0 R_0 A^{1/3}$$

or $E_{th}/4\pi R_0^2TA^{2/3} = 0.260 - 0.215\ \chi$ ...(6)

For an uncharged droplet $\chi = 0$ and $f(0) = 0.260$, hence the critical energy is just the work done against surface tension in separating into two drops. For $\chi \approx 1$, a small deformation from the spherical shape causes the drop to reach the critical shape and separate.

If the critical energy is compared with the excitation energy, it becomes possible to predict fission probability. The excitation energy E_e, contributed to the resultant compound nucleus by the capture of a neutron, is equal to the binding energy of neutron in the compound nucleus and can be calculated by the relation.

$$E_e = B(A + 1, Z) - B(A, Z) = {}_ZM^A + M_n - {}_ZM^{A+1}$$

The values of the excitation energy calculated in this way for a number of heavy nuclei are listed in the table 1 and compared with the corresponding values of the critical energy. In reviewing the results, it is seen that for uranium—238 a critical deformation energy of 6.5 MeV is necessary for fission but it acquires only 5.9 MeV when it takes up a neutron of zero K. E. Thus no fission is possible with thermal neutrons with 0.03 eV energy. If the neutrons have a K. E. of 0.6 MeV fission becomes possible.

Experiments indicate that neutrons of about 1 MeV energy are required. The fission cross section increases rapidly with neutron energy. The situation is quite different with U^{235}. Here the excitation energy or the energy available by the capture of a slow neutron is greater than the threshold energy. It is evident that in this case thermal neutrons should be capable of causing fission of U^{235} nucleus. The reason for the difference in the excitation energy for U^{236} and U^{239} lies in the pairing term which appears in the semi-empirical mass formula. The term makes a positive contribution of about 0.5 MeV to the binding energy of U^{236} but it is zero for U^{235} On the other hand the contribution is zero for U^{239} but is about 0.5 MeV for the U^{238}. Thus the odd-even effect is responsible for approximately 1 MeV of the difference in the energies gained by these isotopes upon the addition of a neutron.

Table 1

Compound Nucleus	E_e (MeV)	E_{th} (MeV)	$E_e - E_{th}$ (MeV)
Pa^{232}	5.4	5.0	0.4
Th^{233}	5.1	6.5	–1.4
U^{236}	6.6	5.5	1.1
Np^{238}	6.0	4.2	1.8
U^{239}	5.9	6.5	–0.6
Pu^{240}	6.4	.4.0	2.4

A general review of the changes in various types of nuclei, after neutron capture, shows that the liberation of energy is greater or the nuclei are likely to undergo fission with slow neutrons if the original nucleus contains an even number of protons and an odd number of neutrons or an odd numbers of both. Whereas fast neutrons would be required for the nuclei containing odd-even or even-even numbers in the same mass region.

Quantum Effects

The values shown in table 13.2 do not agree well with measured values. This disagreement may be the result of two quantum mechanical effects.

(1) The fission may take place for excitation energies below the threshold due to the tunneling effect.

(2) The vibration of the drop in the distorted mode will have a zero point energy.

The tunneling effect is the origin of spontaneous fission. The fission problem is more difficult than the α-decay problem as it is very difficult to have a clear picture of the exact shape of the fission potential barrier. The fission barrier penetration probability

$$P \propto \exp\left[-\frac{2}{\hbar}\int_0^b \{2M(V-E)\}^{1/2}\,dr\right], \qquad \text{...(5)}$$

where M is the reduced mass of the two fragment system, (V – E) is the –ve K. E. in the barrier of width b. For simplicity, let the barrier be of a parabolic form, given as

$$V = 1/2\,K\,(r-R)^2 \qquad \text{...(6)}$$

where r is the separation of the two fission fragments and R is the separation at the top of the barrier. The width of the barrier b is given by the relation

$$E = 1/2\,K(b\ 1/2b)^2 \text{ or } 1/2\,b = \sqrt{(2E/K)} \qquad \text{...(7)}$$

Thus using parabolic function, we get

$$P \propto \exp\left[-\left(\frac{2\pi}{\hbar}\right)\left(\frac{M}{K}\right)^{1/2} E\right]$$

$$\propto \exp\left[-\left(\frac{b\pi}{2\hbar}\right)(2ME)^{1/2}\right] \qquad \text{...(8)}$$

For U^{238}, $E \sim 6$ MeV, $b = 1.5 \times 10^{-14}$ m and $M = 240/4 = 60$ mu

$$\therefore \quad P \propto \exp(-100) \qquad ...(9)$$

Frankel and Metropolis obtained an expression for the life time for spontaneous fission using the idea of barrier penetration as

$$t = 10^{-21} \times 10^{7.85}{}_{th} \text{ sec} \qquad ...(10)$$

Scaborg calculated E_{th} for various heavy elements and found that the fission rate could be determined by the formula

$$t = 10^{178-3.75(Z^2/A)} \text{ sec} \qquad ...(11)$$

After comparing with eqn (10), he obtained the relation

$$E_{th} = 19.0 - 0.36\, Z^2/A \qquad ...(12)$$

which is in close agreement with experiment.

CONTROLLED THERMONUCLEAR REACTIONS

In stars, nuclear fusion reactions release great amounts of energy. The rate depends upon the density and temperature of the gas and upon the cross sections or lifetimes of the reactions involved. Bearing in mind the masses of materials available on earth, it is certain that the reactions of the carbon cycle and the p-p chain would occur extremely slowly. There are thermonuclear reactions which occur much more rapidly and depend on an most abundant material. Among the nuclei of the hydrogen isotopes, ${}_1H^2$ and ${}_1H^3$ reaction are:

${}_1H^2 + {}_1H^2 \rightarrow {}_1H^3 + {}_1H^1 + 4.0$ Mev D-D reaction

${}_1H^2 + {}_1H^2 \rightarrow {}_2He^3 + {}_0n^1 + 3.2$ Mev D-T reaction

${}_1H^2 + {}_1H^3 \rightarrow {}_2He^4 + {}_0n^1 + 17.6$ Mev

The D-D reaction can go in two equally probable ways, the first of which produces ${}_1H^3$ while the second produces ${}_2He^3$. When reactions take place in a chamber the deuterium can react with ${}_1H^3$ to give an α-particle and a neutron. 14.1 MeV energies are carried off by the neutron and 3.5 MeV by the α-particles. In these fusion reactions the energy released is much less than that released in fission reaction but the energy yield per unit mass of material is greater slightly. Deuterium occurs in nature with an abundance of about one part in 6500 hydrogen and can be separated from the lighter isotope quite cheaply It has been calculated that the energy equivalent of the deuterium in one gallon of water is the same as that obtained from the combustion of 300 gallons of gasoline. The more than 10^{20} gallons of water present in the oceans could thus supply the world's power requirement for several million years, if all the deuterium could be utilized to provide energy by fusion processes.

Thermonuclear (hydrogen) Bomb

Even before the achievement of atomic bomb, it was realized by J. Robert Oppenheimer and others that the explosion of an atomic bomb might create the high temperature needed to start a thermonuclear reaction and thus an atomic bomb might serve as the *fuse or trigger* to set off a hydrogen bomb. Hydrogen bomb can be made with deuteron or tritium or the combination. It is possible to produce tritium (as it occurs in nature only in infinitesimal amounts) by the neutron bombardment of lithium in a nuclear reactor. Calculations show that a given weight of tritium would release 7 times as much energy as an equal weight of Pu. One kg of tritium would release as much energy as 140,000 tons of T. N. T. (the recognized abbreviation for the common chemical explosive 2, 4, 6-trinitrotoluene, assuming that the explosive energy of 1 ton of T. N. T. is 10^9 calories).

The energy produced in a hydrogen bomb is in an uncontrolled manner. To control thermonuclear reactions some of the essential conditions must be satisfied. The process must be self sustaining and the balance must be obtained between the energy released in fusion and the amount lost by radiation. The temperature corresponding to this condition of balance is known as critical ignition temperature.

The relation $E = kT$ implies that it is of the order of 10^8 °K. At very high temperatures the atoms are fully ionized and these ions and the free electrons are moving about very rapidly. The result is a completely ionized gas, called a *plasma*. The plasma is electrically neutral so that in the absence of electric or magnetic fields, there are no external forces acting on it except gravity. Because of the internal pressure, the plasma would expand in a vacuum to fill the container in which it is kept. When it comes in contact with the walls, cheats up the walls.

The power produced in fusion reactor will depend on the plasma density and temperature. Because of the very low density of the plasma, used in controlled thermonuclear reactions, and the relatively small volume of the system, it does not behave as a black body. The radiation should consist mainly of bremsstrahlung accompanying the deflection of the rapidly moving electrons in the plasma by the electrostatic fields of the positively charged nuclei. The amounts of energies released per unit time per unit volume in the D-D reactions and in the D-T reactions and of energy lost as radiation at various temperatures are shown in Fig. 1.

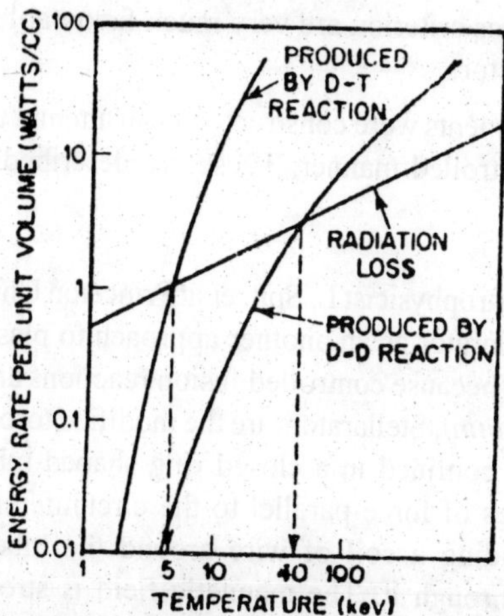

Fig. 1 : Critical ignition temperalure for the D-T and D-D reactions.

In common practice the temperatures are expressed in keV, which is equivalent to 1.16×10^7 °K It is clear from the figure that the bremsstrahlung loss equals to power produced at plasma temperature about 5 keV for a D-T and about 40 keV for D-D reactions This does not necessarily mean that the plasma temperature for a D-D reactor has to be higher than 40 keV to produce a net power gain. The reason is that the bremsstrahlung is not lost but absorbed by the walls of the plasma container and converted therefore into heat in the way similar to the K-E. of the reaction products. The minimum operating temperature for a given reactor will therefore depend on the efficiency with which heat can be converted into other forms of energy. A part of this energy should be taken back by the plasma again to keep it at the operating voltage.

J.D. Lawson in England in 1957 pointed out another necessary condition, known as the *Lawson Criterion*, for a self-sustaining thermonuclear system. According to this criterion for a fusion reactor the total useful recoverable energy shall be at least sufficient to maintain the temperature of the reacting species. It can be expressed in terms of the product nt, where n is the number of reacting nuclei per unit volume and/

is the time in seconds during which thermonuclear reaction takes place or the time during which the high temperature plasma can be confined. 6×10^{19} and 2×10^{21} are the calculated minimum values for nt for D-T system and D-D system respectively. Thus we see that critical ignition temperature and this criterion are very much favourable for D-T reactor than for D-D reactor.

Several equipments were constructed in an attempt to produce fusion reactions in a controlled manner. These are described as given below.

Stellarator

In 1951, the astrophysicist L. Spitzer at Princeton University described a principle of stellarator as an another approach to plasma confinement. This name is used because controlled fusion reactions are similar to those in stars *(stella in Latin)*. Stellarators are the modified torous type machines. Here a plasma is confined in a closed ring shaped tube. The magnetic field with the lines of force parallel to the circumference of the ring is produced by winding a coil of wire around the tube and passing an electric current through it. The magnetic field is stronger at the inner perimeter of the torus than at the outer perimeter, thus causes the plasma as a whole to move to the outer walls. The confinement is thus impossible. Spitzer suggested to twist the toroidal tube into a space like. The difficulty of instability which causes a escape of particles towards the walls is removed in this shape.

The ten-dency of the plasma to drift in a particular direction in one bend of the tube is compensated by the opposite tendency in the other bend. In each circuit of the stellarator tube, a line of force is somewhat displaced from its original position. A magnetic field of this type is said to possess a rotational transform. The earliest stellarators were of the figure of eight type.

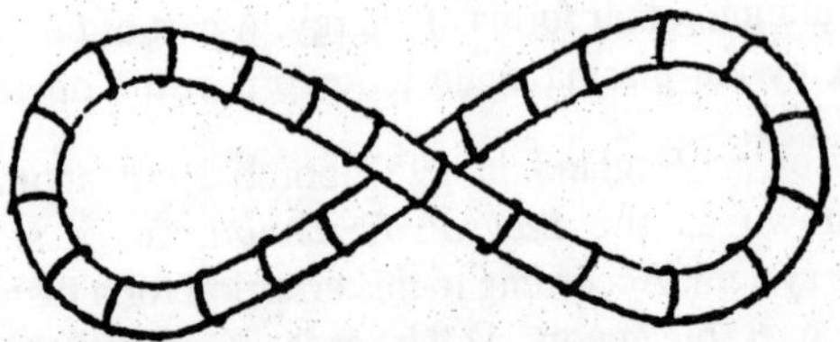

Fig. 2 : Stellarator tube of figure eight.

The most recent stellarator is the Model C stellarator, which was completed in 1961 at Princeton. It is a large and complex but has a flexible facility for the study of plasma confinement and heating. The maximum strength of the axial magnetic field is about 5 tesla which is held constant for periods upto 1 sec. or so.

Magnetic Mirror Systems

In the stellarator and other systems using closed containing tubes, the plasma can escape in the axial direction and so confinement is required only in radial direction. In magnetic mirror systems the plasma is confined in a *bottle shaped magnetic field.* This system consists of a straight tube with magnetic coils wound around it in such a way as to provide a field that is considerably stronger at the ends than in the middle. The ions move in spiral orbits with radii of curvature inversely proportional to the field strength and under certain conditions are reflected when move into regions of higher field strengths. In this way the magnetic mirrors inhibit but do not prevent entirely the escape of plasma from the ends of the tube. For a particle to be reflected it must have a significant component of velocity perpendicular to the magnetic field lines in the central region between the mirrors. The greater the value of B_{max}/B_{min}, the smaller the value of v_1 relative to actual "velocity v for which reflection is possible. The heating of the plasma in this system is accomplished by increasing the magnetic field relatively slowly such that the plasma is compressed. This compression can be carried out in several stages and the compressed plasma is transferred after each stage from one magnetic bottle to a smaller one in which it is further compressed.

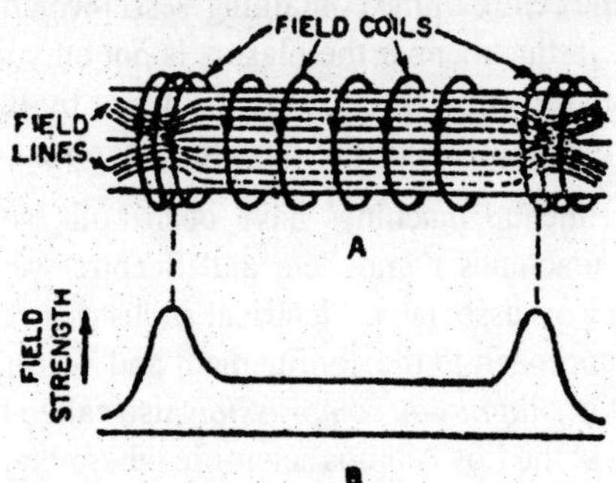

Fig. 3 : (A) Field coils and lines of force in a magnetic mirror system; (B) Variation in the magnetic field strength.

Beam-injection Method

The injection of charged particles with a significant velocity component along the radial direction, directly across the magnetic field presents a difficulty. To solve this problem, the use of electrically neutral atoms of high energy was suggested, but it was very difficult to accelerate neutron atoms to a very high energy. Indirect process is the use of deuterium ions (~20 kV) which are allowed to pass into a chamber containing neutral deuterium gas. In this chamber a charge exchange occurs and the stream of high energy neutral deuterium atoms are produced, which then enter the magnetic field where they are ionized. Similar injection process used at the Oak Ridge National Laboratory starts with a high energy beam of molecular deuterium ions (D_2^+). These ions dissociate as $D_2^+ \rightarrow D^+ \rightarrow D^0$. Since the mass of deuterium is half that of the molecular ions, hence the resulting D^+ ions can be trapped but the undissociated D_2^+ ions are not. The neutral atoms are not affected by the magnetic field and hence escape. As a result of collisions, the trapped deuterons acquire random motion equivalent to a high temperature.

Pinched Discharge

In principle the simplest way of heating and containing the plasma is to make use of a phenomenon called the *pinch, utilized* in U.S.A., U.K., and in the U.S.S.R. simultaneously and independently in the early 1950. In it a very strong electric current amounting to a million amperes is passed through a deuterium like gas at low pressure. This current heats the gas and ionizes it, thus converting it into a *plasma.* At the same time the current flow produces a magnetic field with its lines of force encircling the plasma. The pressure of this field then pinches the plasma into a narrow region in the centre of the containing vessel which may be a torus or a straight tube. In this manner the plasma is not only kept away from the walls but heated also by the current itself and by the compression produced by the magnetic field, the current generates.

Several experimental machines have been built, some on quite a large scale. Two machines name Zeta and Sceptre were designed by Harwell group and by associated electrical industries respectively. An entirely different approach to the confinement and heating of plasmas is by the method of *fast magnitude compression* also called the *theta pinch,* originated in 1957 at the Los Alamos scientific laboratory. The apparatus consists of an open ended tube containing deuterium gas at low pressure surrounded by a single turn coil. An oscillating current of several million amperes is passed through the coil for a very short time by the discharge of a large bank of capacitors. The first half cycle of the oscillating magnetic field produces an ionized sheath of plasma which is rapidly compressed and partially heated. In the second half cycle the direction of the magnetic field is reversed and the mutual annihilation of the oppositely directed fields releases a large amount of energy that heats the deuterium plasma. Scientists at Los Alamos were able to obtain plasmas with temperatures in excess of 5 keV and densities of about 4 × 10^{22} deuterons per m^3 by fast magnetic compression of low density preionized deuterium.

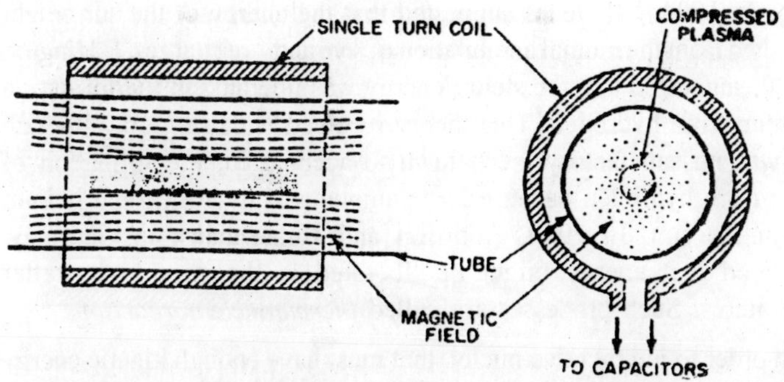

Fig. 4 : Schematic diagram of theta pinch apparatus.

THERMONUCLEAR REACTIONS AND NUCLEAR FUSION

Power from nuclear fission is now a reality both on land and sea and in those countries where coal or oil is costly. An alternative to the fission reaction as a source of energy is its reverse process known us *fusion process* in which the lighter nuclei fuse together and produce a heavier nucleus. The sum of the masses of the Individual light nuclei is more than would be the mass of the nucleus formed by their fusion, and thus the fusion process should result in a liberation of energy.

Some indication of how this might be achieved can be obtained by considering the source of energy produced continuously in the stars, including the sun. It has been calculated that the sun (our nearest star) emits electromagnetic energy at a rate of about (our joules per second. Astronomical and geological evidences show that the sun has been radiating energy at about its present rate for several billion years. Chemical reactions cannot possibly be the source of this energy because even if the sun is supposed to be consisted of pure carbon, its complete combustion would supply energy to maintain these radiations only for few thousand years. Now question arises how can the sun have maintained this energy output for so long and what is the source of all *stellar energy*? Helmholtz in

1853, suggested that the contraction was taking place and thus the gravitational energy was being converted into heat energy in sun. This conversion is analogous to the production of electricity from the falling of water. It has been shown that if contraction were taking place it could supply not more than 1% of the total energy output needed and it led to an estimate of the age of the sun which was much too short.

With the discovery of radioactivity at the end of the 19th century, it appeared possible that atomic energy might be contributing to the sun's energy. In 1904, J.II. Jeans suggested that the energy of the sun might be resulted from the mutual annihilation of +ve and –ve charges. Eddington in 1920, suggested that the stellar energy was liberated in the formation of helium from hydrogen. This theory received wide support, although there was no satisfactory mechanism to account for the formation of helium from hydrogen because large amounts of hydrogen and helium exist in the sun. In 1929, Atkinson and Houtermans, in Germany, considered that energy might be liberated in the very high stellar temperatures. Such processes are called *thermonuclear reactions*.

In order to interact two nuclei, that must have enough kinetic energy to permit them to overcome the electrostatic repulsion barrier which tends to keep them apart. It can be shown from calculations that the energy requited make the nuclear reactions occur at a detectable rate is about 0.1 MeV for nuclei of the lowest atomic number (*e.g.*, isotopes of hydrogen), the larger energies are needed for nuclei of higher atomic number. This energy can be resulted from a sufficient increase in temperature (1000×10^6 °K if the average energy of the particles is to be 0.1 MeV). Such temperatures are considerably higher than those existing in stars from stars, like sun where the central temperatures are less than (50×10^6 °K, the fusion reaction takes place with a finite rate and releases enough energy to keep up the heat and light of the star.

H.A. Bethe in the United States suggested in 1939 that the *production of stellar energy is by thermonuclear reactions* in which helium-4 nuclei are synthesized from four protons (the nucleus of hydrogen, the most abundant element in the universe). A few years ago, it was held that the major portion of the sun energy was derived from the *carbon nitrogen cycle*. Recent modification of the estimates of the central temperature of the sun now favour the *proton-proton chain*. In the carbon-nitrogen cycle carbon acts as a short of catalyst in facilitating the combination of four protons to form a helium nucleus. In this cycle a proton first interacts with a C^{12} nucleus with a release effusion energy. The product nucleus

N^{13} decays in a very short time. The stable nucleus of C^{13}, thus formed then reacts with another proton. The more energy being liberated by this process. The stable nucleus of N^{14} combines with a third proton. The product nucleus O^{15} is a positive β emitter, which decays into N^{15}. This nucleus finally interacts with a fourth proton and regenerates C^{12} nucleus. These reactions can be written as:

$$_6C^{12} + {}_1H^1 \rightarrow ({}_7N^{13}) \rightarrow {}_7N^{13} + 1.94 \text{ MeV } (10^6 \text{ y})$$

$$_7N^{13} \rightarrow ({}_6C^{13}) + {}_{+1}e^0 + v + 1.20 + 1.02 \text{ MeV } (10 \text{ m})$$

$$_6C^{13} + {}_1H^1 \rightarrow ({}_7N^{14}) \rightarrow {}_7N^{14} + 7.55 \text{ MeV } (2 \times 10^5 \text{ y})$$

$$_7N^{14} + {}_1H^1 \rightarrow ({}_8O^{15}) \rightarrow {}_8O^{15} + 7.29 \text{ MeV } (3 \times 10^7 \text{ y})$$

$$_8O^{15} \rightarrow {}_7N^{15} \rightarrow {}_{+1}e^0 + v + 1.74 + 1.02 \text{ MeV } (2\text{m})$$

$$_7N^{15} + {}_1H^1 \rightarrow ({}_8O^{16}) \rightarrow {}_6O^{12} + {}_2He^4 + 4.96 \text{ MeV } (10^4 \text{ y})$$

$$4\ {}_1H^1 \rightarrow {}_2He^4 + 2\ {}_{+1}e^0 + 2v + 26.7 \text{ MeV}$$

It will be noted this chain of reactions can start with either carbon or nitrogen, since each one is reproduced in the reaction. The four protons are associated with four elections to maintain electrical neutrality, whereas only two are required for the helium nucleus, and rest two electrons combine readily with positrons resulting in the formation of γ-rays.

The mass difference released as energy in this chain of reactions is simply the difference between the masses of four protons and one helium nucleus. A small amount of this energy is carried away by the neutrons that are emitted during the e^+-decay. Bethe estimates this to be about 1.84 MeV leaving rest for each α-particle formed.

In the p-p chain, two protons first fuse to produce a deuterium nucleus which combines with an another proton to yield He^3. Two He^3 nuclei interact and form He^4 and two protons. These reactions can be represented by the equations

$$_1H^1 + {}_1H^1 \rightarrow ({}_2He^2) \rightarrow {}_1H^2 + {}_{+1}e^0 + v + 0.42 \text{ MeV } (7 \times 10^9 \text{ y})$$

$$_1H^2 + {}_1H^1 \rightarrow ({}_2He^3) \rightarrow {}_2He^3 + \gamma + 5.5 \text{ MeV } (10 \text{ sec})$$

$$_2He^3 + {}_2He^3 \rightarrow ({}_4Be^6) \rightarrow {}_2He^4 + {}_1H^1 + {}_1H^1 + 12.8 \text{ MeV } (3\times10^5 \text{ y})$$

$$4\ {}_1H^1 \rightarrow {}_2He^4 + 2\ {}_{+1}e^0 + 2v + 2\gamma + 26.7 \text{ MeV}$$

The positrons emitted are annihilated by free electrons with the production of gamma rays. The energy released in p-p chain is the same as in the C-N cycle (26.7 MeV for each helium nucleus)

Which of the two hydrogen-helium fusion processes plays the major role in energy production in star? The answer is based on our knowledge of the stellar temperature. H. Bondi and E.E. Salpeter (1952) developed empirical equations for the rate at which energy is liberated in each of the above chains of reaction. The p-p chain reaction rate varies more slowly with temperature, roughly as T^4 and is much more important at lower temperatures. The carbon cycle rate predominates as the temperature is in the vicinity of 18×10^6 °K.

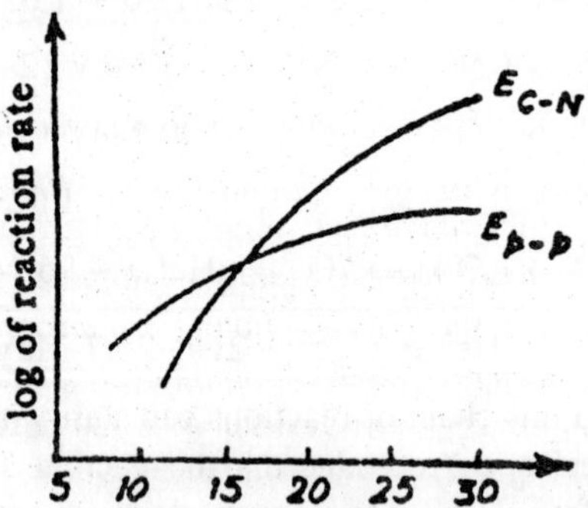

Fig. 5 : Variation in the rate of energy produced with the central teftiperature.

For the case of the sun, they estimated that the rate of generation of energy in the p p chain was about the same as that in the carbon cycle. Actually the p-p chain predominates (96%) because the interior temperature of the sun is about 15×10^6 °K.

Solar Future

When all the hydrogen is used up in the above thermonuclear reactions, the star will consist mostly of helium. At this stage gravitational contraction will occur once again until a temperature of about 10^8 °K is reached and the density of the star is about 10^7 kg/m^3. Under these conditions three helium nuclei combine and form a C^{12} nucleus with the release of about 7.3 MeV. Hoyle estimates that such processes provide energy for an additional 10^7 years. Further gravitational contraction of the star will occur, when all the helium is used up, and will produce a further rise in the temperature of the star. The atoms formed in this way are most stable. Any appreciable combination of these atoms to form heavier ones will lead to endothermic rather than exothermic reactions. Bondi and Salpeter suggested that the endothermic reactions might account for the sudden collapse of a star identified as the sudden appearance of a supernova.

NUCLEAR FISSION

When uranium was bombarded with neutrons it was found that a β-active product resulted, followed by a whole chain of β-active products such as

$$_{92}U^{38} + {_0}n^1 \rightarrow {_{92}}U^{239} \rightarrow {_{93}}X^{239} \rightarrow {_{94}}Y^{239} \rightarrow \text{etc.}$$

It was recognised that more than one neutron was produced. Hahn and Strassman in 1939 proved that the product might be consisted of two large fragments, which they identified as barium and krypton. Frisch and Meitner in 1939 used the *wordfission to describe the process which takes place when a heavy nucleus is caused to break down into two roughly equal parts, known as fission fragments*. In this process neutrons are also emitted with the release of considerable energy. This process has been also observed to occur when heavy nuclides are bombarded with protons, deuterons, α-particles and even electrons and gamma rays. Further work showed that lighter elements could also be fissioned by high energy particles, as for example in the case of copper

$$_{29}Cu^{63} + {_1}p^1 \rightarrow {_{11}}Na^{24} \rightarrow {_{19}}K^{39} \rightarrow {_0}n^1 \quad \text{...(1)}$$

(A) Types of Fission

(a) **Charged-particle Fission :** Elements with Z > 90 show fission process with protons, deuterons and α-particles. High energy charged particles induce fission in elements even in the middle of the periodic table.

(b) **Thermal Fission :** Since a 'thermal neutron adds negligible energy to the fissionable nucleus, it is clear from semi-empirical mass formula that the fission of the nuclei in which the compound nuclei are of even-even structure take place even with the thermal neutrons. Fission of U^{235} and Pu^{231} by thermal neutrons are the most important reactions.

(c) **Fast Fission :** Other isotopes of Uranium and other elements which form compound nuclei of even-odd structure enter into (n, f) reactions with fast neutrons (>I MeV). The example is U^{238}.

(d) **Photo Fission :** High energy photons induce fission in the heavier elements. 5.1 MeV y-rays can produce fission with U^{238}.

In all cases very large disintegration energies were released and fast neutrons are emitted. Meitner and Frisch indicated that, because of their exceptionally high neutrons to protons ratio, the fission fragments should be unstable, undergoing a chain of β^-distintegrations. For example

$$\begin{array}{lcr} & {}_{40}Zr^{98} \rightarrow {}_{41}Cb^{98} \rightarrow {}_{42}Mo^{98} & \\ {}_{92}U^{235} + {}_0n^1 \rightarrow {}_{92}U^{236} & & + \; {}_{20}n^1 \qquad ...(2) \\ & {}_{52}Te^{136} \rightarrow {}_{53}I^{136} \rightarrow {}_{54}Xe^{136} & \end{array}$$

A typical beta decay chain is

$$ {}_{54}Xe^{140} \rightarrow {}_{55}Cs^{140} \rightarrow {}_{56}Ba^{140} \; {}_{57}La^{140} \rightarrow {}_{58}Ce^{140} \text{ (Stable)} \qquad ...(3)$$

For the nuclei near the middle of the periodic table (E_B = 8.5 MeV/nucleon), the fission process is expected to release a large amount of energy. Assuming fission of a ucleus of mass number 240, for which $E_B \simeq 7.6$, into two similar fragments with A ~ 120, the energy released in fission is $2 \times 120 \times 8.5 - 240 \times 7.6 = 216$ MeV. Assuming their separation equal to the sum of their nuclear radii, the Coulomb repulsion energy is given by

$$E_c = \frac{Z_1 Z_e e^2}{4\pi \epsilon_0 d} = \frac{52 \times 40 \times (1.6 \times 10^{-19})^2 \times 9 \times 10^9}{1.5 \times 10^{-14} \times 1.6 \times 10^{-13}} \text{ MeV.}$$

$$\sim 200 \text{ MeV.}$$

Thus we see that the Coulomb repulsion energy is nearly equal to the energy released in fission process. We can, therefore, regard this fission process as a result of Coulomb repulsion. The Q value of fission reaction (2) is also calculated from the exact mass difference of the two sides of the equation. The combined isotopic masses before and after fission are $\Sigma m_i = m(U^{235}) + m({}_0n^1) = 235.0439 + 1.0087 = 236.0526$ mu and $\Sigma m_f = m(Mo^{98}) + m(Xe^{136}) + m(2{}_0n^1) = 97.9054 + 135.9072 + 2.0154$ = 235.83 mu.

Hence $Q = \Delta mc^2 = (\Sigma m_i - \Sigma m_f)c^2 = 210$ MeV, which is comparable with the value calculated by the binding energy method.

(B) Distribution of Fission Products

We have seen that a fissionable nucleus gives only two fission fragments which decay by β^-emission to a *stable end product*. Although, the sum of two large fragments always adds up to 234, there is a wide distribution in possible products. The mass distribution of the fission products is shown most conveniently in the form of a *fission yield curve*, in which the percentage yields (in log scale) of the different products are plotted against mass number. There is a tendency for masses to concentrate respectively round 90 and round 140.

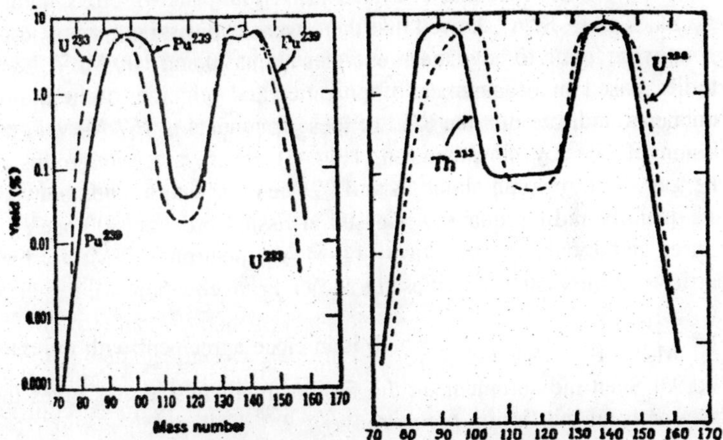

Fig. 6 : (a) Thermal fission of U^{233} and Pu^{239},
(b) Fast fission of Th^{232} and U^{238}.

Actually about 97% of the total fission products fall within the narrow range 85—104 for the lighter fragments and 130-149 for the heavier. In the case of U^{235} the maxima lie near mass numbers 95 and 140; fission produced by slow neutrons is highly asymmetric process and division into two equal fragments occurs in only about 0.01% of the fissions. Symmetric fission becomes increasingly more probable with increasing neutron energy.

Single peak appears for high energy neutrons. The trough in the curve also tends to fill, when the nuclei are fissioned by the panicles other than neutrons, A comparison of the U^{233} and Pu^{239} fission yield curves shows that the distribution of the heavier products are practically the same, whereas the portions of the curves for the lighter fragment yields are displaced by six mass units with respect to other.

The mass distribution of the fission fragments can also be obtained from the distribution of their kinetic energies. The nucleus undergoing fission can be considered to be at rest initially and, if the neutrons emitted are neglected, the law of conservation of momentum gives $M_1V_1 = M_2V_2$.

$$\therefore \quad \frac{E_1}{E_2} = \frac{\frac{1}{2}M_1V_1^2}{\frac{1}{2}M_2V_2^2} = \frac{M_2}{M_1} \quad \text{...(4)}$$

Thus we see that the masses are inversely proportional to the kinetic energies. W. Jentschke and F. Prankl in Germany and Booth, Dunning and Slack in U. S. A. showed that there were two distinct groups, each group were found to have mean energies about 70 and 100 MeV. Later studies, based on measurement of ionization and velocities of the fission fragments, indicate that the K.E. of these fragments is 167 MeV in the fission of U^{235} by slow neutrons (Fig. 7). There are two groups of energies, with maxima about 68 and 99 MeV. The difference between this quantity and the quantity released in fission process (200 MeV) is carried out by the gamma rays (11 MeV), neutrons (5 MeV), beta particles (7 MeV) and neutrinos (11 MeV). From equation (4), it is clear that $\frac{M_1}{M_2} = \frac{E_2}{E_1} = \frac{99}{68} \simeq 1.5$, which is in close agreement with the ratio 140/95, Such measurement on the fission fragment energies gives the asymmetry of the fission process.

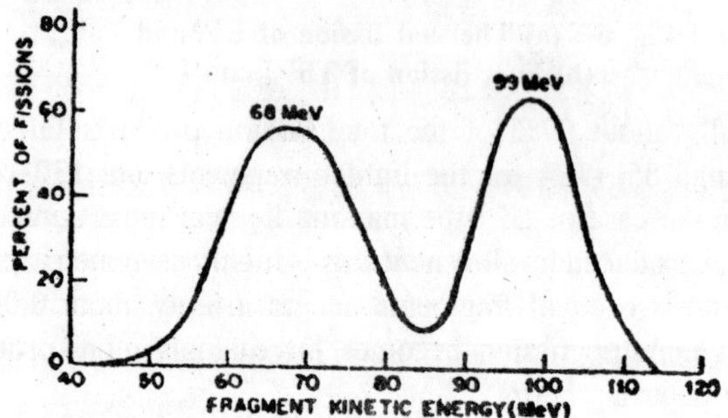

Fig. 7 : Kinetic energy distribution of fission fragments in U^{235}.

(C) Neutron Emission in Fission

An accurate knowledge of the average number of neutrons emitted per second is of great importance to the scientists. Actually the number of neutrons released in any one fission is an integer, but average value, v_{av} of the number of neutrons in one fission is not an integer, because the fissionable nucleus can divide in at least 30 different ways, and can be obtained by the relation

$$v_{av} = \Sigma v_0 \, n \, v_0 / \Sigma n v_0 \quad ...(5)$$

Most of these neutrons are emitted within possibly 10^{-13} sec. and are called prompt neutrons. A smaller number of neutrons is also emitted with a time lag of several seconds to more than a minute after the fission. These trons arc called *delayed neutrons.*

For sufficiently large piece of fissionable substance, the neutrons that are released in a first fission process will be absorbed by the other nuclei and produce new processes which in turn emit new neutrons. Act lly some of the compound nuclei decay to the ground state by gamma emission, rather than fission. The ratio of the radiative capture cross-section to the fusion cross-section is usually denoted by a given by

$$\alpha = \frac{\sigma_r}{\sigma_f} \quad ...(6)$$

and the number of fusion neutrons released per neutron absorbed in the fissionable nuclide is denoted by η and is given by

$$\eta = \frac{v_{ab}}{(1+\alpha)} \quad ...(7)$$

The relative probability that a compound nucleus decays by fission is thus be $(1 + \alpha)^{-1}$. Values of v_{av}, α and η for thermal neutrons are listed in the Table 2.

Table 2

	U^{233}	U^{235}	Natural uranium	Pu^{239}
σ_f	525 ± 4	577 ± 5	4.18 ± 0.06	742 ± 4
σ_r	53 ± 2	101 ± 5	3.50	286 ± 4
v_{av}	2.51 ± 0.02	2.44 ± 0.02	2.47	2.89 ± 0.03
α	0.101 ± 0.004	0.18 ± 0.01	0.83	0.39 ± 0.03
η	2.28 ± 0.02	2.07 ± 0.01	1.34 ± 0.02	2.08 ± 0.02

(D) Fissile and Fertile Materials

For a number of reasons, isotopes such as U^{233}, which only fission with energetic neutrons, can not alone be used to fuel nuclear reactors. The only neutrally occurring nuclide that can be fissioned with thermal neutrons is U^{235}, which is 0.71% of the naturally occuring uranium. The only other nuclides that can undergo fission with thermal neutrons are

U^{233} and Pu^{239}. These do not occur in nature but can be produced by the interaction of neutrons with Th^{232} and U^{238} respectively and are called *fissile materials*. Th^{232} and U^{238} are not fissile materials but can be used as raw material for the production of fissile isotopes and are called *fertile or fissionable materials*. The nuclear reactions which convert these fertile materials into fissile materials are called *breeding reactions*. They are neutrons capture processes with subsequent β-decay.

$$
\begin{aligned}
{}_{92}U^{238} + {}_{0}n^{1} &\rightarrow {}_{92}U^{230} + \gamma \\
{}_{92}U^{239} &\rightarrow {}_{93}Np^{239} + {}_{-1}e^{0} \qquad \text{...(8)} \\
{}_{93}Np^{239} &\rightarrow {}_{94}Pu^{239} + {}_{-1}e^{0}
\end{aligned}
$$

Similarly

$$
\begin{aligned}
{}_{90}Th^{232} + {}_{0}n^{1} &\rightarrow {}_{90}Th^{233} + \gamma \\
{}_{90}Th^{239} &\rightarrow {}_{91}Pa^{233} + {}_{-1}e^{0} \qquad \text{...(9)} \\
{}_{91}Pa^{233} &\rightarrow {}_{92}U^{233} + {}_{-1}e^{0}
\end{aligned}
$$

(E) Spontaneous Fission

Most heavy nuclides undergo spontaneous fission in competition with the a-emission. Spontaneous fission is predicted by the empirical nuclear mass equation. Consider the special case, when the nucleus splits into equal parts. Neglecting the pairing term δ, we have the Q-value for the fission reaction

$$E_f = [{}_{Z}M^{A} = 2\,({}_{Z/2}M^{A/2})]c^2 \qquad \text{...(10)}$$

Consider Weizsaker's semi empirical binding energy equation

$$M(Z, A) = ZM_p + A- Z)\,M_n - a_v A + a_s A^{2/3} + a_c Z^2 A^{1/3} + a_a (A - 2Z)^2 A^{-1} \qquad \text{...(11)}$$

$$M\left(\frac{1}{2}Z, \frac{1}{2}A\right) = \frac{1}{2}ZM_P \mp \frac{1}{2}(A - Z)M_n - a_c\left(\frac{1}{2}A\right) + a_s\left(\frac{1}{2}A\right)^{2/3} + a_c\left(\frac{1}{2}Z\right)^2\left(\frac{1}{2}A\right)^{-1/3} + \frac{1}{2}a_c(A - 2Z)^2 A^{-1} \qquad \text{...(12)}$$

Substituting value of ${}_{Z}M^{A}$ and ${}_{Z/2}M^{A/2}$ in eqn. (10), we have

$$E_f = \left[a_s\left\{A^{2/3} - 2\left(\frac{1}{2}A\right)^{2/3}\right\} + a_c\left\{Z^2/A^{1/3} - 2\left(\frac{1}{2}Z\right)^2/\left(\frac{1}{2}A\right)^{1/3}\right\}\right]c^2$$

$$= -3.42\,A^{2/3} + 0.22\,\frac{Z^2}{A^{1/3}}\ \text{MeV} \qquad \text{...(13)}$$

This eqn shows that the splitting of a nucleus affects Coulomb energy and surface energy in such a way that the change in one and that in other tend to cancel one another partially. This is reasonable and to be expected, since the division of the nucleus increases (1) the separation between proton groups, thus reducing their Coulomb potential energy, (2) the total nuclear surface which increases the surface energy. Thus for spontaneous fission

$$-3.42\ A^{2/3} + 0.22\ Z^2/A^{1/3} \geq 0 \quad \text{or} \quad Z^2/A \geq 15 \qquad ...(14)$$

This relation shows that the fission should be energetically possible for nuclei with mass number $A \geq 85$. However, the slow neutron fission does not take place even with many of the heavy nuclei. In order to explain this discrepancy Bohr and Wheeler considered the Coulomb's potential barrier of the two fragments at the instant of separation. The existence of this barrier prevents the immediate breaking of these two. If we denote the height of the Coulomb barrier by E_b, we can say that the nucleus will be unstable and break apart into two fragments if $E_f > E_b$. The barrier height corresponding to the Coulomb potential between the two symmetric fragments when they are just in contact with each other is given by

$$E_b = \left(\frac{1}{2}Z\right)^2 \frac{e^2}{A\pi \in_0 (2R)} = \frac{Z^2e^2}{32\pi \in_0 R_0}\left(\frac{1}{2}A\right)^{1/2}$$

$$= 0.15\ Z^2/A^{1/3}\ \text{MeV} \qquad ...(15)$$

$$\therefore\ E_b - E_f = 0.15\ Z^2/A^{1/3} - [-3.42\ A^{2/3} + 0.22\ Z^2/A^{1/3}]$$

$$= 3.42\ A^{2/3} - 0.07\ Z^2/A^{1/3} \qquad (16)$$

Thus the condition for stability gives

$$E_b - E_f \geq 0 \text{ or } Z^2/A \leq 49 \qquad ...(17)$$

For a particular nucleus, the closer the value of Z^2/A to 50, the shorter should be the half life for spontaneous fission. It is clear that Z^2/A may have value 50 for a nucleus of mass number 250, hence a nuclide $(A > 250)$ would be too unstable to exist for more than 10^{-12} sec or less. Computing E_b and E_f for various values of A for symmetric fission And plotting on the same graph, we get Fig. 8. The graph shows that for A $\simeq$ 250. E_b, becomes equal to E_f and $E_f >$ Eb for nuclei with $A > 250$. This indicates that we do not expect nuclei with $A > 250$ to be found in nature. Tins graph also shows that fission begins to become exoergic in the neighborhood of $A \simeq 85$ in agreement with the earlier occurring result.

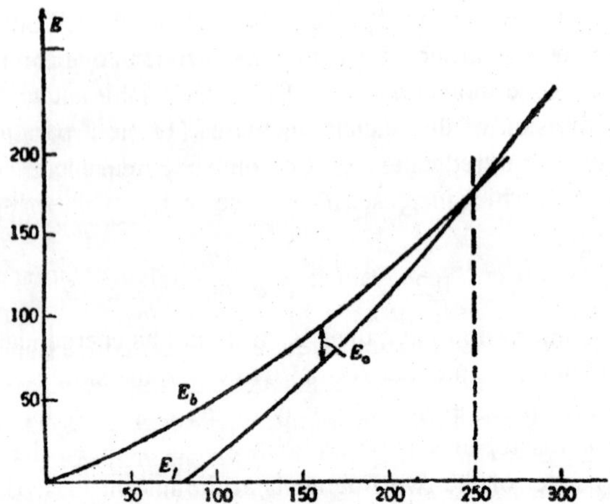

Fig. 8 : Variation of E_b and E_f with A.

(F) Deformation of Liquid Drop

The fission process can be explained with the help of liquid drop model. The incident neutron combines with the nucleus to form highly energetic compound nucleus. Its extra energy is partly the kinetic energy of the neutron but largely the added binding energy of the incident neutron. This energy appears to initiate a series of rapid oscillations in the drop, which tend to distort the spherical shape so that the drop may become ellipsoidal in shape. *The surface tension forces lend to make the drop return to its original spherical shape, while the excitation energy fends to distort the shape still further*. If the exciation energy is sufficiently large, the drop may attain the shape of a *dumb-bell*. If the oscillations become so violent that stage fourth is reached then the final fission into stage fifth is inevitable.

Thus there is a *threshold energy* or a *critical energy* required to produce stage fourth after which the nucleus can not return to stage first. When the distortion produced is not pronounced enough to get the nucleus beyond the critical point, the ellipsoid will return to the spherical shape with the excitation energy being liberated in the form of γ-rays and we have a radiative copture rather than fission. The potential energy of the drop in the different stages can be calculated as a function of the degree of deformation of the drop. It is plotted against r, the separation of the centres of two fission fragments. The curve is supposed to be divided into three regions.

In region I the fragments are completely separated and their potential energy E is simply the electrostatic Coulomb energy resulting from the mutual repulsion of the two positively charged nuclear fragments If distance r = 2R, when the drops just touch each other, energy E at that point if less than the corresponding Coulomb potential by an amount CD. This amount is equal to the potential of the surface forces which are just beginning to come into play at this point. As we pass through region II we reach the critical distance r_c, where the potential energy curve has a maximum value En. This corresponds to barrier height and explains why fission does not take place spontaneously in all cases where $E_f > 0$. An additional amount of energy $E_a = R_b - E_f$, the activation energy is required by the nuclear system before the potential barrier can be surmounted and fission can take place. In the III region, the fragments have coalesced and the short range nuclear forces have become predominant.

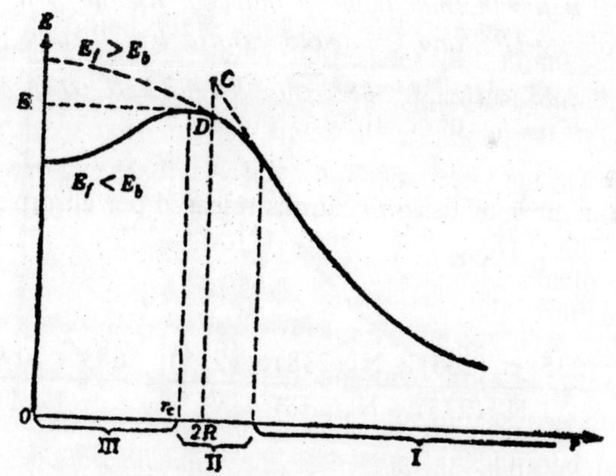

Fig. 9 : Potential energy curve for fission.

SOLVED EXAMPLES

Example 1:

Calculate the excitation energy for U^{236*} and for U^{239*}. Estimate the rare of spontaneous fissioning of 1 gm of U^{235}, given its half life ~3 × 70^{17} years.

Solution:

The excitation energy $E_e = B(Z, A + 1) - B(Z, A)$.

For U^{236}

$E_e = a_v(236 - 235) - a_s(236^{2/3} - 235^{2/3}) - a_c(92^2/236^{1/2} - 92^2/235^{1/2})$
$- a_a(52^2/236 - 51^2/235) + a_v(236^{-3/4} - 0)$

$= a_v + - 0.1\ a_s - 22\ a_c - 0.4\ a_a + 0.0167\ a_p = 6.8$ MeV

Similarly for U^{239}, $E_e = 5.9$ MeV.

Rate of spontaneous fissioning $\frac{dN}{dt} = \lambda N$, where

$N = 6.023 \times 10^{23}/235$ and $\lambda = \log_e 2/T_{1/2}$

$= 0.693/3 \times 10^{17} \times 3.15 \times 10^7$

$\therefore\ dN/dt = 2 \times 10^{-4}\ sec^{-1} = 0.7\ hr^{-1}$.

Example 2:

Calculate η for thermal neutron induced, fission of a uranium mixture containing U^{235} and U^{238} isotopes in a 1 : 20 ratio.

Given $v = 2.43$, $\sigma_a(U^{235}) = 683$, $\sigma_a(U^{238}) = 2.73b$, $\sigma_f(U^{235}) = 583b$

Solution:

Average number of fission neutrons released per absorption

$$\eta = v(\sigma_a/\sigma_a)$$

For a mixture

$$\sigma_a = \frac{N_0(235)\,\sigma_a(235) + N_0(238)\,\sigma_a(238)}{N_0\,(235) + N_0(238)} = \frac{683 + 20 \times 2.73}{1 + 20}$$

$= 35.1$ barns.

and $\sigma_f = \frac{N_0(235)\,\sigma_f(235)}{N_0\,(235) + N_0(238)} = \frac{283}{1 + 20} = 27.8$ barns.

Hence $\eta = 2.43 \times (27.8/35.1) = 1.924$.

Example 3:

A thermonuclear device consists of a torus of diameter 3 m with a tube of diameter 1 m, containing deuterium gas at 10^{-2} mm mercury pressure and at room temperature. A bank of capacitors of 1200 μf is dencharged through the tube at 40 kV.

If only 70% of the electrical energy is transformed to plasma kinetic energy, what is the maximum temperature attained ? Assuming that the energy is equally shared between the deuterons and electrons in the plasma.

Solution:

Cross-sectional area of the torus $1/4\pi$ meter^2

Circumference = 3π meter.

$\therefore$ Volume of Torus $= \dfrac{3\pi^2}{4}$ 7.4 m^3

Pressure of the gas $10^{-5} \times 13.6 \times 10^3 \times 9.81 = 1.34$ N/m^2.

From gas eqn. PV = NkT, we have Nk = $1.34 \times 7.4/293 = 0.0338$.

The energy obtained from the discharge

$= 1/2\ CV^2 = 1/2 \times 1200 \times 10^{-6} (4 \times 10^4)^2 = 9.6 \times 10^6$ joules.

$\therefore$ Energy transformed to plasma K.E. $= 9.6 \times 10^4$ joules.

We know that the average kinetic energy associated with the gas molecule is given as $E_r = 3/2\ NKT_k$. Since each deuterium molecule produces two ions and two electrons, hence

$$4\frac{3}{2} NkT_k = 9.6 \times 10^4 \text{ or } T_k = 4.75 \times 10^5 \text{ °K}$$

Example 4:

Calculate the amount of energy released if all the deuterium atoms in the water in lake of area about 10^5 sq. miles and of depth $1/20^{th}$ mile are used up in fission,

Solution:

The volume of water = $10^5 \times 1/20 = 5000$ cubic miles.

$\therefore$ Mass of water $= 5000 \times (1.6 \times 10^3)^3 \times 10^3$ kg.

or No. of molecules of water $= 2.13 \times 10^{16} \times 6.02 \times 10^{26}/18$

$= 7.13 \times 10^{41}$ molecules.

As the abundance of deuterium is 0.0156% so that the total number of deuterium atoms $= 7.13 \times 10^{41} \times 2 \times 0.0156 \times 10^{-2} = 2.22 \times 10^{38}$. M the fusion of 6 deuterium atoms gives an energy release of 43 MeV, hence the total energy released $= 2.22 \times 10^{33} \times (43/6)$

Example 5:

Calculate the fission rate for U^{235} required to produce 2 wait and the amount of energy that is released In the complete fissioning of 1/ 2 kg of U^{235}.

Solution:

As we know that 200 MeV energy is released per fission of U^{235}

$\therefore$ Fission rate = 2 watt/200 MeV per fission

$= 6.15 \times 10^{10}$ fission/sec.

No. of U^{235} nuclei in 1/2 kg. of U^{235} = (0.5/235) × 6.0247 × 1026

On fissioning this number of U^{235} nuclei, the energy release

will be $= (0.5/235) \times 6.0247 \times 10^{26} \times 200$ MeV

$= 2.57 \times 10^{26}$ MeV $= 10^{10}$ kilocalories.

EXERCISES

1. In the fission reaction $Sn^{118} + p \rightarrow Na^{24} + Zr^{94} + n$, determine (a) the energy equivalent of the mass difference between reactants and products, (b) the excitation energy necessary for passage over the potential barrier and (c) the threshold energy for the reaction.
2. How much U^{235} and U^{238} will be lost in 1 year from spontaneous fission in a stockpile of 1000 kg of natural uranium? How much will be lost from α-decay?
3. The energy released in a nuclear reaction is usually expressed in kT of TAT, which releases 103 cal/gm. How many fissions are involved in 50 kT explosion? How many gms of uranium are fissioned 1 How much mass is converted ?
4. The total volume of water in the oceans is 1.34×10^{9}. What is the total mass of deuterium contained in the water? What will be the total energy available from allot this deuterium through reaction $H^2 - H^2 \rightarrow H^3 + p + 4.04$ MeV? How much mass will be lost in this energy release?

5

BASIC PROPERTIES OF THE NEUTRONS

INTRODUCTION

The neutron is usually assumed to have no net charge. Dee investigated the ionization proceed in air in a cloud chamber irradiated by fast neutrons and found one ion pair per 3 metre of the path and thus concluded that the charge of the one neutron is less than 1/700 of the proton charge. Rabi and his coworkers have found the neutron charges less than 10^{-12} electron charge. Fermi and Marshall set an upper limit of ~ 10^{-18} electron charge for the charge on the neutron.

Neutron Mass

Chadwick gave a rough estimation of the mass of a neutron from the information obtained in collisions of neutrons with protons on the one hand and with nitrogen nuclei on the other.

From the equations for the conservation of energy and for the conservation of momentum for a head on collision, and from the lengths of accompanying tracks in the cloud chamber, Chadwick found that neutron mass M_n = 1.15 amu, roughly the same as that of the proton.

A more precise value for the mass of the neutron was obtained by Chadwick from the study of the nuclear reaction

$$_5B^{11} + {_2He^4} \rightarrow {_7N^{14}} + {_0n^1} + Q.$$

The Q-value was found from the kinetic energies of B^{11}, N^{14}, $_0n^1$ and α-particles. Substituting all the known values in terms of atomic mass units in the above equation Chadwick obtained 1.0067 mu. for the mass of the neutron.

J. Chadwick and M. Goldhaber in 1934 investigated the photo-nuclear reaction ($_1H^2 + h\nu \rightarrow {}_1H^1 + {}_0n^1$) using gamma rays of ThC", with $h\nu = 2.62 \times 10^6$ eV. Since the mass of the deuteron and of the proton were known, the only required measurement was the kinetic energy of the proton and of the neutron.

The proton energy was measured by an ionization method and found to be almost 1.05 MeV. If it is assumed that the neutron and proton mass are nearly equal than the disintegration energy is nearly equal to twice the recoil energy of the proton. Using the best mass-spectrographic atomic masses of hydrogen and deuterium and substituting the mass equivalents for the γ-ray energy and disintegration energy (proton energy + neutron energy), they calculated

$$\text{Mass of neutron} = 1.0087 \pm 0.0003 \text{ amu.}$$

The most precise determination of the mass of the neutron was later made by Bell and Elliott from the reaction

$$_0n^1 + {}_1H^1 \rightarrow {}_1H^2 + h\nu.$$

They carefully measured the energy of the γ–ραψσ and obtained a value of 2.230 ± 0.007 MeV. Therefore, the mass of the neutron from above equation becomes 1.008672 mu.

NEUTRON STATISTICS AND SPIN

Since the deuteron is a loose combination of proton and neutron and is supposed to have a spin 1. From the study of band spectra of deuterium and the proton spin (1/2) the neutron spin comes out to 1/2. Neutrons have been found to obey Fermi statistics. The experimental fact that the magnetic moment of the deuteron is nearly equal to the algebraic sum of the magnetic moments of the neutron, and proton strongly supports this value. This value of 1/2 for the spin of the neutron has been confirmed by reflection of neutrons from magnetised mirrors by Hughes and Burgy.

DECAY OF THE NEUTRON

The first accurate determination of the mass of neutron by Chandwick and Goldhaber showed that the neutron mass was greater than that of a proton. It was also evident that the neutron outside the nucleus is unstable and could be expected to decay with the emission of a beta-particle plus a neutrino, leaving a proton. It was also assumed that a similar process accounted for the β-emission of β-unstable radioactive nuclei.

$$n \rightarrow p + e^- + \overline{\nu} + 782 \text{ keV}.$$

Using the neutrons from the Oak Ridge pile, A.H. Snell and his coworkers (1948) were able to detect proton coming out from the beam of neutrons. The presence of electrons were also shown by them after some time. J.M. Robson (1951), using the much more intense source of neutrons available from the Chalk River Pile, not only observed the decay of the neutrons but also succeeded in measuring the energy spectrum of the decay electrons.

A collimated beam of pile neutrons from a nuclear reactor was allowed to pass into a vacuum chamber. Some of the neutrons decayed in their fight through the vacuum chamber. The decay protons, with a small fraction of the disintegration energy, were deflected by the deflector (high voltage electrode), maintained at a potential of + 13 kV with respect to the earthed vacuum chamber, into a proton spectrometer and detected with an electron multiplier. The β-rays emitted during this disintegration were deflected to the left into a β-rays spectrometer and were detected with a scintillation counter using anthracene crystals as the phosphor. These two detectors were connected to a coincidence circuit which would register only when pulses from the two detectors arrived there simultaneously. Owing to its greater mass the proton would move more slowly through its spectrometer system and would be delayed with respect to the electrons. Calculations gave this delay as about 0.9 micro second. The rate of proton-beta particle coincidences is plotted against the energy of the β-particles, measured in the β-spectrometer. Extrapolation of the curve gives a value for the maximum energy of the spectrum as 782 ± 13 keV, which is the close approximation of the mass difference between the neutron and the proton.

HALF LIFE OF NEUTRON

The estimate of the half life of the neutron was obtained from a determination of the density of the neutron beam and the number of neutrons decaying per unit time per unit volume. The neutron density was obtained by exposing manganese foils to the neutron beam and measuring their activities against the activities of the calibrated standard foils of identical thickness. At the centre of the beam the value so obtained was 1.6 × 1010 neutrons/m^3. The number of neutrons decaying per unit time per unit volume was estimated from the number of protons per unit time striking the first electrode of the electron multiplier and the volume of the beam where the protons originate. The value so obtained was 6.3 ×

108 neutrons/min m^3 of the neutrons beam. Therefore, the half life T becomes

$$T = 1.16 \times 10^{10} \times 0.693/6.3 \times 10^8 = 12.8 \text{ min.}$$

Robson's observations gave a value for the half life T as 12.8 ± 2.5 min, Recently, a more accurate determination of T by a Russian group by a similar method gave the value T = 11.3 ± 0.3 min. An alternative method was used by D. Angelo at the Argonne National Laboratory in the same year. His result is

$$T = 12.7 \pm 1.9 \text{ min.}$$

Wave Properties

For non-relativistic velocities (v << c), the wavelength associated with neutrons is given by

$$\lambda = \frac{h}{p} = \frac{h}{(2ME)^{1/2}} = \frac{2.86 \times 10^{-11}}{E^{1/2}} \text{ metre,}$$

where E is in eV. Thus for thermal neutrons A = 1.82 A° which is comparable with atomic dimensions. For E = 1 MeV, $\lambda = 2.86 \times 10^{-14}$ metre approaching nuclear dimensions. The relativistic correction is important above energy of 100 MeV. The wave properties are of primary importance in determining the nature of the interaction between neutrons and nuclei.

Magnetic Moment of the Neutron

It is sometimes said that the neutron should not have a magnetic moment because it has zero charge. Actually magnetic moments depend upon currents rather than charges. For example the hydrogen atom has no net charge but does have a magnetic moment. Thus neutron can be supposed to have a magnetic moment if it contains equal +ve and –ve charges which are in motion. According to Yukawa's meson field theory neutron transforms into a proton and a negative π-meson for a fraction of time. Since this meson is lighter than the proton, its motion will be faster and the resultant magnetic moment is that of revolving a negative charge. Hence it is negative. The proton and π-meson also possess spin and, therefore, have intrinsic magnetic moments, that of the meson being large due its small mass. This situation also contributes to a negative moment of the neutron.

Nuclear resonance method in the original form cannot be applied here because it is not possible to have neutrons in large densities just like protons. The value of magnetic moment of neutron was measured with considerable accuracy by Alvarez and Bloch in 1940. They used essentially the magnetic resonance beam method employing solid magnetized steel plates as polarizer and analyzer in place of the magnets.

In an experiment used by Bloch and others, neutrons produced in a beryllium target by cyclotron accelerated deuterons are slowed down in a paraffin block, then emerged through a hole in a 9-in-shielding wall and moved as a collimated beam through a sequence of three magnets (all with magnetic fields directed upward). The neutrons in the beam, after passing through the highly saturated iron piece present in the gap of the first magnet, are partly polarized (down) because more spin up than spin down neutrons have been scattered out.

In the absence of second magnet, some of the neutrons in the beam when passing through the iron piece present in the gap of the third magnet, are scattered out of it. The amount of reduction of the beam intensity depends on the magnitude of the polarization previously attained. The relative loss in the third magnet is less than in the first since more than half of the neutrons in the beam hitting the scatterer have spin down. An R.F. field with field direction perpendicular to the main field is produced in the gap of the second magnet. When this field equals the Larmor frequency of the neutrons in this magnet such that the neutron spins can be flipped, the effect is to destroy fully or partly the beam polarization and the result is a reduction in intensity observed at the counter. For a constant field in the precession magnet, the frequency of the R.F. field was varied and the quantity

$$E_T = \frac{(I_{on} - I_{off})}{(I_{off} - I_{bg})} \quad \text{...(1)}$$

was measured. Here I_{on} and I_{off} are the counting rates with the R.F. field on and off respectively and I_{bg} is the background counting rate.

Instead of attempting to measure directly the field induction B in the precession magnet and thus determining μ, Bloch and others measured the ratio $|\mu_n|\mu_n$ $(= \nu_n/\nu_p)$. Here ν_n and ν_p are the larmor frequencies in the same field for neutron and proton respectively. They made use of the fact that the ratio $\nu_p \; \nu_n \sim 2/3$. The signal used in the proton resonance apparatus was the sixth harmonic of a master oscillator of frequency

$\nu_p' = \nu_p/6$. Similarly the frequency of another mass oscillator was $\nu_n' = \nu_n/4$. We thus have

$$\frac{|\mu_0|}{m_p} = \frac{\nu_n}{\nu_p} = \frac{4\nu_n'}{6\nu_p'} = \frac{2}{3}\left(1 + \frac{\Delta\nu_n'}{\nu_p'}\right) \quad ...(2)$$

Here $\Delta\nu_n'\left(= \nu_n' - \nu_p'\right)$ is the difference between the two master oscillators. It was measured very accurately. A typical neutron resonance observed by Bloch. For this curve we have maximum R.F., field BRF = 11.5 gauss. $\nu_p' = 7.674$ Mc/sec and $\Delta\nu_n' = 210.5$ kc/sec. at resonance. We thus have

$$\frac{|\mu_n|}{\mu_p} = \frac{2}{3}\left(1 + \frac{0.2105}{7.674}\right) = 0.6850 \quad ...(3)$$

Using the results for μ_p, obtained in chapter 1, we have

$$|\mu_n| \simeq 1.91303 \text{ nuclear magneton.}$$

A better method, achieving a much higher degree of polarization is to reflect the beam at a very small glancing angle from a magnetized mirror. Ramsay and others used this technique and were able to measure the neutron magnetic moment accurately. Neutron beams from cyclotron induced reaction (d, n) or high flux reactors are convenient sources. The neutrons are reduced to thermal velocities by a hydrogeneous moderator before they are collimated by cadmium sheets. This collimated beam of thermal neutrons (700,000 counts/rein.) is incident at a glancing angle of 17' on a magnetic mirror used as a polarizer. An alloy of 93% Co and 7% Fe is a suitable material for the mirror. The reflected beam has a polarization as high as 90%. The beam of polarized neutrons passes through a 5 ft long uniform magnetic field H_0 (~ 8500 gauss). The beam is then reflected at a small angle from a second magnetic mirror used as a analyzer (magnetized in the same direction as first mirror). The intensity of the doubly reflected beam (~ 7500 counts/min) is reflected by BF_3 proportional counter.

Ramsay has previously shown that the method of using two separated current loops produces narrower resonance lines. The neutron spins in the long magnetic field are flipped by two radio-frequency current loops mounted 42 inches apart. Resonance flipping will occur when the angular frequency of the radio-frequency supply coincides with the Larmor angular frequency of the neutron spins. The spin flipping reverses the polarization of the beam. Thus the resonance is revealed by a sharp decrease in the number of neutrons detected by the counter. Similar to the molecular beam resonance experiments the resonance condition for spin half particles is given by

$$2m_nB_0 = \hbar\omega \text{ or } \mu_n = \frac{\hbar\omega}{2B_0}. \quad ...(4)$$

The magnetic field B_0 is conveniently measured using a proton resonance probe. The value of the neutron magnetic moment obtained by Ramsay, Corngold and Cohen is

$$\mu_n = -1.913148 \pm 0.000066 \text{ nm.}$$

The – ve sign indicates that the magnetic moment of the neutron is similar to that generated by rotating – ve charge with its spin parallel to that of the neutron.

Since the deuteron is believed to consist of a proton and a neutron with parallel spins, the magnetic moment should be equal to the sum of the separate nucleon moments. But this sum differs from deuteron moment by 0.022 nuclear magneton. This discrepancy may be due to following reasons:

(a) The deuteron spends part of its time in a state of higher orbital momentum, in which its magnetic moment would be definitely less than the ground state.

(b) There would be a difference in the magnetic moments of the nucleons in the free and combined conditions.

CLASSIFICATION OF NEUTRONS AS TO ENERGY

The neutron energy can be determined by measuring the ranges projected by them in an ionisation chamber. For more detailed examination of the results of the interactions of neutrons with matter it is also desirable to classify them according to their kinetic energy.

1. Ultra Fast Neutrons

Neutrons with energies beyond 50 MeV are called *ultra high energy neutrons*. They are produced in the target by p-n interactions induced in nuclei by high energy protons. The cosmic radiation is also a source of neutrons with energies well above those which are likely to be producer by accelerations.

2. Slow Neutrons

Neutrons with energies from zero to about 1000 eV are usually included in this category. In this range sub-classifications occur. The most important are given below.

(a) Epithermal Neutrons

Consider an arrangement in which a fast neutron source is placed inside a moderator which slows the neutrons down until they are in equilibrium with the molecules of moderator. Before establishing the thermal equilibrium, the distribution of velocities will contain velocities which exceed any permitted by a Maxwell distribution for the temperature of the moderator. Such a distribution is called *epithermal* and the neutrons in it are called *epithermal neutrons.*

(b) Cold Neutrons

Neutrons having their average energy less than thermal neutrons are called cold neutrons. These neutrons are produced by a device depending on the coherent scattering. From the Bragg law ($n\lambda = 2d \sin \theta$) we have no reflection if the neutron wavelength λ exceeds 2d, where d is the grating spacing of the crystal. Thus the neutrons with the lower energies are transmitted. Since 2d for graphite is 6.7 A° the computed maximum energy of the transmitted neutrons is 0.002 eV, which is well below the average energy of thermal neutrons.

(c) Resonance Neutrons

Since the various nuclei exhibit strong absorption of neutrons at fairly well defined energies in the range of energies between 1 to 100 eV. These absorptions are known as resonance absorptions and the neutrons having the corresponding energies are known as resonance neutrons.

(d) Thermal Neutrons

When fast neutrons have been slowed down until the average energy of the neutrons is equal to the average thermal energy of the atoms around them, the neutrons are called *thermal neutrons.* At each collision with these atoms, the neutron may gain or loss energy. They are in thermal equilibrium with their surroundings and they may be assumed to have a velocity distribution analogous to Maxwell distribution for the molecules of a gas.

$$dn(v) = Av^2 e^{-Mv^2/2kT} dv, \quad ...(5)$$

where v is the neutron velocity, M its mass and T the absolute temperature. The maximum number of neutrons will have the energy kT, which is about 0.025 eV at 20°C.

3. Very Fast Neutrons

This energy interval (10-50 MeV) is distinguished from the proceeding by the appearance of nuclear reactions involving the emission of more than one product, such as the (n, 2n) reaction.

4. Intermediate Neutrons

Intermediate neutrons in the energy region between 1000 eV and 0.5 MeV are obtained by the deceleration of fast neutrons. We have less informations about intermediate neutrons than about slow neutrons because of the difficulty of finding efficient detectors. In this energy range elastic scattering process is dominant. Recently a number of techniques have been developed for the study of this energy range.

5. Fast Neutrons

Neutrons with energies having range between 0.5 to 10 MeV are called fast neutrons. This energy region is characterised by the appearance of many nuclear reactions which are energetically impossible at lower neutron energies of which the most important is inelastic scattering.

NEUTRON DETECTION

Neutrons are uncharged particles, therefore, produce very little direct ionization about one ion pair per metre of path in passing through matter. They cannot be detected in any instrument whose action depends on the ionization due to the entering particle. For reasons of practical efficiency, the secondary particles must be generated at energies which can produce ionization conveniently detectable. The secondary charged particles may be

(a) the protons released by collisions of neutrons with hydrogen nuclei,

(b) the direct result of nuclear disintegration produced by neutrons,

(c) the radioactive disintegrations from product nuclei which become radioactive as a result of neutron capture or

(d) the strongly ionizing fission fragments produced by the absorption of neutrons in fission reactions.

Two general principles have been employed in the detection of neutrons, first, use is made of the charged particles produced by the interaction of neutrons with various substances introduced into the counter, and second, advantage is taken of the recoil of nuclei of light elements after being struck by neutrons. For the detection of slow neutrons the first method is used while for fast neutrons we use second method.

1. Slow neutron detection

(a) Boron detectors

Detection process is based on the reaction.

$$_{5}B^{10} + {}_{0}n^{1} \rightarrow {}_{3}Li^{7} + {}_{1}He^{4} + 2.78 \text{ MeV}.$$

For thermal neutrons (0.025 eV) the cross-section for this reaction is 3770 barns. The 2.78 MeV reaction energy being carried off by the resulting lithium nucleus and alpha particles which produce considerable ionization in their tracks. However, if the target is natural boron the effective cross-section is reduced to 710 barns as the abundance of the B^{10} isotope is only 18 83%. Better results are obtained if the boron compound is enriched in B^{10} isotope. The cross-section for the above reaction follows the 1/v law and falls off with increasing neutron energy. Hence these counters are useful for slow neutrons. Ionization chambers with boron coated electrodes or proportional counters filled with boron trifluoride vapour (enriched in B^{10}) are in common use. Full advantage of the proportional counter is taken to discriminate neutron induces pulses from smaller background gamma pulses.

The BF_s proportional counter tube has been modified by Hanson and Mc Kibben (1947) to make it sensitive to fast neutrons in the energy range 10 keV to 3 MeV. This counter is called *long counter*. The long (10") thin (diameter 1/2") cylindrical BF_3 proportional counter tube is embedded in a cylinder of paraffin (dia 8"). Around this paraffin is a layer of an absorber for slow neutrons which is covered with an outer layer of paraffin. Fast neutrons are moderated (*i.e.*, they are rapidly reduced to thermal energies by elastic collisions with the nuclei of the moderator) in the paraffin wax. The neutrons incident at an energy of about 100 keV have a slightly better chance of detecting than those at lower energy. Experimental tests have verified the expectation that the long counter has an efficiency for the detection of neutrons which is practically constant over a considerable range of neutron energies. The detector efficiency for neutrons incident on the 8 inch circle on the front face is less than 1 per cent.

(b) Fission Chamber

The relatively high cross-sections for fission by thermal neutrons of U^{235}, Np^{237} and Pu^{239} has led to the development of thermal neutron detectors. The fission fragments have large energies (> 50 MeV) and if one of them escapes from the thin surface layer of the fissionable material into a gas of the chamber, a large, easily distinguished, ionization pulse is generated. A simple type of fission chamber for observing neutrons thus consists of an ionization chamber for which one electrode is coated with uranium oxide, preferably enriched in the uranium 235 isotope. There are following limitations on the design of fission counters :

(i) One is imposed by strong a-particle background-always present in fission chambers.

(ii) Another involves the short range of the fission fragments.

To have the pulses produced by the fission fragments much larger than the pulses from the α-particles, the sensitive layer of fissionable material introduced into the counting chamber must be quite thin (~ 100 micrograms/cm^2), also the resolution of the recording system must be high to reduce the effect of coincident α-particle pulses.

(c) Activation Method

The important discovery made by Fermi, that many elements exposed to a flux of the thermal neutrons become radioactive, has led to a useful method of detecting allow neutrons. The feasibility of detecting the induced radioactivity depends on its life time, which cannot be appreciably shorter than the time which must elapse between exposure to neutrons and measurement of the induced activity, on the other hand, the life time must not be so long that the decay rate is negligible. A number of elements have a large activation cross-section for the (n, γ) reaction. Thin foils of a suitable element (manganese, gold, indium and rhodium are often used) are exposed to a neutron flux for a time interval greater than six or seven times the half life of the induced activity. They are then removed and the extent of the induced activity is determined by counting the emitted radiations with an appropriate Geiger counter, ionization chamber, scintillation counter or other radiation detector. Indium has a large cross-section for the reaction at low energy range, therefore, is a very efficient detector for slow neutrons at the energy. One method of estimating the thermal component of a neutron flux is to measure the induced activity in a bare indium foil and other wrapped in cadmium foil (~ 1 mm. thick). For low energy neutrons (0.3 eV} cadmium is nearly opaque (the cross-section for the reaction is several thousand barns but it does not yield a radioactive isotope) whereas for epithermal neutrons the Cd foil is nearly transparent.

Szilard and Chalmers in 1934 devised a technique for separating radio-nuclei, induced by n, γ reaction, from their isotopic environment. The γ-rays emitted from product atoms in the (n, γ) reaction can provide recoil energy E_γ sufficient to break the chemical bond between the radioactive atoms and thereby to change the chemical state of the product nucleus as compared to the normal nucleus in the medium. The recoil energy in eV from the emission of a γ-ray of energy E_γ in MeV is given as

$$E_r = 536\frac{E_r^2}{A} \quad ...(6)$$

where A is the mass number of the recoiling atom. For E_γ = 7.5 MeV and A = 100, the recoil energy is 300 eV which is fairly large than necessary to disrupt chemical bonds (1 – 5 eV). The above reaction known as *Szilard-Chalmers* reaction is used to obtain radioactive samples of high specific activity, especially when only relatively weak neutron source were available.

(d) Scintillation Counter

Considerable work has been done by various investigators to develop scintillation counters for slow neutrons. Because neutrons cannot produce scintillation directly, the methods used are based on the addition to the scintillator material of an element which interacts with neutrons to yield particles capable of producing scintillations directly. One of the early methods involved the preparation of alternate thin layers of a boron compound and ZnS. The α-particles released by the neutrons from the boron were detected by the well known ability of activated ZnS. The α-particles were relatively insensitive to γ-rays, they did not have a very satisfactory efficiency for the detection of neutrons. Not all ionizing particles from the disintegration of the boron succeeded in producing scintillations and only a fraction of the light from the scintillations reached the photo-multiplier, much of the light being lost by reflection.

Both these difficulties can be corrected to some extent by fusing the ZnS in a glassy layer of boric anhydride. The resulting water is applied directly to the window of the photomultiplier tube. When large volume detectors are required, liquid scintillators containing boron compounds are often used. Methyl borate dissolved in terphenyl and phenyl-cyclohexane (main solvent) is an efficient detector of slow neutrons.

2. Intermediate Neutron Detection

Cross-sections m this range of neutron energies are generally so low that to yield adequate intensities of ionizing radiations thick layers of the sensitive elements are required. Sensitive elements which emit γ-rays are not subject to the same limitation. Rae and Bowey (1953) have designed a detector which utilizes the soft gamma rays from the reaction

$$_5B^{10}\ (n, \alpha)\ _3Li^{7*} \rightarrow\ _3Li^7 + \gamma(480\ \text{keV})$$

for measuring neutrons of intermediate energies. Layers of $_5B^{10}$ which are opaque to neutrons with energies upto about 1 keV may then be used as the detector. By using a NaI phosphor with a rapid rate of rise of the luminous response in a single, channel scintillation spectrometer, the disturbing background from other radioactive sources and the cosmic radiation was materially reduced. In addition a thorough system of lead screening and neutron shields were provided. The detector consists of two photomultipliers, each with NaI phosphor, faced the boron detector from opposite sides. As the amount of boron used is no longer opaque to neutrons above 1 keV, a drop of ~ 15% in the efficiency is expected at 5 keV and of ~ 1% at 50 keV.

3. Fast Neutron Detection

(a) Proton-recoil Detector

This method for detecting fast neutrons is based on the observation of the ionization produced by the recoiling protons in the elastic scattering of neutrons by hydrogeneous materials. For this purpose a proportional counter may be filled with hydrogen or deuterium, but a better procedure is to use argon, or one of the heavier inert gases as a filling gas and to place a thin sheet of a hydrogeneous material such as paraffin at one end of the chamber. Fast neutrons striking the target (paraffin) cause the ejection of protons, which produce ionization in their paths through the counter and so can be detected. If neutron's energy is not large, the recoil pulses will be masked by the small pulses caused by γ-rays or cosmic rays. In 1954 Sun and Richardson designed a multiple-wire proportional counter contained 1.2 atmospheres of methane and operated at 3400 volts. This has beer found to be almost completely insensitive to γ-radiations when operated at an over all efficiency of 0.17 percent for detecting neutrons of energies ranging from 0.3 to 10 MeV.

(b) Fission Process

The isotope U^{235} has a very large cross section for slow neutrons, where as U^{288} has a large cross section for fast neutrons. Hence a fission chamber for the detection of fast neutrons can be designed using a uranium compound which has been depleted in the lighter isotope. This device is useful for fast neutron detection only. For both slow and fast neutrons a mixture of U^{236} and U^{238} is used. The slow neutron component can again be cut out if desired, by shielding the counter either with cadmium or boron.

(c) Scintillation Counter

Hornyak (1952) designed scintillators which were effective for fast neutrons by molding a scintillating phosphor which was sensitive to recoiling protons, such as ZnS, into a plastic serving as a fairly dense radiator. Another approach to this problem has been suggested by McCrary, Taylor and Bonner (1955).

They have found that isolated spheres of anthracene will yield brighter scintillation from recoil protons than from the electrons released by γ-rays if the diameter of the anthra-cene spheres is in the range from 3 to 8 mm. The same secondary electron with energy of the order of 5 MeV can be prevented from exciting more than one sphere if several millimeters of glass or quartz separate the individual spheres. Taylor.

Lonsjo and Bonner in 1954, used an ingenious scintillation detector of fast neutrons. A number of small plastic scintillating spheres is immersed in a non-hydrogeneous fluid ($C_6Cl_3F_3$) of nearly the same refractive index. The liquid is a non-scintillator. It optically couples the plastic spheres via a quartz rod to the 4 cm diameter cathode of a photomultiplier tube. Since γ-rays produce small pulses, the neutron pulses are easily distinguished from γ-pulses.

The optimum diameters of the plastic spheres for detecting neutrons of kinetic energy 3.5, 4.5, 7.0 and 14 MeV are 1.5, 2.5, 6.0 and 10 mm respectively.

(d) Photographic process

The nuclear track emulsion is an excellent detector for the protons which recoil from neutrons in the neutron-proton-scattering process when the neutron energy exceeds about 2 MeV. Fast neutrons can be observed by photo graphing the tracks of recoil protons in a hydrogen water vapour filled cloud chamber.

4. Ultra-High-Energy Detection

The efficiencies of detectors of neutrons with energies greater than 50 MeV are usually low. Proton recoils from a thin hydrogeneous target may be measured with telescope consisting of three proportional counter tubes in line. The telescope can be made insensitive to protons generated by neutrons of energies lower than some arbitrary value by introducing suitable absorbers in between two telescopes.

The advantage of this telescope is that the total range of the recoil protons can be measured and the energy of the neutrons can be deduced from this measurement. Because neutrons of energies greater than 50 MeV can induce fission in some of the heavy stable elements (Bi), fission chambers made with such elements as electrodes may be used to detect these ultra-high-energy neutrons. Liquid scintillators are also used for the detection of ultra-high-energy neutrons.

5. Threshold Detectors

Numerous reactions of neutrons with nuclei have definite thresholds for the energies of the neutrons below which the reactions do not occur. These thresholds extent from a few tenths of 1 MeV to 20 MeV and higher involve either (n, 2n) or charged particle reactions. For example the reaction ($_6C^{12} + {_0n^1} \rightarrow {_6C^{11}} + 2{_0n^1}$) does not occur unless the neutron energy is greater than 22 MeV, the threshold energy of the reaction. Thus we see that $_6C^{12}$ can be used as a threshold detector for neutrons whose energies exceed 22 MeV.

SLOWING DOWN NEUTRONS

The neutrons emitted in nuclear reactions have an energy of a few MeV. Since neutrons are not electrically charged they cannot loose energy by ionization but are slowed down by successive scattering in matter until they reach in thermal equilibrium with their surroundings. After this they diffuse like gaseous molecules until they are captured by nuclei.

A. Energy Loss

In 1934, Fermi and his colleagues showed that the neutrons are slowed down in the hydrogeneous materials and the slower neutrons have a greater probability of inducing radioactivity than do more energetic neutrons. The amount of energy that a neutron losses in a single collision can be calculated by using the equations of conservation of energy and momentum. Another method, which is very simpler involves the use of two reference systems. The first is the L-system, in which the target nucleus is initially at rest and is approached by the incident neutron. The second is the C. M. system in which the centre of mass of the colliding particles is at rest initially and throughout the collision. *In the L-system,* before the collision, the neutron of mass m moves towards the nucleus of mass M, which is assumed to be at rest, with speed ν_0. The total linear momentum of the colliding particles is zero in C. M. system. If ν_c be the velocity of incident particle in this system and V that of the target nucleus, then

$$m\nu_c - MV = 0. \quad ...(1)$$

The velocity of the target nucleus in the C.M.-system is equal and opposite to the velocity of the C. M.-system in the L-system. Thus we can write

$$\nu_0 = \nu_c + V. \quad ...(2)$$

$$\therefore \quad V = \frac{m\nu_0}{(m+M)}$$

and
$$\nu_c = \frac{M\nu_0}{(m+M)} \quad ...(3)$$

If the ratio of moderator mass to neutron mass M/m is called A, the mass number of the moderator nucleus, the above relations can be written as

$$V = \frac{\nu_0}{(1+A)} \text{ and } \nu_c = \nu_0 \frac{A}{(1+A)}. \quad ...(4)$$

Principle of conservation of energy follows that the speeds of the colliding particles are not changed after the collision in C. M.-system. If the neutron moves at an angle ϕ with its initial direction after the collision, to conserve total momentum, the nucleus must move off at an angle $(180 + \phi)$ with the direction of the incident neutron. Thus the total effect in the C.M.-system is a rotation of the particle system by an angle

ϕ only. In the L-system, the neutron is scattered through an angle θ, has a velocity v_1 and the nucleus has a velocity v_2. The v_1 is the vector sum of the neutron velocity in C.M.-system and the velocity of C.M.

$$v_1^2 = v_c^2 + V^2 + 2V_{vc}\cos\phi = v_0^2 \frac{A^2 + 2A\cos\phi + 1}{(1+A)^2}.$$

If E_0 is the neutron energy in the L-system before the collision and E_1 the neutron energy in the same system after the collision. Then we can write

$$\frac{E_1}{E_0} = \frac{v_1^2}{v_0^2} = \frac{A^2 + 2A\cos\phi + 1}{(1+A)^2} \quad \text{...(5)}$$

This relation indicates that $E_1 = E_0$ for a grazing collision ($\phi = 0$), which means that no energy is transferred from the neutron to the moderator nucleus and $E_1 = \alpha E_0$ for a head on collision ($\phi = \pi$), which means that maximum energy is transferred from neutron to the moderator nucleus. Here

$$\alpha = \frac{(A-1)^2}{(A+1)^2}. \quad \text{...(6)}$$

Thus in a single collision maximum possible energy loss by the neutron

$$(\Delta E)_{max} = E_0 - (E_1)_{min} = E_0(1 - \alpha).$$

$$\therefore \text{ Maximum fractional energy loss } \left(\frac{\Delta E}{E_0}\right)_{max} = 1 - \alpha \quad \text{...(7)}$$

Since α depends on A, the maximum loss of K. E. depends on the atomic weight of nucleus encountered. For example, for hydrogen A = 1 and $\alpha = 0$ and the maximum loss of kinetic energy is E_0.

For larger values of A, α can be written as

$$\alpha = \left(\frac{1-1}{A^2}\right)\left(\frac{1+1}{A}\right)^{-2} = \left(\frac{1-2}{A}\right)^2 \simeq \frac{1-4}{A}.$$

$$\text{or Maximum fractional energy loss} = I - \alpha = \frac{4}{A}. \quad \text{...(8)}$$

Thus we see that the light nuclei are more effective moderators than heavy nuclei and the maximum loss of energy is proportional to the initial energy.

B. Scattering Angles in L-system and C. M. System

The scattering angle ϕ in the C.M.-system can be related to the scattering angle θ in the laboratory system as :

$$v_1 \cos \theta = v_c \cos \phi + V. \qquad ...(9)$$

Substituting here values of v_1, v_c and V, we have

$$\cos \theta = \frac{1 + A \cos \phi}{\left(A^2 + 2A \cos \phi + 1\right)^{1/2}}. \qquad ...(10)$$

When the collision is against a very heavy nucleus A >> 1, the second and third term in the denominator of this equation can be neglected compared with A^2. Thus we can write

$$\cos q = \cos \phi + \frac{1}{A}. \qquad ...(11)$$

Thus the scattering angle is same in two systems if A + 1.

C. Average Energy Loss Per Collision

Before we proceed to calculate the average energy that a neutron is going to loss per collision we should remember the experimental fact that in the C.M. system all scattering angles are equally probable. From this empirical law we can now deduce the probability of occurrence of a final kinetic energy E_1. To find the probability of a neutron being scattered through an angle lying between ϕ and $\phi + d\phi$, we describe a unit sphere about the scattering centre O and evaluate the corresponding space angle $d\omega_\phi$, as

$$d\omega_\phi = 2\pi \sin \phi \, d\phi.$$

Since the probability of a particle being scattered in such a way that it passes through area dS is equal to the fraction of the projected area on to the unit sphere as compared to the total surface area of this sphere. Hence.

$$dW = \frac{d\omega_\phi}{4\pi} = \frac{1}{2} \sin \phi \, d\phi$$

$$= -\frac{1}{2} d\left(\cos \phi\right). \qquad ...(12)$$

The eqn. can be written in terms of a as

$$\cos\phi = \frac{1}{1-\alpha}\left[\frac{2E}{E_0} - (1+\alpha)\right]$$

$$\therefore \qquad dW = -\frac{dE}{E_0(10-\alpha)}. \qquad ...(13)$$

Thus we see that the probability that the neutron K.E. after the collision will lie in an interval dE centered around a kinetic energy E after collision is independent of this energy. In other words we can say that *all values of E are equally probable after a collision.* In eqn. (13) dW appears negative because dE is itself negative, representing, an energy loss. The probability per unit energy loss

$$P(E) = -\frac{dW}{dE} = \frac{1}{E_0}(1-\alpha). \qquad ...(14)$$

The average energy loss of the scattered neutrons after one collision

$$\Delta E = \frac{\int_{\alpha E_0}^{E_0} (E_0 - E)P(E)dE}{\int_{\alpha E_0}^{E_0} P(E)dE}.$$

Since the sum of the probabilities over all values of the energy is equal to unity, hence

$$\Delta E = \int_{\alpha E_0}^{E_0} (E_0 - E)P(E)dE = \frac{1}{2}E_0(1-\alpha).$$

$$\therefore \qquad \text{Average fractional energy loss } = \frac{1}{2}(1-\alpha). \qquad ...(15)$$

Thus we see that the same fraction of neutron energy is transferred to the moderator nucleus in each successive collision.

D. Transport Mean Free Path and Scattering Cross Section

The scattering angles in the L-system are smaller than the scattering angle in the C.M.-system.

It means in effect that the neutrons show a preferential forward scattering in the L-system, although the scattering is spherically symmetrical (isotopic) in the C.M. system.

The deviation from spherical symmetry is expressed in terms of the average value of cosines of all possible angles.

$$\cos\theta = \frac{\int_{\alpha E_0}^{E_0} \cos\theta \, P(E)dE}{\int_{\alpha E_0}^{E_0} P(E)dE} = \frac{\int_1^{-1} \cos\theta \left(-\frac{1}{2}\right) d(\cos\phi)}{\int_1^{-1} \left(-\frac{1}{2}\right) d(\cos\phi)} = \frac{2}{3A}.$$

Thus we see that for large A the average forward component becomes very small and scattering is almost isotropic. The forward scattering is predominant in the collisions with lighter nuclei. The predominant forward scattering in the L-system also affects the mean free path in the sense that the average distance travelled by the neutron before it is scattered through an angle 90° is greater than the corresponding average distance for isotropic scattering. The increased effective mean free path for nonisotropic scattering is called the *transport mean free path* λ_{ir} and is related to the *scattering mean free path* λ_s by the relation

$$\lambda_{tr} = \frac{\lambda_s}{1-\cos\theta} = \frac{\lambda_s}{1-\frac{2}{3A}}. \qquad ...(16)$$

$$\therefore \text{ Transport cross section } \sigma_{tr} = \frac{1}{N_0\lambda_{tr}} = \sigma_s\left(1-\frac{2}{3A}\right) \qquad ...(17)$$

E. Average Logarithmic Energy Decrement

We shall now find the average value of decrease of the natural logarithm of neutron energy in a collision or *average logarithmic energy decrement,* usually denoted by ξ. According to the definition

$$\xi = \Delta \log E = \log E_0 - \log E = \log \frac{E_0}{E}. \qquad ...(18)$$

This is done by dividing the probability of occurrence of a certain event by the probability of occurrence of all events of that type. If P(E) dE be the probability for a neutron to be scattered, in one collision, into an energy range lying between E and E + dE, then the probability that the logarithmic decrement log (E_0/E) be the one corresponding to a final energy E, when the initial energy was E_0, is given by the product log (E_0/E) P(E) dE. If we assume the energy E to vary between its minimum (αE_0) and maximum (E_0) values, then

$$\xi = \frac{\int_{\alpha E_0}^{E_0} \log\left(\frac{E_0}{E}\right) P(E)\, dE}{\int_{\alpha E_0}^{E_0} P(E)dE}$$

$$= -\int_{\alpha E_0}^{E_0} \log\left(\frac{E}{E_0}\right) \frac{d\left(\frac{E}{E_0}\right)}{(1-\alpha)} = 1 + \frac{\alpha \log \alpha}{1-\alpha}.$$

In terms of the atomic mass number A, this becomes

$$\xi = 1 + \frac{(A-1)^2}{2A} \log \frac{A-1}{A+1}. \qquad ...(19)$$

For large values of A (> 10), it is easy to write

$$\xi = \frac{2}{\left(A + \frac{2}{3}\right)}. \qquad ...(20)$$

Thus we see that for very large A, ξ approaches zero. This is in line with our previous finding that heavy elements are poor moderators. We find a very good agreement if we compare the theoretical value with the experimental results.

When ξ is known, *the average number of collisions needed to bring about a given decrease in neutron energy*, can easily be calculated. If the neutrons start out with a average energy of E_1 and are slowed down to E_2 the total logarithmic energy loss is $\log\left(\frac{E_1}{E_2}\right)$. Then the average number of collisions is given by

$$n = \log \frac{\left(\frac{E_1}{E_2}\right)}{\xi}. \qquad ...(21)$$

F. Slowing-down Power and Moderating Ratio

For efficient slowing down we not only require a large value of ξ and a large probability of cross section for scattering and large N, the number of atoms per unit volume, but the *absorption cross section must be small* otherwise too many neutrons would be lost by absorption. The three quantities ξ, σ_s and N jointly determine the slowing down ability of a moderator and their product is called the *slowing down power (sdp).*

$$\text{sdp} = \xi\sigma_s N = = \xi\Sigma_s = \frac{\xi}{\lambda_s}. \qquad ...(22)$$

Hence sdp can be interpreted as the average loss in logarithm of the energy per unit distance of travel in the moderator. Above eqn. indicates that a good moderator is one which has large values of Σ and of macroscopic scattering cross section Σ_s. The material chosen for the moderator should have a large sdp and should also have a low cross section for absorption. The ratio of sdp to macroscopic absorption cross section is the quantity which will indicate the better moderator, and is given the name of moderating ratio, represented by mr.

$$mr = \frac{sdp}{\Sigma_a} = \frac{\xi \Sigma_s}{\Sigma_\alpha}. \qquad ...(23)$$

When a moderator is not a one element substance, but consists of a combination of elements, the approximate value of ξ is given by

$$\xi = \frac{\xi_1 \Sigma_s^1 + \xi_2 \Sigma_s^2 + ...}{\Sigma_s^1 + \Sigma_s^2 + ...} \qquad ...(24)$$

$$\therefore \text{ For a compound, } sdp = \xi_1 \Sigma_s^1 + \xi_2 \Sigma_s^2 + ... \qquad ...(25)$$

$$\text{and } mr = \frac{\xi_1 \Sigma_s^1 + \xi_2 \Sigma_s^2 + ...}{\Sigma_a^1 + \Sigma_a^2 + ...}. \qquad ...(26)$$

G. Slowing-down Density

The rate at which the neutrons per unit volume of a moderator slow down past a particular energy E is called the slowing down density and is denoted by q(E). Owing to the discontinuous nature of the slowing down process and of the presence of absorption resonances for neutrons of discrete energies, the variation of slowing down density with energy is not expected to be simple. We can simplify the problem by treating the slowing down as a virtually continuous process, which is possible with sufficiently heavy moderator nuclei for which average energy loss per collision is very small.

If the rate of production of neutrons with an initial high energy E_0 is constant and equal to Q neutrons per unit volume per sec. If no neutrons are lost either by escaping (leakage) or by absorption before they have reached thermal energies, the slowing-down density q(E) will remain constant and will be equal to the source density Q. This is the case of an ideal moderator as we can only reduce the neutron escaping to negligible proportions by using a moderator of thick walls and the absorption is not zero, even if the moderator is infinite in extent. The number of neutrons per unit volume whose energies lie between E and E + ΔE is equal to n(E) Δ E. If dt is the time needed by the neutrons to transit through the energy interval ΔE, the number of neutrons that are within this energy interval at a given time will be given by

$$n(E)\ \Delta E = -\ q(E)\ \Delta t. \qquad ...(27)$$

The negative sign is introduced because energy decreases with the increase of time. Here Δ(E) A E represents the neutron density crossing E per collision. The number of collisions this neutron group makes per second $= v/\lambda_s = v\Sigma_s$. Hence, the neutron density that crosses the energy E per second

$$q(E) = \{n(E)\ \Delta\ E\} \times v\Sigma_s. \qquad ...(28)$$

In the event of small energy changes in collision we can write

$$\frac{\Delta E}{E} = \Delta(\log E) = \xi \text{ or } \Delta E = E\xi.$$

$$\therefore \qquad q(E) = n(E)\ E\ \xi_v \Sigma_s, \qquad ...(29)$$

Thus in the absence of neutron absorption [q(E) = Q], the neutron flux per unit energy is given by

$$\phi(E) = n(E)\ v = \frac{Q}{E\xi\Sigma_s}. \qquad ...(30)$$

This expression shows that the slowing down neutron flux per unit energy is inversely proportional to the energy E.

The quantity $\phi(E)\ \Sigma_s$, is called *collision density* and is given by

$$F(E) = \phi(E)\Sigma_s = \frac{Q}{\xi E}. \qquad ...(31)$$

This relation shows that the scattering loss (no. of neutrons scattered out of the energy interval ΔE per sec. per unit volume) is equal to the neutron influx gain (the no. of neutrons scattered into the energy interval ΔE per sec. per unit volume). In the presence of absorption, q(E) will be related to Q as

$$q(E) = Q\ P(E), \qquad ...(32)$$

where P(E) is the probability or fraction of unabsorbed neutrons. The absorption loss $\Delta q = n(E)\ \Delta E\ v\Sigma_a = \phi(E)\ \Delta\ E\Sigma_a$. Hence in the steady state within the energy interval Δ E, Scattering loss + absorption loss = Influx

$$\phi(E)\ \Sigma_s\ \Delta E + \phi(E)\Sigma_a\ \Delta E = q(E)\ \frac{\Delta E}{E\xi}$$

or
$$\frac{\phi(E)\Sigma_a\ \Delta E}{\phi(E)\Sigma_a\ \Delta E + \phi(E)\Sigma_s \Delta E} = \frac{\Delta q E\xi}{q(E)\Delta E}$$

or
$$\left[\frac{\Sigma_a}{(\Sigma_a + \Sigma_s)}\right]\left[\frac{\Delta E}{E\xi}\right] = \frac{\Delta q}{q(E)}. \qquad ...(33)$$

H. Slowing Down Time

The time required for the neutrons to slow down from an initial energy E_0 to a final energy E is known as the slowing down time, can be obtained by eqns. (33) and (34).

$$\therefore \qquad \Delta t = -\frac{\Delta E}{E\xi\ v\Sigma_s}. \qquad ...(34)$$

If we assume that ξ and Σ_s remain constant over the slowing down range of energy, the above expression can be integrated as

$$\text{Slowing down time } T = \int_0^T dt = -\frac{1}{\xi\Sigma_s}\int_{E_0}^{E_f} \frac{dE}{Ev}$$

$$= -\frac{\left(\frac{1}{2}m\right)^{1/2}}{\xi\Sigma_s}\int_{E_0}^{E_f} E^{-3/2} dE$$

$$= \frac{(2m)^{1/3}}{\xi\Sigma_s}\left[\frac{1}{E_f^{1/2}} - \frac{1}{E_0^{1/2}}\right] \qquad ...(35)$$

With the help of this relation we can also calculate the average slowing-down time of neutrons for various moderators.

ENERGY DISTRIBUTION OF THERMAL NEUTRONS

The slowing-down process is due to an elastic scattering process as emphasised earlier. The moderating material, used for this slowing down process, should have a large scattering cross-section and small absorption cross-section for neutrons. As a result of the slowing-down process, neutrons reach the state in which their energies are in equilibrium with those of the atoms or molecules of the moderator in which they are moving. At this time their energy distribution will be approximately

Maxwellian, corresponding to the temperature of the surrounding medium. In thermal equilibrium the velocity distribution of neutrons will be given by the expression

$$dn = n(v)\ dv = 4\pi n \left(\frac{m}{2\pi kT}\right)^{3/2} v^2 e^{-mv^2/2kT} dv, \qquad ...(1)$$

where n is the total number of neutrons per unit volume, m is the mass of the neutron, T the absolute temperature and k the Boltzmann constant. The number of neutrons whose velocities lie between v and v + dv is given by dn = n(v) dv. The velocity distribution function has a maximum for a value of the velocity v_p, which is called the most *probable velocity*. It can be determined by the usual procedure of differentiating the right hand side of eqn. (42) with respect to v and setting equal to zero. In this way we have $v_p = (2kT/m)^{1/2}$. The energy corresponding to this velocity is denoted by E_p and is given by

$$E_v = \frac{1}{2} m v_p^2 = kT. \qquad ...(2)$$

The energy distribution of the neutrons is given by

$$dn = n(E)\ dE = 2\pi n\ (\pi kT)^{-3/2}\ E^{1/2}\ e^{(-E/KT)}\ dE, \qquad ...(3)$$

this distribution leads to a most probable energy E_0, which can be found by differentiating eqn. (3) and equating this to zero. The value thus obtained is 1/2kT, which is one half of the energy E_p, the energy corresponding to the most probable velocity. The velocity corresponding to this energy is known as average velocity and is given as

$$\bar{v} = \left(\frac{2}{\pi^{1/2}}\right) v_p = 1.128\, v_p.$$

The average energy E is given by kinetic theory as

$$\bar{E} = \frac{3}{2} kT = \frac{3}{2} E_0. \qquad ...(4)$$

This average energy does not correspond to the average velocity but to the root mean square velocity v_{rms}, given by

$$\bar{E} = \frac{1}{2} m v^2 = \frac{1}{2} m v_{rms}^2$$

or

$$v_{rms} = \left(\frac{3kT}{m}\right)^{1/2} = \left(\frac{3}{2}\right)^{1/2} v_p. \qquad ...(5)$$

The velocity distributions of thermal neutrons from different sources have been measured experimentally. For various experimental reasons, highly accurate measurements are hard to make, but the measured velocities do follow Maxwell distribution within the experimental accuracy of about 10%. This discrepancy corresponds to a temperature which is slightly greater than that of the moderator material. The steady influx of high energy neutrons and at the same time a steady absorption of the low energy neutrons raises the high energy end of the Maxwell distribution and depresses the low energy portion of the distribution.

NEUTRON DIFFUSION

Diffusion is the process that takes place when particles of one type move between particles of another type into which they are penetrating. In the previous article we focused our attention on the energy loss suffered by the neutrons as a consequence of collisions with the nuclei of the moderator. It is often necessary to know the spatial distribution of the neutrons, *i.e.,* the dependence of the neutron density on position. The problem can be treated with the aid of the theory of diffusion. This is not new in physics but is similar to the problem of the diffusion of a gas through another gas or to that of the diffusion of electrons though a gas. For the sake of simplicity we make several assumptions (i). The neutron flow does not change with (ii) We assume that neutron absorption is negligibly small (iii) In lab. system, the probability of a particular scattering angle is a constant. (iv) The neutrons are monoenergetic and the scattering is entirely due to elastic collisions. The result of calculations based on these assumptions are in fairly good agreement with experiment.

(a) Neutrons Current Density

If the neutron density is not constant throughout a given volume, neutrons move from regions of higher density to regions of tower density. If we consider a unit area in a volume occupied by a neutron gas. There will be an excess of neutrons passing through this area in one direction as compared to the number of neutrons crossing it in the opposite direction. When the neutron density becomes uniform in the neighbourhood of our unit area, the net flow stops. The net number of neutrons travelling per second through the unit area which is normal to the direction of flow is known as *neutron current density*. It is a vector quantity and can be started in cartesian coordinates as :

$$J = J_x i + J_y j + J_x k. \quad ...(1)$$

According to Fick, neutron current density J and the space rate of change of the neutron density dn/ds are proportional to each other,

$$J = -D\ dn/ds. \qquad ...(2)$$

Here D is the *diffusion coefficient* and negative sign indicates that the neutron current is in the direction of decreasing neutron density. In standard vector notation dn/ds = grad n = ∇n = density gradient. For monoenergetic neutrons flux $\phi = n\nu$, hence

$$J = -\frac{D}{\nu}\frac{d\phi}{ds} = -D_0\,\nabla\phi. \qquad ...(3)$$

Hence D_0 is the diffusion constant. Since grad $\phi = \left(\frac{\partial\phi}{\partial x}\right)i + \left(\frac{\partial\phi}{\partial y}\right) i + \left(\frac{\partial\phi}{\partial z}\right)k$, hence after substituting this value of grad ϕ in eqn. (49) and comparing it with eqn. (1), we have

$$J_x = -D_0\frac{\partial\phi}{\partial x},\ J_y = -D_0\frac{\partial\phi}{\partial y} \text{ and } J_x = -D_0\frac{\partial\phi}{\partial z}. \qquad ...(4)$$

Each of these components represents the net current density along the respective axis, which is the difference between a current density component in the positive direction and a current density component in the opposite direction. Hence

$$J_x = J_x^+ - J_x^-;\ J_v = J_v^+ - J_v^-;\ J_z^+ - J_z^-. \qquad ...(5)$$

It can be shown that current density components are related with neutron flux ϕ and macroscopic scattering cross-section as

$$J_{x_+} = \frac{\phi}{4} - \frac{1}{6\Sigma_s}\frac{\partial\phi}{\partial x},\ J_{x_-} = \frac{\phi}{4} + \frac{1}{6\Sigma_s}\frac{\partial\phi}{\partial x}. \qquad ...(6)$$

$$\therefore \qquad J_x = -\frac{1}{3\Sigma_s}\frac{\partial\phi}{\partial x} = -\frac{\lambda_{tr}}{3}\frac{\partial\phi}{\partial x}. \qquad ...(7)$$

Thus we have $D_0 = 1/2\ \lambda_{tr}$. ...(8)

(b) Neutron Leakage Rate

Let us consider a small volume element dV = dxdydz in a rectangular frame of reference. The neutron leakage rate can now be calculated as the difference between the number of particles entering and the number of particles exiting. Let us first consider a direction parallel to the x-axis. The number of particles crossing the face dy dz is J_x dy dz. After entering they continue to move in the same direction and at a distance dx they encounter the opposite face of volume element. If $\frac{\partial J_x}{\partial x}$ is the rate of change of the current density per unit length along the x-axis, the total number of particles crossing the opposite face of volume element dV will thus be

$$\left[J_x + \left(\frac{\partial J_x}{\partial x}\right) dx\right] dy\, dz.$$

Hence the net leakage rate of neutrons from the volume element in x-direction $= \left(\frac{\partial J_x}{\partial x}\right) dx\, dy\, dz.$

The same calculations lead to the leakage from the volume element in y- and z-directions. Thus

$$\text{Total leakage rate} = \left(\frac{\partial J_x}{\partial x} + \frac{\partial J_y}{\partial y} + \frac{\partial J_x}{\partial z}\right) dx\, dy\, dz$$

$$= - D_0(\partial^2\phi\ \partial x^2 + \partial^2\phi/\partial y^2 + \partial^2\phi/\partial z^2)\ dx\ dy\ dz$$

$$= - D_0\ \nabla^2\phi\ dV = - D\nabla^2 n\ dV. \qquad ...(9)$$

The ∇^2 is used for the sum of second derivatives of a function and is known as the Laplacian. In spherical and cylindrical coordinates respectively it is given by

$$\nabla^2 = \frac{1}{r^2}\frac{\partial}{\partial r}\left(r^2 \frac{\partial}{\partial r}\right) + \frac{1}{r^2 \sin^2\theta}\left[\frac{\partial^2}{\partial \phi^2} + \frac{\partial}{\partial \theta}\left(\sin\theta \frac{\partial}{\partial \theta}\right)\right] \qquad ...(10)$$

$$\nabla^2 = \frac{1}{r}\frac{\partial}{\partial r}\left(r \frac{\partial}{\partial r}\right) + \frac{1}{r^2}\frac{\partial^2}{\partial \theta^2} + \frac{\partial^2}{\partial z^2}. \qquad ...(11)$$

(c) Thermal Neutron Diffusion

To simplify the problem, let us treat thermal neutrons as a monoenergetic group. All the neutrons of the group are assumed to have same energy. On the average net energy change remains zero when colliding with the nuclei of the moderator. At a given point r (x, y, z) of the moderator, the neutron density n(r) will depend upon following three factors : (a) The rate of production of thermal neutrons per unit volume, Q; (b) The rate of absorption of thermal neutrons per unit volume, $nv\Sigma_a$; (c) The rate of diffusion or leakage per unit volume, $-D\nabla^2 n$.

Hence the rate of change of neutron concentration with time $\partial n/\partial t$ = rate of production–rate of absorption–rate of leakage

$$= Q - n\nu\Sigma_a - \left(-D\nabla^2 n\right) = Q - \frac{n\nu}{\lambda_a} + \frac{1}{3}\lambda_{trv}\nabla^2 n \qquad \text{....(12)}$$

This is the diffusion equation for neutrons that has to be solved for each particular case according to the requirements of the problem. When a steady state has been established the neutron density n at any given point inside the moderator will be independent of time t. This is in fact reached after enough time has elapsed, so that the production rate of neutron and the leakage and absorption have adjusted themselves to an equilibrium value. Since production of thermal neutrons is mainly due to slowing down of fast neutrons to thermal energies, hence Q can be replaced by slowing down density q. Thus *steady state equation* can be written as

$$\nabla^2 n - \frac{3}{\lambda_{tr}\lambda_a} n + \frac{3q}{\nu\lambda_{tr}} = 0. \qquad \text{...(13)}$$

In terms of neutron flux ϕ we can write

$$\nabla^2\phi - \frac{3}{\lambda_{tr}\lambda_a}\phi + \frac{3q}{\lambda_{tr}} = 0. \qquad \text{...(14)}$$

Let us imagine that the source is a point. We can imagine surrounding the source point a sphere of infinitesimal radius so that in all the space outside it, the q = 0. Thus the diffusion equation for these regions reduces to

$$\nabla^2\phi - \frac{3}{\lambda_{tr}\lambda_a}\phi = 0 \text{ or } \nabla^2\phi - \frac{1}{L^2}\phi = 0, \qquad \text{...(15)}$$

where $L^2 = \lambda_{tr}\lambda_a/3$. This term L is known as *thermal diffusion length.* To get a solution of eqn. (15), the ∇^3 is expressed in coordinates most appropriate to the geometry of the problem.

We shall write the diffusion equation for the spherical reference system, making use of the Laplacian in spherical form for the region outside the point source. The equation is hence

$$\frac{1}{r^2}\frac{\partial}{\partial r}\left(r^2\frac{\partial\phi}{\partial r}\right) - \frac{1}{L^2}\phi = 0. \qquad \text{...(16)}$$

After substituting $R = r\phi$, we get

$$\frac{1}{r^2}\frac{d}{dr}\left(r\frac{dR}{dr} - R\right) - \frac{1}{L^2}\frac{R}{r} = 0, \text{ or } \frac{d^2R}{dr^2} - \frac{R}{L^2} = 0.$$

This equation has the solution

$$R = Ae^{r/L} + Be^{-r/L} \quad ...(17)$$

or
$$\phi = \frac{A}{r}e^{r/L} + \frac{B}{r}e^{-r/L}. \quad ...(18)$$

These arbitrary constants A and B, can be determined from the boundary conditions. Since the neutron flux is finite for all values of the variable r, hence the constant A must be zero. To determine B we can make use of the physical condition that the total neutron absorption rate for the medium must be equal to the source strength Q. Therefore

$$Q = \int_0^{\infty} \phi\Sigma_a \, dV = \int_0^{\infty} \phi\Sigma_a \, 4\pi r^2 dr$$

$$= \int_0^{\infty} \frac{B}{r}e^{-r/L}\Sigma_a \, 4\pi r^2 dr = 4\pi\Sigma_a BL^2$$

or
$$B = \frac{Q}{4\pi\Sigma_a L^2} = \frac{3Q}{4\pi\lambda_{tr}} \quad ...(19)$$

$\therefore$
$$\phi = \frac{3Q}{4\pi\,\lambda_{tr}}\frac{1}{r}e^{-r/L}. \quad ...(20)$$

Similarly if the *neutron source is in the form of an infinite plane* that emits Q thermal neutrons per unit area per second. We shall consider a slab of material infinite in the y- and z-directions but of finite thickness in the x-direction. Thus the steady state equation becomes

$$\frac{d^2\phi}{dx^2} - \frac{1}{L^2}\phi = 0. \quad ...(21)$$

The general solution of this equation can be written as

$$\phi = Ae^{x/L} + Be^{-x/L}.$$

Again A = 0 if ϕ is to remain finite for all values of x. B can be evaluated, we can make use of the condition that the neutron current density J_x parallel to the source plane is

$$J_x = -\frac{1}{3}\lambda_{tr}\frac{d\phi}{dx} = \frac{1}{3}\lambda_{tr}\frac{B}{L}. \quad ...(22)$$

Since the neutron source emits Q neutrons per sec per unit area in both +ve and –ve directions, hence by virtue of symmetry of the conditions on both sides of the source one can write

$$J_x = \frac{1}{2}Q = \frac{1}{3}\lambda_{tr}\frac{B}{L} \text{ or } B = \frac{3QL}{2\lambda_{tr}}.$$

and the solution thus comes out to be

$$\phi(x) = \frac{3QL}{2\lambda_{tr}} \dot{e}^{-x/L}. \qquad ...(23)$$

$$\text{If } x = L, \; \phi = \frac{3QL}{2\lambda_{tr}} \frac{1}{e}. \qquad ...(24)$$

Thus for an infinite plane as neutron source, L is the distance from the source at which the neutron flux is reduced by a factor 1/e and is equal to the *relaxation length* which is more generally known as the *thermal diffusion length.* This is a measure of the air line distance travelled by neutron between the point of its origin as a thermal neutron A and the point of its absorption B.

For a point source, a relationship between the mean square distance from the source to the point of neutron absorption $\overline{r^2}$ and L^2 can be established by the following procedure. Let the flux of neutrons at a distance r from the point source emitting neutrons at a given rate be ϕ neutrons per unit area per second. If Σ_a is the macroscopic absorption cross section, the rate of absorption of neutrons will be $\Sigma_a \phi$ neutrons per unit volume per second. Thus the number of neutrons absorbed per second in a spherical shell of thickness dr at a distance r from the neutron source is $4\pi r^2 dr\, \Sigma_a \phi$. This is the probability of absorption of neutron when it is within the spherical shell of thickness dr. Therefore the mean square distance travelled by a neutron from the source to the point where it is absorbed is defined by

$$r^2 = \frac{\int_0^\infty r^2\, 4\pi r^2 dr\, \Sigma_a \phi}{\int_0^\infty 4\pi r^2 dr\, \Sigma_a \phi} = \frac{6L^4}{L^2} = 6L^2$$

or

$$L^2 = \frac{1}{6}\overline{r^2} \qquad ...(25)$$

In other words, the square of the diffusion length is equal to one sixth the average of the square of the shortest distance that a neutron travels from the point where the neutron is emitted to where it is absorbed. The parameter L^2 is called the *diffusion area.*

Similarly we have an expression for the slowing down length $L_s^2 = \frac{1}{6}\overline{r_s^2}$, where r_s is the distance measured in a straight line from the point where a fission neutron is produced to the point where it becomes a thermal neutron.

(d) The Exponential Pile

It is used for measuring diffusion length for a moderator experimentally. It is a rectangular column built from the moderator material with the neutron source at one end. The neutron flux in such a column falls off exponentially as $\phi \sim e^{-x/L_1}$ where L_1 is a combination of the diffusion length L for the moderator and the geometry of the column and can be given as

$$\frac{1}{L_1^2} = \frac{1}{L^2} + \frac{\pi^2}{a^2} + \frac{\pi^2}{b^2} \qquad ...(26)$$

Here a and b are the linear dimensions of the base of the column.

(e) Fast Neutron Diffusion and Fermi Age Equation

Let us now consider the diffusion of neutrons during the pre-thermal or slowing down stage. By virtue of their slowing down, neutrons undergo considerable energy changes while diffusing. The neutrons change their average energy and diffuse until they reach thermal energy. In the slowing down region neutron density per energy interval n(E) depends on the difference between the slowing down density q(E+dE] into the energy interval dE and the slowing down density q(E) out of it. This difference is zero for thermalized neutrons. In order to simplify the calculations, a continuous loss of energy for a slowing down neutron is assumed instead of actual discontinuous energy loss.

Let us investigate the neutron balance for a moderator in which there is no neutron absorption, $\Sigma_a = 0$. and in which neutrons are not being produced but diffuse through it before getting thermalized. If n is the initial number of neutrons per unit volume with energies between E and E + Δ E, the only physical processes that can cause neutron n to change are : *(i) diffusion of neutron into or out of the unit volume and (ii) slowing down of neutron into the energy interval ΔE and out of it.* The neutron flow and neutron balance in a space energy diagram. Neutron leakage by diffusion along the space co-ordinate is compensated by a neutron excess flowing along the energy co-ordinate into the energy interval dE.

In terms of neutron density the rate of diffusion is given by

$$-D\nabla^2 n = -\frac{1}{3}\lambda_{tr}\, v\nabla^2\, n. \qquad ...(27)$$

The number of neutrons slowing down into the energy interval dE and remaining will be equal to the difference of the inflow and the outflow, that is

$$q(E+dE)-q(E)-\left(\frac{\partial q}{\partial E}\right)dE \qquad ...(28)$$

Hence
$$-\frac{1}{3}\lambda_{trv}\nabla^2 n=\left(\frac{\partial q}{\partial E}\right)dE. \qquad ...(29)$$

By successive differentiation, we have

$$\nabla^2 q = E\xi\nu\Sigma_s\nabla^2 n \text{ or } \nabla^2 n=\frac{\nabla^2 q}{E\xi\nu\Sigma s}. \qquad ...(30)$$

Substituting this value of $\nabla^2 n$ in equation. we get

$$\nabla^2 q=\frac{\partial q}{-\left(\frac{\lambda_S\lambda_{tr}}{2\xi E}\right)dE}. \qquad ...(31)$$

Let us introduce a new variable τ, such that

$$d\tau=-\frac{\lambda_s\lambda_{yt}}{3\xi}\frac{\partial E}{E} \qquad ...(32)$$

or
$$\tau=\int_{E_0}^{E}d_\tau=\int_{E}^{E_0}\frac{\lambda_S\lambda_{tr}}{3\xi}\frac{\partial E}{E}=\frac{\lambda_{tr}\lambda_S}{3\xi}\log\frac{E_0}{E}. \qquad ...(33)$$

Substituting this new variable in equation (32), we have

$$\nabla^2 q - \partial q/\partial\tau = 0. \qquad ...(34)$$

This is the general equation of diffusion including all energies and is known as the *Fermi age equation* and the variable τ as the *Fermi age* or as the *neutron age*. Despite its name it must be emphasized that dimensions of τ are not of time as can readily be seen from equation (34), but are those of $(\text{length})^2$. It plays the same role in the slowing down equation as does the time in the heat conduction equation and is more strictly analogous to the variable kt/cp in heat conduction.

As (1/ξ) log E_0/E represents the average number of collisions a neutron, undergoes with the moderator nuclei to change energy from E_0 to E. If C is used for this number, equation (33) becomes

$$\tau=\frac{1}{3}\lambda_{tr}\lambda_S\,C=\frac{1}{8}\lambda_{tr}\,\Lambda_S. \qquad ...(35)$$

where As represents the total zig-zag path length of a neutron between the moment of the beginning of its slowing down and the moment of its arrival at energy E, which is usually thermal energy. Here we see that

Λs is quite analogous to λ_a in relation for thermal diffusion length. We can, therefore, define a quantity τ_0 such that

$$\tau_0 = \frac{1}{3}\lambda_{tr}\,\Lambda S = L_f^2. \qquad \text{...(36)}$$

The quantity L_f defined in this manner is called the *fast diffusion length.* For a point source emitting fast neutrons the solution of the Fermi age equation comes out to be

$$q(r) \propto e^{-r^2/4t/t^{3/2}} \qquad \text{...(37)}$$

This is a neutron distribution for a given τ. Its form is similar to the Gaussian distribution. At the origin ($r = 0$) it has its maximum value. The maximum is higher, the smaller the value of τ. K is clear from equation (33) that the value of τ is determined by the choice of the lower limit for to energy in the integral. The more advanced the degree of federation of the neutron, the lower its energy and the greater the corresponding value of τ. Thus we see that *the neutron age is a direct measure of the degree of moderation.* The slowing down density distribution represented by the *thermal neutrons have the larger range and a large fermi age where as the faster neutrons have smalier range and small fermi age.* The root mean square distance for a given neutron age can be obtained by the similar procedure adopted for thermal neutrons, as

$$\overline{r^2} = \int_0^{\infty} r^2 (q) 4\pi^2\,dr \Big/ \int_0^{\infty} q(r) 4\pi r^2\,dr$$

$$= 6\tau. \qquad \text{...(38)}$$

If $\tau = \tau_0$, the neutron age that corresponds to the termination of the moderation process and the attainment of thermal energies of the neutron, then we have

$$\overline{r^2} = 6\tau_0 = 6L_f^2. \qquad \text{...(39)}$$

Thus the neutron age τ is defined as the one sixth of the mean square distance from the point of creation to the point where their energy has been reduced to a value E which corresponds to that τ. This is analogous to that of L^2 in thermal diffusion. The sum of τ_0 and L^2 is called the *migration area.*

$$\tau = -\int_t^{t_0} \frac{\lambda_S \lambda_{tr}}{3\xi}\frac{dE}{E} = -\int_t^{t_0} \frac{\lambda_{trv}}{3} dt = -\int_t^{t_0} D dt$$

$$= -(t_0 - t).\frac{1}{(t_0 - t)}\int_t^{t_0} D dt = -(t_0 - t)\overline{D} = \overline{D} \times (t - t_0)$$

= Average diffusion coefficient × Slowing down time. ...(40)

Experimental Determination of Fermi's Age : A neutron source, for example polonium beryllium mixture emitting 10^{11} neutrons/m^3 sec approximately, is placed inside a pile of graphite. Indium foils are placed in cadmium boxes so as to exclude thermal neutrons. Because of the strong resonance of indium of 1.46, the values of saturation activity I_∞ are proportional to slowing down density q at this energy.

$$I_\infty = Ke^{-r^2/4\tau} \quad ...(41)$$

where K is a constant. Logarithm of this gives

$$\log I_\infty = \log K - \frac{r^2}{4\pi} \quad ...(42)$$

A plot I_∞ verses r^2 then given a straight line whose negative slope is $1/4\tau$. Thus fermi age can be determined.

NEUTRON VELOCITY SELECTOR

Even accelerator sources do not provide monoenergetic neutrons for accurate cross-section analysis. The problem has been solved with the aid of devices for selecting monoenergetic neutrons from heterogeneous neutron sources. These are known as velocity selectors. We now consider some of these devices.

1. Crystal Spectrometer

For a thermal and epithermal neutrons (E < 10 eV) a moroenergetic neutron beam can be selected by diffraction from a single crystal. The design is based on the wave properties cf neutrons. It can be seen from relation $A = h/mv = h/\sqrt{(2mE)}$ that neutrons with energies in the range 0.01 eV to about 10 eV have wavelengths of the order of magnitude as the wavelengths of X-rays and of the distance between crystal lattice planes. When a beam of slow neutrons of wavelength λ is reflected from the surface of a crystal, the diffraction maxima are observed at angles given by the Bragg relation

$$2d \sin \theta = n\lambda, \quad ...(1)$$

where n is an integer which represents the order of the reflection, d is the spacing of the lattice planes and θ is the glancing angle of the incident neutrons for with order reflection. If the de Broglie relation is used, the velocity of the neutrons at the nth maxima is given by

$$v = \frac{h}{m\lambda} = \frac{nh}{2md} \sin\theta \qquad ...(2)$$

$$\therefore \qquad E = \frac{n^2h^2}{8md^2} \sin^2\theta. \qquad ...(3)$$

A systematic diagram of a crystal spectrometer. A well collimated beam of thermal neutrons from an intense source, *e.g.*, a nuclear fission reactor, is diffracted by a prominent set of lattice planes of a large well shielded single crystal. The diffracted neutrons enter a second long collimator before entering a well shielded BF_3 cylindrical proportional counter. Neutrons of various speeds are present in the beam but for a particular value of θ, the great majority of the neutrons reaching the detector will have a velocity given by equation (2). If the glancing angle θ is changed, neutrons with a different velocity will now reach the detector. As the glancing angle diminishes with increasing neutron velocity, the angle ultimately becomes too small to be measured. This sets an upper limit to the speed of the neutron which can be studied in the crystal spectrometer. The energy resolving power dE/E decreases with increasing neutron energy. The differentiation of equation (3) gives

$$dE = \frac{n^2h^2}{8md^2} \frac{2}{\sin^3\theta} \cos\theta \, d\theta$$

$$\therefore \qquad \frac{dE}{E} = -2\cot\theta \, d\theta.$$

For small θ, dE/E = –2 dθ/θ. ...(4)

For 1 eV neutrons reflected from the crystal (d = 2.32 Å) in the first order, 0=3.5°. If the sharpness of collimation determines dθ = 0.1°, the resolution dE/E ~ 6%. The main drawback of the crystal diffraction method is the occurrence of second order reflection of neutrons of wavelength λ/2 at the same angle as first order neutrons of wavelength **λ.**

2. Mechanical Velocity Selectors

The design of this type of chopper is based on the fact that cadmium is a strong absorber of neutrons with energies less than 0.3 eV, while other metals such as aluminium are weak absorbers in this energy region. A cylinder, made up of alternate layers of cadmium and aluminium running parallel to the axis, is rotated before a strong source of heterogeneous neutrons. This device acts as a shutter and passes neutrons through it only when the laminations are parallel to the neutron direction.

Two bursts of neutrons are released at each revolution. Neutrons then arrive at a detector after travelling a distance/and their time of flight t depends on their velocity $v = l/t$. The detector must be sensitive for a short period at a definite interval after the neutrons have been transmitted. The timing is achieved by means of a double sided mirror fixed to the axis of the rotating cylinder. This reflects a beam of light from a fixed lamp into a photocell at an instant between successive bursts given by the angular setting of the cell.

The time t is determined by the speed of rotation of the cylinder and also by the relative positions of source and cell. In the slow neutron chopper of Fermi, J. Marshall and L. Marshall, the BF_3 neutron counter was placed 1.46 metres beyond the rotating cylinder. The arrangement is shown in Fig 4.18. The neutrons detected are those with a speed of l/t. The cross-section can be obtained by introducing the absorbing material at A and can be determined as a function of energy.

Above about 0.3 eV the Cd sheets become too transparent and a more satisfactory chopper uses materials with a large scattering cross-section, *e.g.,* nickel steel or nickel coated cadmium. Fast choppers are more complicated and difficult to operate. In one form of fast neutron chopper two steel cylinders, about 16" long and 4" diameter are mounted horizontally end to end in front of the neutron source. Each cylinder has a number of narrow slits on the outside, running parallel to the axis. One cylinder is stationary and the other is rotated at high speed. This type of fast chopper can be used with neutrons having energies up to about 5 keV.

3. Time of flight Velocity Selector

It is necessary to measure cross-sections at neutron energies greater than those covered by the slow chopper and crystal spectrometer. The time of flight method for measuring the neutron velocity can be used with a cyclotron or linear accelerator which produces a neutron beam. Fast neutrons are slowed down in paraffin block and then allowed to pass through a thin slab of material. The time required for the neutrons to reach a detector about 10 metres away from the source is electronically measured. The timing is usually done by distributing the counter pulses to a series of channels according to time. In this way neutrons of certain velocities are selected from a non-homogeneous beam. Transmission curve obtained with an instrument used at Columbia University. The dips in the transmission curve correspond to the resonances. This instrument can be used for neutron energies from 0.003 to 5000 eV.

One of the best methods of producing short intense neutron bursts in the pulsed electron accelerator. The latest machine of this type installed at Harwell (1960) accelerates pulses of electrons (~ 0.25 μ sec) to an energy of 30 MeV. They strike a high atomic number target which serves the dual purpose of producing energetic bremestrahlung, which in turn liberates neutrons through the (γ, n) reaction.

SOURCES OF THE NEUTRON

Rutherford in his Bakerian Lecture to the Royal Society in 1920 said, "*Under some conditions, ..if may be possible for an electron to combine much more closely with the hydrogen nucleus (i.e., a proton), forming a kind of neutral doublet. Such an atom would have novel properties. Its external field would be practically zero...and consequently it should be able to move freely through matter. Its presence would probably he difficult to detect.*" In 1930, Bothe and Becker bombarded lithium, beryllium and boron with α-particles from polonium and found a very penetrating but nonionizing radiation, they assumed that the radiation was of the γ-ray type because of its tremendous penetration. While repeating these experiments in 1932, Dr. H. Joilot and his wife, Dr. Irene Curie Joliot, daughter of Madam Curie found that which a sheet of hydrogen containing material, particularly paraffin, was interposed in the path of these radiations, protons were ejected with a considerable velocity. From the ranges of these recoil protons, the maximum proton energy E proved to be about 5.3 MeV. Assuming that the protons were produced as the result of elastic collisions with the *γ-ray photons*, calculations showed that each photon must have possessed an amount of energy of about 52 MeV. These results were entirely inconsistent with the results from the experiments on the absorption of these rays in lead (about 7 MeV). For an α-particle of 5 MeV energy :

$$_4Be^2 + {}_2He^4 + E_a \rightarrow ({}_6C^{13}) \rightarrow {}_6C^{12} + h\nu$$

$$9.01503 + 4.00388 + .00536 = 13.00751 + h\nu$$

or $$h\nu = 0.01676 \text{ amu}, = 15.5 \text{ MeV}.$$

This is much shorter than the required energy in order to explain the proton scattering. At Cambridge, Webster observed that the energy of the photons emitted by the beryllium varied according to their direction of emission, heaving least value if emitted in the direction opposite to that of the incident α-particles. The situation was resolved by Chadwick, in England in 1932, after performing a series of measurements of the energies of recoil of protons ejected from thin targets by the penetrating "Be-radiation" with a pulse ionizing chamber and amplifier. He said, *"These results are very difficult to explain on the assumption that the radiation from beryllium is a quantum radiation, if energy and momentum are to be conserved in the collisions. The difficulties disappear, however, if it be assumed the radiation consists of particles of mass* 1 *and charge 0, or neutrons."* Chadwick's theory made sense of Webster's puzzling discovery that the neutron energy varied with the direction of emission. These neutrons were formed as a result of the highly exoergic nuclear reaction

$$_4Be^9 + {}_2He^4 \rightarrow ({}_6C^{13}) \rightarrow {}_6C^{12} + {}_0n^1 + 5.65 \text{ MeV}.$$

Thus we see that *comparatively late date of the discovery of the neutron is partly due to the similarity in the properties of the neutron and high energy photons and the confusing effect that γ-rays as well as neutrons are associated usually with neutron sources.*

PHOTO-NEUTRON SOURCES

When γ-rays of sufficient energy (greater than the neutron binding energy) are incident upon light elements a probability exists that neutrons will be emitted by the energized nuclei. The majority of nuclei have a neutron binding energy of 5 MeV or more, except $_4Be^9$ (1.67 MeV) and $_1H^2$ (2.23 MeV). g-rays from artificially produced radio-nuclei can serve in photo disintegration sources. A standard source consisting of a *MsTh* or Na^{24} gamma source is placed at the centre of a spherical vessel filled with heavy water. Yield, the number of neutrons per sec due to 1 gm. of Be or D_2O at a distance of 1 cm. from 1 curie, of various gamma sources.

Table 1 : Radioactive (γ, n) sources.

γ-Source	Target	Half Life	E_γ (MeV)	E_n (MeV)	Yield
Na^{24}	Be	14.8 hr.	2.76	0.83	13×10^4
Na^{24}	D_2O	14.8 hr.	2.76	0.22	27×10^4
Y^{88}	Be	87 d.	1.9, 2.8	0.16	10×10^4
Y^{88}	D_2O	87 d.	2.8	0.31	0.3×10^4
Sb^{12}	Be	60 d.	1.7	0.025	19×10^{24}

The advantages of this type of source are :

(a) The reproducible neutron strength,

(b) The choice of E_n by choosing a suitable value of E_γ.

(c) The wide variety of artificial γ-sources available from atomic reactors,

(d) The neutrons produced are more mono-energetic than those obtained from (Be – α) sources.

Wattenberg gave the equation for the energy E_γ of the emitted neutrons as

$$E_n = \frac{A-1}{A}\left[E_\gamma - Q - \frac{E_\gamma^2}{1862(A-1)}\right] + E_\gamma \cos\theta \left[\frac{2(A-1)\left(E_\gamma - Q\right)}{931\,A^3}\right]^{1/2}$$

where A is the mass of the target nucleus, E_γ the energy of γ-ray in MeV, the threshold energy in MeV for the (γ, n) reaction in the target nucleus and θ is the angle between the path of the γ-ray and the direction of the emitted neutron. As Na^{24}-source is of short half life and hence Sb^{124} is very convenient for intermediate energy neutrons with a relatively long half life.

2. Radioactive Sources

The majority of the neutrons is produced by the (a, n) reaction given above. If the radium is in secular equilibrium with its decay products, extra deuterons are released according to the reaction

$$_4Be^9 + h\nu \rightarrow {}_4Be^8 + {}_9n^1.$$

The product nucleus of this reaction Be^8, is unstable and thus decays into two a-particles within 10^{-15} sec. This neutron source is isotropic and produces neutrons over a wide energy range (0 → 13 MeV). The neutrons are not mono-energetic due to following reasons :

(a) The α-particles do not all have the same initial energy.

(b) Many α-particles lose part of their initial energy by ionization before nuclear capture,

(c) The neutron energy varies with the direction of emission having least value if emitted in the direction opposite to that of the incident α-particles.

(d) $_6C^{12}$ can be left in an excited state.

We can have slow neutrons after putting the Ra – α – Be source in water, paraffin, or some other hydrogen-containing material. For a 2-MeV neutron only 25 elastic collisions with protons are needed. Radon is some times used instead of radium, as it is a gas and can be made into a more compact source. But it has the disadvantage of decaying rapidly because of the snort half life of 3.8 days. Polonium is used instead of radium when a neutron source with relatively few g-rays is desired. This has several disadvantages :

(i) the neutron output is much lower (3×10^6 neutrons/sec per curie of Po while it is $10 - 18 \times 10^6$ neutrons/sec per curie of Ra for radium),

(ii) special chemical facilities are needed to prepare Po,

(iii) the half life is short (140 days) — compared with that of Ra – Be source (1600 years).

3. Reactor Sources

The most powerful source of neutrons available in these days is a nuclear reactor or chain reacting pile on which uranium undergoes fission. The fission of each heavy nucleus by slow neutron capture is accompanied by the release on the average of about 2.5 neutrons. The neutrons produced from the fission on of uranium atoms are fast and are slowed down in the moderator. The disadvantages of reactor sources are the wide spread in neutron energy and the impossibility of directly pulsing the neutron source.

4. X-rays Sources

Through the (γ, n) reaction, high energy X-ray machines can produce the high energy neutrons. A flux of high energy photons is produced when the electrons accelerated to high energies in linear accelerators, betatrons, synchrotrons etc. impinge on target elements of high atomic number. A large number of neutrons within the target material are released by the high flux of photons. The disadvantage of this type of source is the wide spread in neutron energy.

5. Ultrafast Neutrons

As a deuteron behaves as a relatively loosely bound system consisting of a neutron and a proton which are frequently outside the range of their mutual forces, hence, deuterons accelerated to about 200 MeV in a synchro-cyclotron can readily be "stripped" into protons and neutrons by directing the deuteron beam to pass through a target nucleus. Recently neutrons of energy > 300 MeV have been produced by stripping 650 MeV deuterons accelerated in the Birmingham synchrotron. Almost any element may be used as the target material for stripping out the yield varies with its nature.

6. Neutrons from Accelerated Particle Reactions

Neutrons are produced by (d, n) and (p, n) reactions in particle accelerators. These sources give mono-energetic neutrons in the range from a few keV to 20 MeV. In deuteron bombardment the energy of the released neutron is progressively larger for most of the target elements, which is due to the low binding energy of the deuteron. The C^{12} (d, n) N^{13} reaction is an exception to the rule with a Q value of – 0.28 MeV. Those neutrons which are emitted in the direction of the incident beam of deuterons will have a maximum energy, equal to the sum of the incident and the reaction energies approximately. Some (d, n) reactions are:

$$_1H^1 + d \rightarrow {}_2He^3 + {}_0n^1 + 3.265 \text{ MeV}$$

$$_1H^3 + d \rightarrow {}_2He^4 + {}_0n^1 + 17.6 \text{ MeV}$$

$$_3Li^7 + d \rightarrow {}_4Be^8 + {}_0n^1 + 15.0 \text{ MeV}$$

$$_4Be^9 + d \rightarrow {}_5B^{10} + {}_0n^1 + 3.79 \text{ MeV.}$$

Several light and moderately heavy elements undergo (p, n) reactions. These endoergic reactions are the convenient sources of neutrons of moderate energy. Some (p, n) reactions are :

$$_1H^3 + p \rightarrow {}_2He^3 + {}_0n^1 - 0.764 \text{ MeV}$$

$$_3Li^7 + p \rightarrow {}_4Be^7 + {}_0n^1 - 1.646 \text{ MeV.}$$

The Li^7 (p, n) Be^7 reaction suggests that accelerated charged particles can be used to produce nearly monoenergetic neutrons.

NEUTRON COLLIMATORS

Neutrons produced by the above sources possess very high energy range. Most of the neutrons are called fast neutrons as they have enormous velocities. Since the ordinary methods used for charged particles and γ-rays can not be used for the collimating a beam of fast neutrons. Dunning (1938) has devised a collimator known as the *neutron howitzer*. He used the property of slowing down of neutron in hydrogeneous materials like paraffin or water. The radium beryllium source S is placed at the bottom of a barrel inside a paraffin block.

A cylindrical slit is cut in this block which is lined from all sides by cadmium sheet. Neutrons are emitted in all directions. The neutrons energy is negligibly small in all directions except through the barrel. The thermal neutrons resulting from the slowing down of the fast neutrons in the paraffin are effectively absorbed by the cadmium coating over the outer surface a paraffin and walls of the barrel. The cadmium layer is covered by a layer of lead to g-rays arising from neutron capture.

SOLVED EXAMPLES

Example 1:

How many collisions are required for neutrons to loss, on the average, 90% of an initial energy of 2 MeV in a Be^9 moderated assembly ($\xi = 0.209$)? Compare this with the total member of collisions required to reach thermal energies.

Solution:

If neutrons of energy E_1 are slowed down to energy E_2, then number of collisions required to reduce this amount of energy

$$n = \frac{1}{\xi}\log\left(\frac{E_1}{E_2}\right) = \frac{1}{0.209}\log_e\frac{100}{1} = 22.$$

Similarly the number of collisions required to reduce the energy to thermal energy (0.025 eV) is

$$n = \frac{0}{0.209}\log_e\left(\frac{2\times10^6}{0.025}\right) = 87.$$

Example 2:

Calculate the average distance travelled by a neutron away from an infinite planer neutron source in a moderator before being absorbed. Compare it with the root mean square of the distance travelled.

Solution:

The average distance is obtained by averaging over all distances

$$\therefore \qquad x = \frac{\int_0^\infty xe^{-X/L}dx}{\int_0^\infty e^{-x/L}dx = L}$$

= Thermal diffusion length.

Similarly we can obtain

$$\overline{x^2} = \frac{\int_0^\infty x^3 e^{-x/L} dx}{\int_0^3 e^{-x/L} dx = 2L^3}.$$

or $$x_{n.m.s} = \sqrt{(2)}L = \sqrt{(2)}\,\overline{x}.$$

Example 3(a):

Calculate the neutron age and slowing down length for fission of neutrons 2.0 MeV average energy to thermal energy 0.025 eV in graphite and beryllium. Given : Σ_s = 33.5 for graphite and 57 for beryllium.

Solution:

We have,

$$\tau_0 = \frac{\lambda_{tr}\lambda_s}{3}\frac{\log\left(\frac{E_0}{E}\right)}{\xi} = \frac{\lambda_s}{\left(1-\frac{2}{3A}\right)}\frac{\lambda_s}{3}\frac{\log\left(\frac{E_0}{E}\right)}{\frac{2}{\left(A+\frac{2}{3}\right)}}$$

$$= \frac{1}{6\Sigma_s^2}\frac{A+\frac{2}{3}}{1-\frac{2}{3A}}\log\left(\frac{E_0}{E}\right).$$

For graphite, A = 12

$$\tau_0 = \frac{1}{6\times(33.5)^2}\frac{12+\frac{2}{3}}{1-\frac{2}{36}}\log\frac{2\times10^6}{0.025}$$

$$= 3.65 \times 10^{-2}\text{m}^2 = 365 \text{ cm}^2.$$

$\therefore$ $L_f = 0.192$ m.

Similarly for beryllium, A = 9

$\tau_0 = 9.7 \times 10^{-3}$ and $L_f = 9.8 \times 10^{-2}$ m.

Example 3(b):

Calculate diffusion length for thermal neutrons in graphite. Given: σ_a = 3.2 milli barns, σ_s = 4.8 barns, and ρ = 1.62 gm/cm^2.

Solution:

Diffusion length L is given by the relation

$$L = \left[\frac{\lambda_{a\Lambda tr}}{3}\right]^{1/2},$$

where $\lambda_a = \frac{1}{\Sigma_a} = \frac{1}{N_0\sigma_a} = \frac{A}{N\rho\sigma_a}$

$= 12/6.02 \times 10^{26} \times 1.62 \times 10^3 \times 3.2 \times 10^{-31}$

$= 38.40$ m.

$$\lambda_{tr} = \frac{\lambda_s}{\frac{1-2}{3A}} = \frac{\frac{1}{\Sigma_s}}{\frac{1-2}{3A}} = \frac{\frac{A}{N\rho\sigma_s}}{\frac{1-2}{A}}$$

$$= \frac{12/6.02\times10^{16}\times1.62\times10^3\times4.8\times10^{-28}}{\frac{1-2}{3\times12}} = 0.0254 \text{ m}.$$

$\therefore$ $L = 0.572$ m.

Example 3(c):

Calculate the energy of the neutrons reflected from the crystal of lattice 2.3 Å, if the glancing angle of the incident neutrons for Ist order reflection is 3.5°.

Solution:

Energy of the neutrons $E = \frac{n^2h^2}{8md^2\sin^2\theta}$

$$= \frac{(6.6256\times10^{-27})}{8\times1.6748\times10^{-24}\times(2.3\times10^{-8})^2(0.061)^2}$$

$= 1.04$ eV.

Example 4:

Calculate the average energy loss for neutrons that have made one collision with carbon nuclei using $\xi = 0.158$ for carbon.

Solution:

Since $\xi = \overline{\Delta \log E} = \overline{\log (E_0/E)}$, hence

$$\left(\overline{\frac{E_0}{E}}\right) = e^{\xi} \text{ and } \left(\overline{\frac{E_0 - E}{E_0}}\right) = 1 - e^{\xi} = 0.1462.$$

Another method of finding average fractional energy loss is the use of eqn. (21). Then, we get

$$\left(\overline{\Delta E/E_0}\right) = \frac{1}{2}(1-\alpha) = \frac{1}{2}\left[1 - \frac{(A-1)^2}{(A+1)^2}\right].$$

This result shows that it is incorrect to use ξ which is the mean value of the log and not the log of the mean value for finding average energy loss per collision.

Example 5:

Calculate the fractional energy loss of a neutron being scattered through an angle of 60° in the C.M. system in collision with 8.3. Hence prove that the maximum fractional energy loss in this is greater than that for collision with U^{238} nuclei.

Solution:

Relation (11) can be written in terms of α, as

$$\frac{E_1}{E_0} = \frac{1}{2}(1+\alpha) + \frac{1}{2}(1-\alpha)\cos\phi.$$

$$\text{As } \alpha = \frac{(A-1)^2}{(A+1)^2} = \left(\frac{8}{10}\right)^2 = 0.64$$

$$\therefore \quad \frac{E_1}{E_0} = \frac{1}{2}(1+0.64) + \frac{1}{2}(1-0.64)\cos 60 = 0.91.$$

Hence $$\frac{\Delta E}{E_0} = 1 - \frac{E_1}{E_0} = 0.09 = 9\%.$$

As $$\left(\frac{\Delta E}{E_0}\right)_{max} = 1 - \frac{(A-1)^2}{(A+1)^2}, \text{ hence}$$

$$\text{for } Bc^2 \left(\frac{\Delta E}{E_0}\right)_{max} = 1 - \left(\frac{8}{10}\right)^2 = 0.36 \text{ or } 36\%.$$

$$\text{and for } \quad U^{238} = -\left(\frac{237}{239}\right)^2 = 0.02 \text{ or } 2\%.$$

Example 6:

A manganese foil of 100 mg is placed in an exponential pile and exposed to the thermal neutron flux for 2 hour. Now activity of the foil is found to be 200 disintegrations/sec. If the cross-section of manganese for neutrons of 0.025 eV is 13.2 barns and the half life of the radioactive isotope is 2.58 hour. Calculate the thermal neutron flux of the pile at the point of insertion of the foil, if the point is assumed at temperature 50°C. Calculate the diffusion length L or thermal neutrons for the material if the base is a rectangular of 1.5 m × 2m and the neutron flux is found to be twice when the experiment is repeated with the foil inserted at a distance of 16 cm. nearer the base of the pile.

Solution:

Neutron energy corresponding to a pile temperature of 50°C

$$E = kT = 1.38 \times 10^{-23} \times 323 = 4.46 \times 10^{-21} \text{ Joule}$$

$$= 0.028 \text{ eV}$$

Absorption cross-section

$$\sigma_a = \sigma_0 \left(\frac{E_0}{E}\right)^{1/2} = 13.2\left(\frac{0.025}{0.028}\right)^{1/2} = 12.5 \text{ barns}$$

Activity $\quad A = \lambda N_0(1 - e^{-\lambda t}) = \Sigma_a \phi V(1 - e^{-\lambda t})$.

Here $\quad \lambda = \frac{0.693}{T_{1/2}} = \frac{0.693}{2.58} \text{ hr} = 0.269 \text{ hr}^{-1}$.

$\therefore \quad (1 - e^{-\lambda t}) = 1 - e^{-0.538} = 0.416.$

Total number of Mn^{55} nuclei in foil

$$N = 6.02 \times 10^{26} \times 0.1 \times 10^{-3}/55 = 1.094 \times 10^{21}$$

$\therefore$ Neutron flux $\quad \phi = 200/1.094 \times 10^{21} \times 12.5 \times 10^{-28} \times 0.416$

$$= 3.5 \times 10^8 \text{ neutrons/m}^2 \text{ sec.}$$

As the neutron flux $\phi \propto e^{-x/L}$

$$\therefore \quad \frac{\phi_1}{\phi_2} = \frac{e^{-x_1/L_2}}{e^{-x_1/L_1}} \text{ or } \log_e\left(\frac{\phi_2}{\phi_1}\right) = \frac{x_2 - x_1}{L_1}$$

or
$$\frac{1}{L_1} = \log_e\left(\frac{\phi_2}{\phi_1}\right)(x_2 - x_1) = \frac{0.693}{0.16} = 4.33 \text{ m}^{-1}.$$

From eqn. (72), we get

$$\frac{1}{L^2} = \left(\frac{1}{4.33}\right)^2 + \left(\frac{3.14}{1.5}\right)^2 + \left(\frac{3.14}{2}\right)^2 = 11.9$$

$$\therefore \quad L = 0.29 \text{ metre.}$$

EXERCISES

1. A certain (1/v) absorber reduces the intensity of a thermalized neutron beam of effective temperature T = 300°K by 10%. What would be the reduction in intensity if the same experiment were done at a temperature of 327°K.
2. Calculate the age difference for neutrons slowing down in graphite between neutrons that have reduced the indium resonance energy of 1.44 eV and neutrons that have reduced thermal energies 0.025 eV.
3. Compute the average logarithmic energy loss per collision, the number of collisions needed to reduce the energy from 2 MeV to 0.025 eV, the slowing down power and the moderating ratio of bismuth, given σ = 30 millibarn and σ_s = 9 barn.
4. Calculate the maximum energy of neutrons released by a (γ, n) reaction from a phosphorus target by 17 MeV γ-rays.
5. What would be the energy, mean velocity and wavelength of thermal neutrons at 20°?
6. A crystal spectrometer has a beryllium crystal with a lattice spacing of 0.7323 Å. What are the Bragg angles for the first order reflection and of neutrons of energy 1, 3, 5 and 30 eV respectively?